Chemistry, Thermodynamics, and Reaction Kinetics for Environmental Engineers

This book aims to be the preeminent university chemistry textbook for environmental engineers. It provides undergraduate and graduate environmental engineering students with basic concepts and practical knowledge about chemistry that they would need in their professional careers. It focuses on the fundamental concepts of chemistry and its practical applications (e.g., understanding fate and transport of chemicals/pollutants in the environmental as well as the chemical/physicochemical processes applied in environmental engineering industry). This book also serves as a valuable resource for entry-level professionals to solidify their fundamental knowledge in environmental engineering chemistry. This book

- Presents the fundamentals of chemistry with focus on the needs of environmental engineers.
- Explains how an understanding of chemistry allows readers a better understanding of the fate and transport of chemicals in the environment as well as various treatment processes.
- Examines the fundamentals of chemical reaction equilibrium from learning the basics of thermodynamics.
- Presents the basic types and designs of reactors as well as reaction kinetics.

Fundamentals of Environmental Engineering
Series Editor: Jeff Kuo

Air Pollution Control Engineering for Environmental Engineers
Jeff Kuo

Site Assessment and Remediation for Environmental Engineers
Cristiane Q. Surbeck and Jeff Kuo

Chemistry, Thermodynamics, and Reaction Kinetics for Environmental Engineers
Jeff (Jih Fen) Kuo

For more information about this series, please visit: https://www.routledge.com/Fundamentals-of-Environmental-Engineering/book-series/CRCFUNOFENV

Chemistry, Thermodynamics, and Reaction Kinetics for Environmental Engineers

Jeff (Jih Fen) Kuo

CRC Press
Taylor & Francis Group
Boca Raton London New York

CRC Press is an imprint of the
Taylor & Francis Group, an **informa** business

Designed cover image: Shutterstock

First edition published 2025
by CRC Press
2385 NW Executive Center Drive, Suite 320, Boca Raton FL 33431

and by CRC Press
4 Park Square, Milton Park, Abingdon, Oxon, OX14 4RN

CRC Press is an imprint of Taylor & Francis Group, LLC

ISBN: 978-1-032-81983-9 (hbk)
ISBN: 978-1-032-82042-2 (pbk)
ISBN: 978-1-003-50266-1 (ebk)

DOI: 10.1201/9781003502661

Typeset in Times
by codeMantra

Contents

Author

Dr. Jeff (Jih Fen) Kuo had worked in the environmental engineering industry for 10 years before joining Department of Civil and Environmental Engineering at California State University, Fullerton (CSUF) in 1995. He gained his industrial experience from working at Groundwater Technology Inc.; Dames and Moore; James M. Montgomery Consulting Engineers, and the Sanitation Districts of Los Angeles County (LACSD). His industrial experience includes design and installation of air strippers, activated carbon adsorbers, soil vapor extraction systems, bioremediation systems, and flare/catalytic incinerators. He is also experienced in site assessment; fate and transport analysis of toxic compounds in the environment; RI/FS work for landfills and superfund sites; design of flanged connections to meet stringent fugitive emission requirements; and development of emission factors for VOC emissions from wastewater treatment. He also participated in application research on various wastewater treatment projects when working at LACSD.

Areas of research in environmental engineering include dechlorination of halogenated aromatics by ultrasound, fines/bacteria migration through porous media, biodegradability of heavy hydrocarbons, surface properties of composite mineral oxides, kinetics of activated carbon adsorption, wastewater filtration, THM formation potential of ion exchange resins, UV disinfection, sequential chlorination, nitrification/denitrification, removal of target compounds using nanoparticles, persulfate oxidation of persistent chemicals, microwave oxidation for wastewater treatment, destruction and removal of PFOAs, landfill gas recovery and utilization, greenhouse gases control technologies, fugitive methane emissions from the gas industry, biogas production from anaerobic digestion of food wastes, and stormwater runoff treatment.

He received a B.S. in Chemical Engineering from National Taiwan University, an M.S. in Chemical Engineering from University of Wyoming, an M.S. in Petroleum Engineering, and an M.S. and a Ph.D. in Environmental Engineering from University of Southern California. He is a professional civil, mechanical, and chemical engineer registered in California.

Preface

My industrial and teaching experience gives me an in-depth understanding of what an environmental engineering student and/or a junior environmental professional should know about chemistry. I had worked at several environmental engineering consulting companies and the Sanitation Districts of Los Angeles County (LACSD) before I started my tenure-track career at California State University, Fullerton (CSUF) in 1995. I have been teaching several undergraduate- and graduate-level environmental engineering courses, and I have also conducted quite a few externally funded research projects as the principal investigator at CSUF.

This book aims to be the preeminent university chemistry textbook for environmental engineers. It provides basic concepts and practical information about chemistry that undergraduate and graduate environmental students would need. It is also useful for entry-level environmental professionals as a tool book to solidify their required knowledge in environmental engineering chemistry. This is not an "environmental chemistry" book or an "environmental engineering chemistry" book; it is a book of "chemistry for environmental engineers". The focus is to provide information/knowledge that typical environmental engineers need to know and to have.

This book is organized into five segments, each with chapters encompassing the themes. The first segment of this book focuses on fundamentals of chemistry for environmental engineers. It includes three chapters: (1) basic concepts related to general chemistry, (2) organic chemistry for environmental engineers, and (3) inorganic chemistry for environmental engineers. The second segment of this book focuses on thermodynamics. It includes two chapters: (4) fundamentals of thermodynamics and (5) chemical-reaction equilibria.

The third segment of this book focuses on solution chemistry. It includes four chapters: (6) solutes, solvents, and solutions; (7) ions in aqueous solutions; (8) acid-base equilibria in aqueous solutions; and (9) oxidation-reduction equilibria. The fourth segment of this book focuses on reaction kinetics. It includes two chapters: (10) fundamentals of chemical reaction kinetics, and (11) types and design of chemical reactors. The fifth/last segment of this book focuses on chemistry in environmental media. It includes three chapters: (12) water quality parameters, (13) chemistry in soils, and (14) chemistry in the atmosphere.

There are many chemistry textbooks available in the market for environmental engineers, and most, if not all, are at least 400 pages long. Although the information contained in those books is valuable and useful, it would be nearly impossible for a lecturer to cover all the content in one semester. In addition, students may not be able to grasp and retain all the information that the authors intend to convey.

This textbook appears to be shorter than the others in pages, but it serves well as a textbook for a 3-semester-units chemistry course for environmental engineering students. In a succinct format, this practical engineering textbook provides well-digested information for both engineering students and junior engineering professionals. My hope and the main objective of this book are that the readers would read this book from cover to cover and gain the common sense needed so that they will be well-equipped and successful in their engineering professions. Though I strongly believe that it is important for an engineer to obtain professional licensures, this book is not a preparation book for the Fundamental of Engineering (FE) exam and/or the Professional Engineering (PE) exam. However, I took these two exams into consideration when I prepared this book.

1 Basic Concepts Related to General Chemistry

1.1 CHEMISTRY FOR ENVIRONMENTAL ENGINEERS

Chemistry is the study of matter and changes it can undergo. *Matter* is anything that has mass and occupies space; it is composed of a basic element or various combinations of some basic elements (i.e., *compounds*). Matter that has a specific composition as well as physical and chemical characteristics is called a *substance*. Chemistry is the basic science that deals with properties, compositions, and structures of substances (i.e., elements and compounds). Environmental engineers need to have sufficient knowledge of chemistry because almost all pollutants, that they may come across at work, are chemical compounds; understanding the properties of pollutants is critical in applying appropriate technologies to remove and/or to destruct them. Although both environmental engineers and chemists need to understand the properties of chemical compounds, their learning focuses could be different. Chemists are often more interested in formulation of chemicals, whereas environmental professionals usually focus more on removal/destruction of chemical pollutants. Fate and transport of *compounds of concern* (COCs) are of great concern to most, if not all, environmental engineers. In addition to knowing the properties of pure substances, environmental engineers also need to know their behaviors in different environmental media (i.e., water, soil, and air) and their interactions with the media in which they are present. These are just a few examples.

1.2 ATOMS AND ELEMENTS

A *chemical element* is a fundamental substance that cannot be broken down into simpler substances by chemical methods. An atom is a particle that defines a chemical element. To this date, 118 elements have been identified.

1.2.1 Atoms

Protons, neutrons, and electrons. An *atom* has a nucleus of *protons* (positively charged) and *neutrons* (neutral) and a cloud of *electrons* (negatively charged) in different shells/orbits surrounding the nucleus. Atoms bond to each other to form an element (e.g., carbon; C); each element is distinguished by the number of its protons, known as *atomic number* (Z). For example, the atomic number of C is 6 and that of oxygen (O) is 8. More details with regard to an atom are described below:

- Each proton carries a positive charge ($+1.602 \times 10^{-19}$ Coulomb), and each electron has a negative charge (-1.602×10^{-19} Coulomb).

DOI: 10.1201/9781003502661-1

- Each atom is neutral because "its number of protons = its number of electrons".
- *Mass number* = number of protons + number of neutrons. Each element has its specific atomic number (Z), but its mass number can be different due to the difference in the number of neutrons in its nucleus. They are called *isotopes*. For example, carbon (Z = 6) has three isotopes; they are carbon-12 (^{12}C), carbon-13 (^{13}C), and carbon-14 (^{14}C) with 6, 7, and 8 neutrons (mass number = 12, 13, and 14), respectively.
- The *atomic mass unit (amu)* is defined as one twelfth of the mass of a ^{12}C atom (with six protons, six neutrons, and six electrons); and "1 $amu \cong 1.67 \times 10^{-24}$ gram (g)".
- The atomic masses of proton, neutron, and electron are approximately 1.00728, 1.00866, and 0.00055 *amu*, respectively. A neutron is slightly heavier than a proton; a proton is approximately 1,837 times heavier than an electron.

  Hydrogen (H) is the lightest element. Each hydrogen atom has only one proton and one electron; it has a mass of ~1.67×10^{-24} g. In practice, hydrogen was assigned a relative mass of unity, and the other elements were assigned "atomic masses" relative to that of hydrogen. For example, atomic masses of C and O are 12 and 16, respectively (they are used to calculate the molar mass and molecular weight of compounds).
- The *atomic mass* of an element is the weighted average of masses of its isotopes (e.g., the atomic mass of magnesium is 24.3 *amu*).

Figure 1.1 illustrates atomic symbols, with atomic and mass numbers, in general form and of four atoms (i.e., carbon-12, carbon-13, hydrogen, and oxygen).

1.2.2 The Periodic Table

The *periodic table of elements* is a systematic catalog of all elements in which they are arranged in the order of the atomic number. The periodic table lists all known 118 elements by the atomic number (see Figure 1.2). Hydrogen is the first element in the periodic table because it has the smallest atomic number of unity. These elements are arranged in periods (i.e., the rows of the table) and in groups (i.e., the columns). There are seven periods, and each is filled with elements sequentially, according to the atomic number from left to right. There are 18 groups; elements in each group have the same electron configuration in their outermost orbital/shell (i.e., the *valence shell*) resulted in similar chemical properties. This arrangement

$$\text{(a)}\ {}^{Mass\ number}_{Atomic\ number}Atomic\ Symbol$$

$$\text{(b)}\ {}^{12}_{6}C;\ (c)\,{}^{13}_{6}C;\ (d)\,{}^{1}_{1}H;\ (e)\ {}^{16}_{8}O$$

FIGURE 1.1 Atomic symbols: (a) the general form, (b) carbon-12, (c) carbon-13, (d) hydrogen, and (d) oxygen.

	1	2	3	4	5	6	7	8	9	10	11	12	13	14	15	16	17	18
1	1 H																1 H	2 He
2	3 Li	4 Be											5 B	6 C	7 N	8 O	9 F	10 Ne
3	11 Na	12 Mg											13 Al	14 Si	15 P	16 S	17 Cl	18 Ar
4	19 K	20 Ca	21 Sc	22 Ti	23 V	24 Cr	25 Mn	26 Fe	27 Co	28 Ni	29 Cu	30 Zn	31 Ga	32 Ge	33 As	34 Se	35 Br	36 Kr
5	37 Rb	38 Sr	39 Y	40 Zr	41 Nb	42 Mo	43 Tc	44 Ru	45 Rh	46 Pd	47 Ag	48 Cd	49 In	50 Sn	51 Sb	52 Te	53 I	54 Xe
6	55 Cs	56 Ba	57-71 La	72 Hf	73 Ta	74 W	75 Re	76 Os	77 Ir	78 Pt	79 Au	80 Hg	81 Tl	82 Pb	83 Bi	84 Po	85 At	86 Rn
7	87 Fr	88 Ra	89-103 Ac	104 Rf	105 Db	106 Sg	107 Bh	108 Hs	109 Mt	110 Ds	111 Rg	112 Cn	113 Nh	114 Fl	115 Mc	116 Lv	117 Ts	118 Og

Lanthanides	57 La	58 Ce	59 Pr	60 Nd	61 Pm	62 Sm	63 Eu	64 Gd	65 Tb	66 Dy	67 Ho	68 Er	69 Tm	70 Yb	71 Lu
Actinides	89 Ac	90 Th	91 Pa	92 U	93 Np	94 Pu	95 Am	96 Cm	97 Bk	98 Cf	99 Es	100 Fm	101 Md	102 No	103 Lr

FIGURE 1.2 Periodic table of elements.

facilitates predictions of many of their chemical properties and their behaviors in chemical reactions. The periodic table is often viewed as three clusters in two ways:

- ***Main group elements, transition elements, and lanthanides & actinides.*** Elements in Groups 1 to 2 and 13 to 18 are the *main group elements* (listed as the "A groups" in older periodic tables). Groups 3–12 are in the middle of the periodic table, and they are the *transition elements* (listed as the "B groups" in older periodic tables). The bottom cluster below the main periodic table are the lanthanides and actinides (they should be in the rows of 6 and 7 of the main periodic table, respectively). In the main group elements, five groups are also known by their names: *alkali metals* (Group 1), *alkaline earth metals* (Group 2), *chalcogens* (Group 16), *halogens* (Group 17), and *noble gases* (Group 18).
- ***Metals, non-metals, and metalloids.*** The heavy zigzag line, running diagonally from the upper left to the lower right through groups 13–16 of the periodic table, divides the elements into (i) metals (to the left of the line), (ii) non-metals (to the right of the line), and (iii) metalloids, which lie along the line (in the shaded boxes). *Metalloids* (also called *semi-metals*) have properties between metals and non-metals; they are boron (B), silicon (Si), germanium (Ge), arsenic (As), antimony (Sb), tellurium (Te), polonium (Po), and astatine (At) [Note: Po and As are not considered as metalloids by some].

Tables 1.1 and 1.2 tabulate the names, atomic numbers, and atomic masses of all elements in the alphabetical order and in the order of atomic mass, respectively. The atomic mass of an element is the average mass of all its isotopes, by taking their natural abundances into account. The parentheses around the atomic mass in these two tables indicate that the atomic mass in the parentheses is a calculated or estimated value, rather than experimentally determined. It is because the isotopes of an element

TABLE 1.1
List of 118 Elements in the Order of Atomic Number

Name	Symbol	Atomic No.	Atomic Mass
Actinium	Ac	89	227.03
Aluminum	Al	13	26.98
Americium	Am	95	(243.06)
Antimony	Sb	51	121.75
Argon	Ar	18	39.95
Arsenic	As	33	74.92
Astatine	At	85	(209.99)
Barium	Ba	56	137.33
Berkelium	Bk	97	(247.07)
Beryllium	Be	4	9.01
Bismuth	Bi	83	208.98
Bohrium	Bh	107	(270)
Boron	B	5	10.81
Bromine	Br	35	79.90
Cadmium	Cd	48	112.41
Calcium	Ca	20	40.08
Californium	Cf	98	(251.08)
Carbon	C	6	12.01
Cerium	Ce	58	140.12
Cesium	Cs	55	132.91
Chlorine	Cl	17	35.45
Chromium	Cr	24	52.00
Cobalt	Co	27	58.93
Copernicium	Cn	112	(285.17)
Copper	Cu	29	63.55
Curium	Cm	96	(247.07)
Darmstadtium	Ds	110	(281.16)
Dubnium	Db	105	(268.13)
Dysprosium	Dy	66	162.50
Einsteinium	Es	99	(252.08)
Erbium	Er	68	167.26
Europium	Eu	63	151.96
Fermium	Fm	100	(257.10)
Flerovium	Fl	114	(289.19)
Fluorine	F	9	19.00
Francium	Fr	87	(223.02)
Gadolinium	Gd	64	157.25
Gallium	Ga	31	69.72
Germanium	Ge	32	72.59
Gold	Au	79	196.97
Hafnium	Hf	72	178.49
Hassium	Hs	108	(277.15)

(Continued)

TABLE 1.1 (*Continued*)
List of 118 Elements in the Order of Atomic Number

Name	Symbol	Atomic No.	Atomic Mass
Helium	He	2	4.00
Holmium	Ho	67	164.93
Hydrogen	H	1	1.01
Indium	In	49	114.82
Iodine	I	53	126.90
Iridium	Ir	77	192.22
Iron	Fe	26	55.85
Krypton	Kr	36	83.80
Lanthanum	La	57	138.91
Lawrencium	Lr	103	(262.11)
Lead	Pb	82	207.20
Lithium	Li	3	6.94
Livermorium	Lv	116	(293)
Lutetium	Lu	71	174.97
Magnesium	Mg	12	24.31
Manganese	Mn	25	54.94
Meitnerium	Mt	109	(276.15)
Mendelevium	Md	101	(258.10)
Mercury	Hg	80	200.59
Molybdenum	Mo	42	95.94
Moscovium	Mc	115	(288.19)
Neodymium	Nd	60	144.24
Neon	Ne	10	20.18
Neptunium	Np	93	237.05
Nickel	Ni	28	58.70
Nihonium	Nh	113	(284.18)
Niobium	Nb	41	92.91
Nitrogen	N	7	14.01
Nobelium	No	102	(259.10)
Oganesson	Og	118	(294)
Osmium	Os	76	190.20
Oxygen	O	8	16.00
Palladium	Pd	46	106.40
Phosphorus	P	15	30.97
Platinum	Pt	78	195.09
Plutonium	Pu	94	(244.06)
Polonium	Po	84	(208.98)
Potassium	K	19	39.10
Praseodymium	Pr	59	140.91
Promethium	Pm	61	(145)
Protactinium	Pa	91	231.04
Radium	Ra	88	226.03

(*Continued*)

TABLE 1.1 (*Continued*)
List of 118 Elements in the Order of Atomic Number

Name	Symbol	Atomic No.	Atomic Mass
Radon	Rn	86	(222.02)
Rhenium	Re	75	186.21
Rhodium	Rh	45	102.91
Roentgenium	Rg	111	(280.16)
Rubidium	Rb	37	85.47
Ruthenium	Ru	44	101.07
Rutherfordium	Rf	104	(265.12)
Samarium	Sm	62	150.40
Scandium	Sc	21	44.96
Seaborgium	Sg	106	(271.13)
Selenium	Se	34	78.96
Silicon	Si	14	28.09
Silver	Ag	47	107.87
Sodium	Na	11	22.99
Strontium	Sr	38	87.62
Sulfur	S	16	32.06
Tantalum	Ta	73	180.95
Technetium	Tc	43	(97.91)
Tellurium	Te	52	127.60
Tennessine	Ts	117	(294)
Terbium	Tb	65	158.93
Thallium	Tl	81	204.37
Thorium	Th	90	232.04
Thulium	Tm	69	168.93
Tin	Sn	50	118.69
Titanium	Ti	22	47.90
Tungsten	W	74	183.85
Uranium	U	92	238.03
Vanadium	V	23	50.94
Xenon	Xe	54	131.30
Ytterbium	Yb	70	173.04
Yttrium	Y	39	88.91
Zinc	Zn	30	65.38
Zirconium	Zr	40	91.22

may be difficult or impossible to measure directly, or properties of some newly found elements have not been well determined/tested.

1.2.3 Valence Electrons and the Octet Rule

The electrons of an atom are distributed over different energy levels. These energy levels are called the *shells* or the *orbitals* (Note: there are *s*, *p*, *d*, and *f* orbitals).

TABLE 1.2
List of 118 Elements in Alphabetical Order

Atomic No.	Name	Symbol	Atomic Mass
1	Hydrogen	H	1.01
2	Helium	He	4.00
3	Lithium	Li	6.94
4	Beryllium	Be	9.01
5	Boron	B	10.81
6	Carbon	C	12.01
7	Nitrogen	N	14.01
8	Oxygen	O	16.00
9	Fluorine	F	19.00
10	Neon	Ne	20.18
11	Sodium	Na	22.99
12	Magnesium	Mg	24.31
13	Aluminum	Al	26.98
14	Silicon	Si	28.09
15	Phosphorus	P	30.97
16	Sulfur	S	32.06
17	Chlorine	Cl	35.45
18	Argon	Ar	39.95
19	Potassium	K	39.10
20	Calcium	Ca	40.08
21	Scandium	Sc	44.96
22	Titanium	Ti	47.90
23	Vanadium	V	50.94
24	Chromium	Cr	52.00
25	Manganese	Mn	54.94
26	Iron	Fe	55.85
27	Cobalt	Co	58.93
28	Nickel	Ni	58.70
29	Copper	Cu	63.55
30	Zinc	Zn	65.38
31	Gallium	Ga	69.72
32	Germanium	Ge	72.59
33	Arsenic	As	74.92
34	Selenium	Se	78.96
35	Bromine	Br	79.90
36	Krypton	Kr	83.80
37	Rubidium	Rb	85.47
38	Strontium	Sr	87.62
39	Yttrium	Y	88.91
40	Zirconium	Zr	91.22
41	Niobium	Nb	92.91
42	Molybdenum	Mo	95.94

(*Continued*)

TABLE 1.2 (*Continued*)
List of 118 Elements in Alphabetical Order

Atomic No.	Name	Symbol	Atomic Mass
43	Technetium	Tc	(97.91)
44	Ruthenium	Ru	101.07
45	Rhodium	Rh	102.91
46	Palladium	Pd	106.40
47	Silver	Ag	107.87
48	Cadmium	Cd	112.41
49	Indium	In	114.82
50	Tin	Sn	118.69
51	Antimony	Sb	121.75
52	Tellurium	Te	127.60
53	Iodine	I	126.90
54	Xenon	Xe	131.30
55	Cesium	Cs	132.91
56	Barium	Ba	137.33
57	Lanthanum	La	138.91
58	Cerium	Ce	140.12
59	Praseodymium	Pr	140.91
60	Neodymium	Nd	144.24
61	Promethium	Pm	(145)
62	Samarium	Sm	150.40
63	Europium	Eu	151.96
64	Gadolinium	Gd	157.25
65	Terbium	Tb	158.93
66	Dysprosium	Dy	162.50
67	Holmium	Ho	164.93
68	Erbium	Er	167.26
69	Thulium	Tm	168.93
70	Ytterbium	Yb	173.04
71	Lutetium	Lu	174.97
72	Hafnium	Hf	178.49
73	Tantalum	Ta	180.95
74	Tungsten	W	183.85
75	Rhenium	Re	186.21
76	Osmium	Os	190.20
77	Iridium	Ir	192.22
78	Platinum	Pt	195.09
79	Gold	Au	196.97
80	Mercury	Hg	200.59
81	Thallium	Tl	204.37
82	Lead	Pb	207.20
83	Bismuth	Bi	208.98
84	Polonium	Po	(208.98)

(*Continued*)

TABLE 1.2 (*Continued*)
List of 118 Elements in Alphabetical Order

Atomic No.	Name	Symbol	Atomic Mass
85	Astatine	At	(209.99)
86	Radon	Rn	(222.02)
87	Francium	Fr	(223.02)
88	Radium	Ra	226.03
89	Actinium	Ac	227.03
90	Thorium	Th	232.04
91	Protactinium	Pa	231.04
92	Uranium	U	238.03
93	Neptunium	Np	237.05
94	Plutonium	Pu	(244.06)
95	Americium	Am	(243.06)
96	Curium	Cm	(247.07)
97	Berkelium	Bk	(247.07)
98	Californium	Cf	(251.08)
99	Einsteinium	Es	(252.08)
100	Fermium	Fm	(257.10)
101	Mendelevium	Md	(258.10)
102	Nobelium	No	(259.10)
103	Lawrencium	Lr	(262.11)
104	Rutherfordium	Rf	(265.12)
105	Dubnium	Db	(268.13)
106	Seaborgium	Sg	(271.13)
107	Bohrium	Bh	(270)
108	Hassium	Hs	(277.15)
109	Meitnerium	Mt	(276.15)
110	Darmstadtium	Ds	(281.16)
111	Roentgenium	Rg	(280.16)
112	Copernicium	Cn	(285.17)
113	Nihonium	Nh	(284.18)
114	Flerovium	Fl	(289.19)
115	Moscovium	Mc	(288.19)
116	Livermorium	Lv	(293)
117	Tennessine	Ts	(294)
118	Oganesson	Og	(294)

The *electron configuration of an atom* describes the arrangement of electrons in these shells/orbitals around the nucleus. *Valence electrons* are electrons that are located in the outermost shell of an atom (i.e., the *valence shell*). All the other electrons are called the *inner electrons*. Due to being the farthermost from the nucleus and thus the most loosely held by the nucleus, they are the electrons that participate in the formation of molecular bonds and in chemical reactions. The number of valence electrons and the shell they are in determine the atom's reactivity, electronegativity, and the number of chemical bonds it can form.

The *octet rule* dictates that atoms when forming molecular bonds or atomic ions, would be most stable when their valence shells are filled with eight electrons (i.e., a "filled" valence shell). When atoms have fewer than eight valence electrons, they tend to react to form more stable compounds. The octet rule is mainly applicable for the main group elements in the periodic table (i.e., elements not in the transition metal group); the elements in the transition metal group have two valence electron shells.

Lewis electron dot symbols (or *Lewis dot symbols*) are used to predict the number of bonds that would be formed by most atoms when they form compounds with other elements. Figure 1.3 illustrates the Lewis dot symbols for the elements in Period 2 of the periodic table. Each Lewis dot symbol consists of the chemical symbol of the element surrounded by dots, representing its valence electrons. In Period 2, the valence electrons increase from one for lithium (Li) to eight for neon (Ne) [Note: The valence electrons also increase similarly from one for sodium (Na) to eight for argon (Ar) for elements in Period 3]. The way to draw the Lewis dot symbols are: (i) if ≤4 valence electrons, usually place the dots sequentially to the right, to the left, above, and below the symbol and (ii) if >4 valence electrons, place one at a time to form a pair with one of the first four.

The elements in the same group of the periodic table have the same number of valence electrons. This does not apply to transition elements, which have more complicated electron configurations [Note: That is the reason why the transition elements are placed in a cluster different from the main group elements in the periodic table]. Several noteworthy observations from Figure 1.3:

- Lithium (Li) in Group 1 (i.e., the alkali metal group) has only one valence electron.
- Beryllium (Be) in Group 2 (i.e., the alkaline earth metal group) has two valence electrons.
- Boron (B) in Group 13 has three valence electrons; carbon in Group 14 has four valence electrons; nitrogen in Group 15 has five; and oxygen in Group 16 has six valence electrons, respectively.
- Fluorine (F) in Group 17 (i.e., the halogen group) has seven valence electrons.
- Neon (Ne) in Group 18 (i.e., the noble gas group) has eight valence electrons. The electron shell is filled, and it implies that the atoms in the noble gas family are stable.

The Lewis dot symbols for elements in Groups 1, 2, 13, 14, 15, 16, 17, and 18 of Periods 3, 4, 5, and 6 are similar to those in Figure 1.3, except the names of the

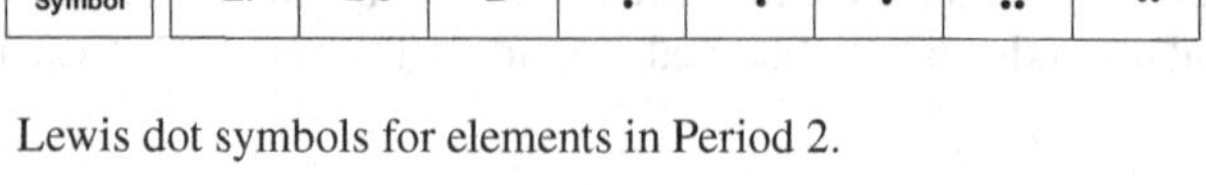

Group #
1
2
13
14
15
16
17
18
Lewis Dot Symbol
Li
Be
B
C
N
O
F
Ne

FIGURE 1.3 Lewis dot symbols for elements in Period 2.

elements. For example, all the elements in the alkali metal group (Group 1) have one valence electron, and all the elements in the halogen family (Group 17) have seven valence electrons.

1.2.4 Ions

When a neutral atom loses or gains electron(s), it becomes an ion. An atom becomes a *cation* (i.e., an ion with a positive charge) by losing negative electron(s); it becomes an *anion* (i.e., an ion with a negative charge) by gaining electron(s). Going back to Figure 1.3, the alkali metals (Group 1) have only one valence electron; they have a great tendency to lose that valence electron so that their new outermost shells will be fully filled to become stable. Consequently, the alkali metals in aqueous solutions are in the form of mono-valent cations (e.g., Li^+, Na^+, and K^+). Similarly, the alkaline earth metals (Group 2) have two valence electrons, and they are commonly found as di-valent cations in aqueous solutions (e.g., Ca^{+2}, Mg^{+2}). On the other hand, the halogens (Group 7) have seven valence electrons, and their valence shells would become filled by gaining one electron. Consequently, they will appear as mono-valent anions in aqueous solutions (e.g., F^-, Cl^-, Br^-, and I^-). Noble gases are stable (with completely filled valence shells); they are neutral, and it is hard to get them ionized.

Ionization energy and electron affinity. Atoms will not lose electron(s) spontaneously, and energy is required to remove valence electron(s) from a neutral atom to form a cation. The *ionization energy* of an element is the amount of energy required to remove an electron from an atom. If an atom possesses more than one valence electron, the amount of energy required to remove successive electrons increases. On the other hand, e*lectron affinity* of an element is the change of energy when an electron is added to an atom or ion.

Below are some general trends with regard to the sizes of atoms and ions as well as ionization energy and electron affinity of elements:

- Atomic radii decrease from left to right across a row in the periodic table (i.e., the same period).
- Atomic radii increase from top to bottom down in a column (i.e., the same group).
- Francium (Fr) at the bottom-left of the periodic table has the largest atomic radius, while helium (He) at the top-right of the periodic table has the smallest atomic radius.
- A cation, having lost electron(s), is always smaller than its parent neutral atom.
- An anion, having gained electron(s), is always larger than its parent neutral atom.
- The ionization energies are always positive (i.e., energy is required to remove an electron from a neutral atom). Opposite to the trends of atomic radii, the ionization energies increase from left to right and decrease from

top to bottom. In other words, the ionization energies increase diagonally from the bottom-left to the top-right of the periodic table.

- The first electron affinities of elements can be negative, zero, or positive; but the second electron affinities are always positive (i.e., energy is required) – see the next section for coverage on electronegativity.

Example 1.1: Atomic Mass, Protons, Neutrons, Electrons, and Ions

Complete the following table:

Symbol	Atomic Number	Mass Number	No. of Protons	No. of Electrons	No. of Neutrons	Net Charge
^{238}U						
$^{27}Al^{+3}$						
	11			10	12	
		19	9			−1

Strategy:
Use the following relationships to complete the table:

- Atomic number = Number of protons = Number of electrons (for a neutral atom)
- Mass number = Number of protons + Number of neutrons
- Cations lost electron(s)
- Anions gained electron(s)

Solution:
The completed table is shown below:

Symbol	Atomic Number	Mass Number	No. of Protons	No. of Electrons	No. of Neutrons	Net Charge
^{238}U	92	238	92	92	146	0
$^{27}Al^{+3}$	13	27	13	10	14	+3
$^{23}Na^{+}$	11	23	11	10	12	+1
$^{19}F^{-}$	9	19	9	10	10	−1

1.3 CHEMICAL BONDING

Metallic solids are solids of metal atoms that are held together in a well-arranged structure [Note: Mercury (Hg) is the only metal that is not in a solid form under ambient conditions]. In metallic solids, the crystal lattice consists of cations and free-flowing electrons. Metals are good conductors of electricity because of having these free-flowing electrons.

Atoms can also join with different atoms to form other types of substances. When the valence shell of an atom is not full, the atom tends to transfer or share electrons with other atom(s) to have a full valence shell to become stable – the "octet rule". A *chemical compound* is a group of atoms held together by chemical bonds. All chemical bonds are due to electrostatic attraction. There are two types of chemical bonds: (i) covalent bonds and (ii) ionic bonds. When atoms are connected by covalent bonds, a *molecule* is formed. On the other hand, if the atoms are connected by ionic bonds, an *ionic compound* is formed.

1.3.1 Covalent Bonds

A *covalent bond* is formed when two atoms share an electron pair. Some elements exist naturally as covalent molecules such as hydrogen gas (H_2), oxygen gas (O_2), nitrogen gas (N_2), and halogen gases (i.e., F_2, Cl_2, I_2, and Br_2). They are diatomic molecules, while "H_2" is the chemical formula of hydrogen gas, for example. Similarly, there are a few pure elements that exist as polyatomic molecules, such as phosphorus (P_4) and sulfur (S_8). Elements of H, N, O, and F have 1, 5, 6, and 7 valence electrons, respectively. Figure 1.4 illustrates the Lewis structures and chemical structural formula of H_2, O_2, N_2, and F_2.

As shown in Figure 1.4, the two hydrogen atoms (H) of hydrogen gas (H_2) share a pair of valence electrons, which forms a covalent bond (a *single bond*), as shown in its chemical structural formula (H–H). The two oxygen atoms of oxygen gas (O_2) share two pairs of valence electrons and form two covalent bonds (a *double bond*) as shown in its chemical structural formula (O=O). With the shared valence electrons, each oxygen atom has eight valence electrons in its outermost shell. The two nitrogen atoms of nitrogen gas (N_2) share three pairs of valence electrons, which form three covalent bonds (a *triple bond*) as shown in its chemical structural formula (N≡N). The fluorine atoms of fluorine gas (F_2) share a pair of valence electrons, which form a covalent bond as shown in its chemical structural formula (F–F). With the shared electrons, each nitrogen atom and fluorine atom in nitrogen gas and fluorine gas also have eight valence electrons in their outermost shells. Below is a summary with regard to covalent bonds of non-metal elements, when forming molecules:

- Hydrogen, with one valence electron, can form a single covalent bond.
- Carbon in Group 14, with four valence electrons, can form four covalent bonds.

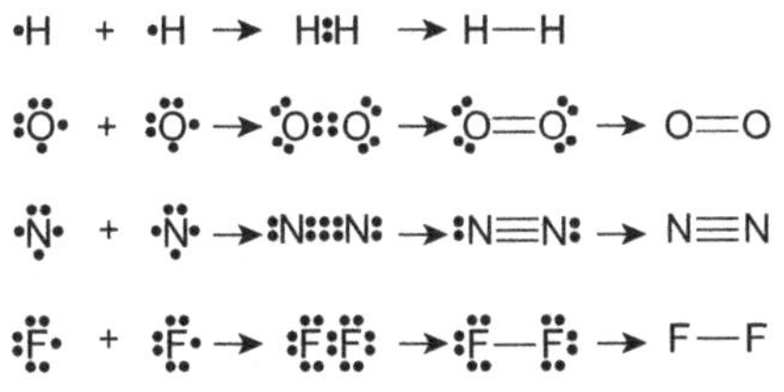

FIGURE 1.4 Lewis structures and chemical structural formula of hydrogen, oxygen, nitrogen, and fluorine gases.

- Nitrogen and phosphorus (P) in Group 15 have five valence electrons. Phosphorus can form three covalent bonds (e.g., phosphorus trichloride, PCl_3) and five covalent bonds (e.g., phosphoric acid, H_3PO_4). On the other hand, nitrogen can only form three covalent bonds (e.g., ammonia, NH_3) because of the relatively small size of the nitrogen atom.
- Oxygen, sulfur (S), and selenium (Se) in Group 16 have six electrons. Oxygen and selenium normally form two covalent bonds, while sulfur can form two (e.g., hydrogen sulfide, H_2S), four (e.g., sulfur dioxide, SO_2), or six covalent bonds (e.g., sulfur trioxide, SO_3; sulfuric acid, H_2SO_4).
- Fluorine (F), chlorine (Cl), bromine (Br), and iodine (I) elements in Group 17 have seven valence electrons; they can form one covalent bond.
- The elements in Group 18 have eight valence electrons, and they typically will not form molecules with other elements.

Equation (1.1) is a "balanced" chemical reaction equation (more in Chapter 5); it shows that one methane molecule (CH_4) reacts with two oxygen molecules (O_2) to form one carbon dioxide molecule (CO_2) and two water molecules (H_2O). Each molecule is shown in its chemical formula.

$$CH_4 + 2O_2 \rightarrow CO_2 + 2H_2O \tag{1.1}$$

Figure 1.5 illustrates the methane combustion, with all the molecules expressed in their chemical structural formula. As shown, a chemical reaction rearranges the combinations of atoms, while the mass is conserved. It should be noted that the actual structures of these molecules may not be planar and/or linear. For example, methane has a shape of tetrahedron; water molecule has a bent or angular shape, and the angle between two O–H bonds is approximately 105°.

1.3.2 Ionic Bonds

In a covalent bond, both atoms share the pair(s) of valence electrons equally or rather equally, whereas in an ionic bond, one atom "donates" valence electron(s) to another atom to fulfill the octet rule. Covalent bonds are formed between two non-metals, whereas ionic bonds are formed between a metal and a non-metal. Atoms that form an ionic bond have a relatively big difference in their electronegativity values. *Electronegativity* is the tendency of an atom in a molecule to attract the shared pair(s) of electrons toward itself. General trends of the electronegativity include:

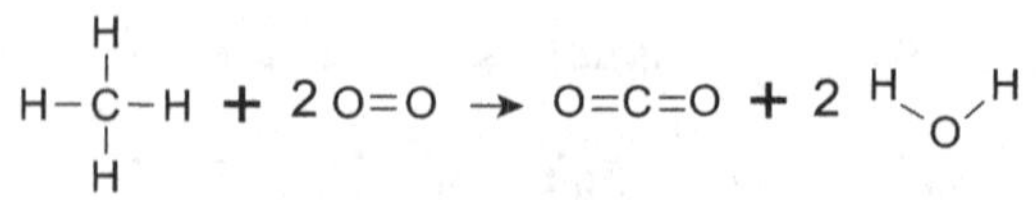

FIGURE 1.5 Reaction equation for methane combustion (Molecules are expressed in their structural formula).

- Values of electronegativity increase from left to right across a row in the periodic table (i.e., the same period).
- Values of electronegativity decrease from top to bottom in a column (i.e., the same group).
- Metals generally have lower values of electronegativity, when compared to non-metals.
- In general, electronegativity values of elements increase from the bottom-left corner to the upper-right corner of the periodic table. Fluorine is the most electronegative element in the periodic table with an electronegativity value of 4.0, while francium (Fr) at the bottom-left of the periodic table has the smallest electronegativity of 0.7. The electronegativity values of metals are typically less than 2.0.
- *Electro-positivity* is the exact opposite of electronegativity. Thus, fluorine is the least electropositive element.

An ionic bond is formed by the attraction between two atoms in which one is highly electronegative, and the other is highly electropositive. The more electronegative atom will gain control over the bond pair of electrons. Figure 1.6 illustrates the Lewis dot symbols of Na and Cl atoms and the Lewis structure of NaCl. Due to the stronger electronegativity of the Cl atom, the Na atom (Group 1) will release its only valence electron to become a cation (i.e., Na^+); the Cl atom (Group 17) will then become an anion (i.e., Cl^-). Both ions achieve a noble gas configuration, and an ionic bond is formed.

If we dissolve salt (NaCl) water, it will dissociate into sodium and chloride ions; the reaction can be written as:

$$NaCl_{(s)} + H_2O \rightarrow Na^+_{(aq)} + Cl^-_{(aq)} + H_2O \quad (1.2)$$

With that, alkali metal ions are single-valent cations (e.g., Li^+, Na^+, and K^+) and ions of the halogen family are single-valent anions (e.g., F^-, Cl^-, Br^-, and I^-). By the same token, ions of the alkaline earth metals (Group 2) are divalent cations (e.g., Be^{+2}, Mg^{+2}, and Ca^{+2}). Consequently, the chemical formula of magnesium chlorine is $MgCl_2$, in which each of magnesium's two valence ions is transferred to a chlorine atom to form an ionic bond.

1.3.3 Polarization of Molecules

The only "pure" covalent bonds occur between identical atoms for diatomic molecules (e.g., H_2, O_2, and N_2), since the electronegativities of the two bonded atoms are the same. Consequently, the bonded pair(s) of electrons are almost equidistant from the two nuclei. On the other hand, the covalent bonds between two atoms of different electronegativities tend to be "polarized". The more electronegative atom pulls the

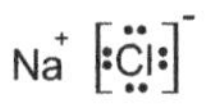

FIGURE 1.6 Lewis dot structure of salt (NaCl).

bond pair(s) of electrons closer to itself, resulting in a partially negative charge (usually denoted by a symbol of δ−), the more electropositive atom develops a partially positive charge (denoted by δ+). These covalent bonds are considered polar and often called *polar covalent bonds.*

Polar compounds are chemical compounds that are held together by polar covalent bonds; they are different from ionic compounds. Atoms in ionic compounds are held together by ionic bonds that arise due to electrostatic forces between ions. A good example of polar compounds is water. Water is polar because the two covalent bonds between hydrogen and oxygen are polar in nature. The more electronegative oxygen pulls each bond pair of electrons closer to itself and develops a partially negative charge, while the two hydrogen atoms carry a partially positive charge. The greater the difference in electronegativity, the more polar the bond is.

A *dipole moment* in physics is a measure of the separation of two opposite electrical charges within a system. The *dipole moment of a chemical bond* is a measure of the bond's polarity between two atoms. The common unit of dipole moment in chemistry is Debye (D), and $1\ D = 3.335 \times 10^{-30}$ Coulomb-meter (C-m). Hydrofluoric acid (HF) is a very polar compound. The electronegativity of fluorine (F) is larger than that of chlorine (Cl); consequently, the dipole moment of hydrofluoric acid is larger than that of hydrochloric acid (1.82 D vs. 1.08 D). The overall dipole moment of a molecule can be determined by the sum of dipole moment of each individual bond. However, it should be noted that the addition of bond dipole moments is not arithmetic because a dipole moment is a vector, not a scalar, quantity. Even if a molecule possesses polar bonds, the entire molecule may not be polar. For example, each of the four covalent C–Cl bonds in carbon tetrachloride (CCl_4) is polar; however, it is a non-polar compound. Similarly, each C=O double bond in carbon dioxide (CO_2) is polar, but its overall dipole moment is zero because of its linear geometry. In addition, the dipole moment of each O–H bond is 1.5 D, and the overall dipole moment of a water molecule is only 1.85 D (not 3.0 D because the angle of two O–H bonds is 104.5°).

1.4 INTRAMOLECULAR AND INTERMOLECULAR FORCES

Intramolecular forces are the forces that hold atoms together within a molecule. As discussed, these atoms are held by one of four types of bonds: (i) metallic, (ii) ionic, (iii) polar covalent, and (iv) covalent.

Intermolecular forces are forces that exist between molecules, and they are generally weaker than those intramolecular forces. However, they are important because they determine many physical properties of molecules (e.g., boiling point, melting point, heat of vaporization, viscosity, and vapor pressure). Major types of intermolecular forces include (i) ionic forces, (ii) dipole-dipole interactions, (iii) hydrogen bonding, and (iv) London dispersion forces. The last three intermolecular forces are often referred to as one group of forces; that is *van der Waals forces*. The intermolecular forces are electrostatic in nature (i.e., attraction and/or repulsion of charged species).

1.4.1 Ionic Forces

Ionic forces are interactions among charged ions. The attractive forces between ions of opposite charges and the repulsive forces between ions of similar charges are described by *Coulomb's law*, in which the forces increase with the charges and decrease with the distance between these ions (i.e., inversely proportional to $(\text{distance})^2$). Due to being highly polarized, ionic compounds typically have higher melting points and higher water solubility.

1.4.2 Dipole–Dipole Interactions

If molecules have net dipole moments, they will tend to align themselves so that the positive end of one polar molecule tends to get near to the negative end of another polar molecule (i.e., an *attractive intermolecular interaction*), or vice versa. This arrangement is more stable than the arrangement in which two positive (or negative) ends are adjacent (i.e., a *repulsive intermolecular interaction*).

Because molecules in an aqueous solution move freely, they are continuously experiencing both attractive and repulsive *dipole-dipole interactions* (the attractive intermolecular interactions typically dominate). The strength of the intermolecular interactions increases as the dipole moments of the molecules increase. The dipole-dipole interactions are significantly weaker than interactions between two ions because each end of a dipole molecule only possesses a fraction of the charge.

1.4.3 Hydrogen Bonding

Hydrogen bonding is a special kind of dipole–dipole interaction that occurs specifically between a hydrogen atom bonded to atoms of high electronegativity such as oxygen, nitrogen, and fluorine (and to a lesser extent to atoms of lower electronegativity such as chlorine and sulfur). Figure 1.7 illustrates the hydrogen bonding among the water molecules, in which the partially positive end of hydrogen (δ+) is attracted to the partially negative end of the oxygen atom (δ−). Hydrogen bonding is a relatively stronger force of attraction between molecules than most of the other dipole–dipole attractive interactions. It would require considerable energy to break the hydrogen bonding; this explains the high boiling points and melting points of compounds such as water, hydrogen fluoride (HF), and hydrogen chloride (HCl). Hydrogen bonding plays an important role in biology; for example, it is responsible for holding pairs of nitrogenous bases in DNA and RNA molecules.

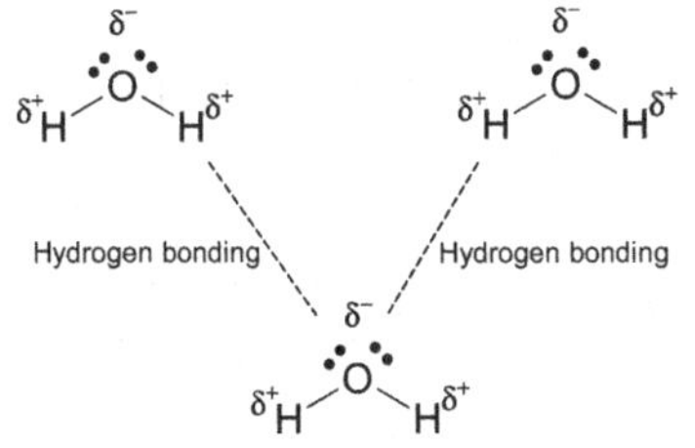

FIGURE 1.7 Hydrogen bonding among water molecules.

1.4.4 London Dispersion Forces

Are there any intermolecular attractive forces between non-polar molecules? When two adjacent atoms have temporary fluctuations in their electron distribution, it may result in short-lived instantaneous dipole moments between two non-polar molecules. This attractive force is temporary, and it is termed as the *London dispersion force*. It should be noted that London dispersion forces can also exist between polar molecules. The London dispersion force is the weakest of all types of intermolecular forces (i.e., it does not require much energy to break the London dispersion forces).

1.5 MOLES, MOLECULE MASS, MOLAR MASS, AND MOLECULAR WEIGHT

1.5.1 Moles

As we know, atoms, ions, and molecules are very tiny in size and mass. A mole (also spelled as mol) in chemistry is a standard scientific unit for measuring large quantities of entities (e.g., atoms, molecules, and ions). The *mole* is defined as the amount of a substance that contains the same number as the number of carbon atoms in 12.0 g of carbon-12. From the most recent experiments, 12.0 g of C-12 contains 6.022×10^{23} C-12 atoms. This number, 6.022×10^{23} is often referred to as the *Avogadro's number.*

One mole of carbon atoms, methane (CH_4) molecules, and eggs contain 6.022×10^{23} carbon atoms, CH_4 molecules, and eggs, respectively. For one mole of methane molecules, they contain one mole of carbon atoms and four moles of hydrogen atoms because each methane molecule contains one carbon atom and four methane atoms. In summary, one mole of anything always has the same number (i.e., 6.022×10^{23}) of that object.

1.5.2 Molecular Mass, Molecular Weight, and Molar Mass

The *molecular mass* (i.e., mass of a molecule) is the sum of the masses of all its atoms. The atomic mass of an element can be found in Table 1.1 or 1.2. The values of atomic mass in those two tables are truncated to two decimal points; they are sufficient for most engineers' uses [Note: Values of atomic mass in literature may carry more significant units]. The units for the atomic mass and the molecular mass are in *amu*. As mentioned, 1 *amu* is a very tiny mass (1 $amu \cong 1.67 \times 10^{-24}$ g).

Engineers consider that "mass" and "weight" are different (i.e., mass is a scalar amount, while weight is a force and it is a vector). However, the terms "mass" and "weight" are often used interchangeably in chemistry. *Molecular weight* (*MW*) can be defined as:

$$MW \text{ of a substance} = \frac{\text{Molecular mass of the substance}}{\text{Atomic mass of C}-12} \times 12 \qquad (1.3)$$

Consequently, the MW has no units. The MW of a compound should have the same numeric value as that of its molecular mass.

Example 1.2: Atomic Mass and Molecular Mass

Determine the molecular mass and MW of calcium carbonate ($CaCO_3$).

Solution:

a. Use Table 1.1 or 1.2 to find the atomic masses:

$$\text{Atomic mass of Ca} = 40.08\,amu$$

$$\text{Atomic mass of carbon} = 12.01\,amu$$

$$\text{Atomic mass of oxygen} = 16.00\,amu$$

b. Molecular mass is the sum of masses of all atoms that the molecule contains:

$$\text{Molecular mass of } CaCO_3 = 40.08 + 12.01 + 16.00 \times 3 = \underline{100.09\,amu}.$$

c. MW of $CaCO_3 = 100.09$ (dimensionless)

Discussion:

In engineering practices, it is often acceptable to truncate the values of atomic mass into integers or into one decimal point. For example, using atomic masses of Mg and Ca as 24.3 and 40 is acceptable. With that, using 100 as the *MW* of $CaCO_3$ is commonly done.

The *molar mass* is defined as the mass of one mole (i.e., 6.022×10^{23}) of that substance (e.g., atoms, ions, and molecules). The standard unit for molar mass in chemistry is g/mole (or g/mol), in which the mole is the gram-mole (or g-mole). However, the SI unit for mass is kg, so the molar mass should be in kg/kg-mole. In addition, lb and "lb-mole" are often used in engineering practices. In this book, we will use the mole in the way as that in chemistry (i.e., g-mole), unless specifically noted. When kg-mole or lb-mole is used, relationships among g-mole, kg-mole, and lb-mole are:

$$1 \text{ kg-mole} = 1{,}000 \text{ g-moles} = 2.2 \text{ lb-moles} \quad (1.4)$$

$$1 \text{ lb-mole} = 454 \text{ g-moles} = 0.454 \text{ kg-mole} \quad (1.5)$$

One mole of C-12 atoms has a mass of 12 g. From Example 1.2, the molecular mass of $CaCO_3$ = 100 amu and its *MW* = 100 (dimensionless); consequently, its molar mass is 100 g/mole. Similarly, the molar mass of water is 18.0 g/mole since its molecular mass is 18.0 amu [Note: The molar mass of 1 kg-mole of water = 18 kg = 18,000 g; the molar mass of 1 lb-mole of water = 18 lbs].

The following relationships relate the mass, moles, and MW of a given substance:

$$\text{Number of moles} = \text{Mass}/MW \quad (1.6)$$

$$\text{Mass} = (\text{Number of moles}) \times MW \quad (1.7)$$

Example 1.3: Mass, Moles, and *MW* Relationships

a. Convert 100 kg of $CaCO_3$ into different units of moles.
b. Find the mass of $CaCO_3$ in 1 g-mole, 1 kg-mole, and 1 lb-mole of $CaCO_3$.

Solution:

a. *MW* of $CaCO_3 = 100$

Number of kg-moles = 100 kg ÷ 100 = <u>1 kg-mole</u>

Number of g-moles (or moles) = 100,000 g ÷ 100 = <u>1,000 g-moles</u>

Number of lb-moles = (100 × 2.2 lb) ÷ 100 = <u>2.2 lb-moles</u>

b. Mass of $CaCO_3$ in 1 g-mole = 1 g-mole × 100 = <u>100 g</u>

Mass of $CaCO_3$ in 1 kg-mole = 1 kg-mole × 100 = <u>100 kg</u>

Mass of $CaCO_3$ in 1 lb-mole = 1 lb-mole × 100 = <u>100 pounds</u>

Discussion:

With *MW* as a dimensionless quantity, the units in the above calculations appear not perfectly matched. However, the bottom line is that the units of mass and mole should be matching (i.e., if the mass is in grams, the moles will be in g-moles).

EXERCISE QUESTIONS

1. Indicate whether each of the following is metal, non-metal, or metalloid: (a) lithium (Li), (b) magnesium (Mg), (c) chromium (Cr), (d) silicon (Si), (e) selenium (Se), (f) phosphorus (P), (g) iodine (I), and (h) neon (Ne).
2. Predict the charge found on the most stable ion formed by (a) sodium (Na), (b) calcium (Ca), and (c) bromine (Br).
3. Complete the following table:

Symbol	Atomic Number	Mass Number	No. of Protons	No. of Electrons	No. of Neutrons	Net Charge
^{222}Rn						
$^{40}Ca^{+2}$	20	40				
	35	80		36		−1
		32	16	18		

4. The following are often referred as USEPA Priority Pollutant Metals: Sb, As, Be, Cd, Cr, Cu, Pb, Hg, Ni, Se, Se, Ag, Tl, and Zn.
   a. Write their full names.
   b. What are their atomic numbers and mass numbers?

c. Which groups of the periodic table are they in?
d. Are they all metals?
e. Which ones are transition metals?
f. Any metalloid?

5. a. Calculate the molecular mass and MW of benzene (C_6H_6) and toluene ($C_6H_5CH_3$).
b. Convert 500 g of benzene and toluene into g-moles, kg-moles, and lb-moles.
c. How many grams, kilograms, and pounds are in 10 lb-moles of benzene and toluene?

2 Organic Chemistry for Environmental Engineers

2.1 INTRODUCTION

2.1.1 Compounds of Concern

One of the main tasks for environmental professionals is to deal with pollution/contamination present in the impacted environmental media (i.e., air, water, and soil). *Contamination* is the presence of a substance at an unwanted location and/or at concentrations above the background level. *Pollution* is contamination that results in, or can result in, adverse impacts on human beings and/or on the environment. A *contaminant* can be considered as a substance present in an unacceptable concentration, at a wrong location, and/or at a wrong time. Everything can be a contaminant (e.g., oxygen is needed for an aerobic environment, but it is unwanted for an anaerobic condition). Environmental engineers need to apply measures to reduce the discharges of compounds of concern (COCs) to the environment as well as to mitigate pollution by destructing/removing contaminants/pollutants from the impacted media. Most, if not all, of the COCs are chemical compounds. One of the goals of this book is to prepare the readers to be more familiar with COCs mentioned in regulations and those they may come across at work. The focus of this chapter is on organic compounds, and inorganic compounds will be discussed in Chapter 3. This section provides a brief coverage on main federal regulations on potable water, ambient water bodies, water discharge, and air quality.

2.1.2 Safe Drinking Water Act

Safe Drinking Water Act (SDWA) was originally passed by the United States Congress in 1974 to protect public health by regulating the nation's public drinking water supply. The law was amended in 1986 and in 1996; the amendments require more actions to protect drinking water and its sources (i.e., rivers, lakes, reservoirs, springs, and groundwater aquifers). The SDWA authorizes the United States Environmental Protection Agency (USEPA) to set national health-based standards for drinking water to protect against both naturally occurring and manmade contaminants that may be found in drinking water. The USEPA, the states, and water systems then work together to make sure that these standards are met.

National Primary Drinking Water Regulations (NPDWR) are legally enforceable primary standards and treatment techniques that apply to public water systems to protect public health by limiting the levels of contaminants in drinking water (USEPA, 2023a). The primary *maximum contaminant levels* (MCLs) include the following [Note: The underlined species are those also on the list of the Clean Water Act's Priority Pollutants– see later]:

DOI: 10.1201/9781003502661-2

a. ***Microorganisms:*** *Cryptosporidium*, *Giardia lambia*, *Legionella*, fecal coliform and *E. coli*, total coliform, *viruses*, heterotrophic plate count (HPC), turbidity (a water quality parameter related to the effectiveness of disinfection)
b. ***Disinfectants:*** chloramines (NH_2Cl and $NHCl_2$), chlorine (Cl_2), and chlorine dioxide (ClO_2)
c. ***Disinfection byproducts:*** bromate (BrO_3^-), chlorite (ClO^-), haloacetic acids (HAA_5), total trihalomethanes (TTHMs) [Note: TTHMs include chloroform ($CHCl_3$), dichlorobromomethane ($CHCl_2Br$), chlorodibromomethane ($CHClBr_2$), and bromoform ($CHBr_3$)].
d. ***Inorganic chemicals:***
   - antimony (Sb), arsenic (As), barium (Ba), beryllium (Be), cadmium (Cd), chromium (Cr), copper (Cu), lead (Pb), mercury (Hg), selenium (Se), and thallium (Tl)
   - cyanide (CN^-), fluoride (F^-)
   - nitrate (NO_3^-), nitrite (NO_2^-)
   - asbestos (fiber >10 μm)
e. ***Organic chemicals***:

   [Note: (i) They are grouped by the author, mainly based on their functional groups, to facilitate later discussion. However, some of them have more than one functional group so that they can be put in groups different from what they are in. (ii) An underline indicates that the compound is also on the list of the Clean Water Act's (CWA) Priority Pollutants. (iii) Some chemical compounds may have more than one name (e.g., a common name, a trade name, and/or another one following the nomenclature of International Union of Pure and Applied Chemistry (IUPAC).]
   - *Simple aromatic hydrocarbons:* benzene (C_6H_6), toluene ($C_6H_5CH_3$), ethylbenzene ($C_6H_5C_2H_5$), total xylenes ($C_6H_4(CH_3)_2$), and styrene (C_8H_8)
   - *Halogenated compounds*:
     - dichloromethane (methylene chloride, CH_2Cl_2), carbon tetrachloride (CCl_4), 1,2-dichloroethane ($C_2H_4Cl_2$), 1,2-dichloropropane ($C_3H_6Cl_2$), 1,1,1-trichloroethane ($C_2H_3Cl_3$), 1,1,2-trichloroethane ($C_2H_3Cl_3$), vinyl chloride (C_2H_3Cl), 1,1-dichloroethylene ($C_2H_2Cl_2$), cis-1,2-dichloroethylene ($C_2H_2Cl_2$), trans-1,2-dichloroethylene ($C_2H_2Cl_2$), tetrachloroethylene (C_2Cl_4), trichloroethylene (C_2HCl_3), and hexachlorocyclopentadiene (C_5Cl_6)
     - chlorobenzene (C_6H_5Cl), o-dichlorobenzene ($C_6H_4Cl_2$), p-dichlorobenzene ($C_6H_4Cl_2$), 1,2,4-trichlorobenzene ($C_6H_3Cl_3$), hexachlorobenzene (C_6Cl_6), and pentachlorophenol (C_6Cl_5OH)
     - ethylene dibromide (dibromomethane; $C_2H_4Br_2$)
   - *Polycyclic aromatic compounds (PAHs):* benzo(a)pyrene ($C_{20}H_{12}$)
   - *Polychlorinated biphenyls (PCBs)*
   - *Dioxins and furans:* 2,3,7,8-TCDD (2,3,7,8-tetrachlorodibenzodioxin; $C_{12}H_4Cl_4O_2$)

- *Other industrial chemicals: acrylamide (C_3H_5NO), di(2-ethylhexyl) adipate (DEHA; $C_{22}H_{42}O_4$),* di(2-ethylhexyl) phthalate (DEHP; $C_{24}H_{38}O_4$), and epichlorohydrin (C_3H_5ClO)
- *Pesticides:*
  - Alachlor (n-(2,6-diethylphenyl)acetamide; $C_{14}H_{20}ClNO_2$)
  - Atrazine (1-chloro-3-ethylamino-5-isopropylamino-2,4,6-triazine; $C_8H_{14}ClN_5$)
  - Carbofuran ($C_{12}H_{15}NO_3$)
  - Chlordane ($C_{10}H_6Cl_8$)
  - 2,4-D (2,4-dichlorophenoxyacetic acid; $C_8H_6Cl_2O_3$)
  - Dalapon (2,2-dichloropropionic acid; $C_3H_4Cl_2O_2$)
  - DBCP (1,2-dibromo-3-chloropropane; $C_3H_5Br_2Cl$)
  - Dinoseb (6-sec-butyl-2,4-dinitrophenol; $C_{10}H_{12}N_2O_5$)
  - Diquat ($C_{12}H_{12}Br_2N_2$)
  - Endothall ($C_8H_{10}O_5$)
  - Endrin ($C_{12}H_8Cl_6O$)
  - Glyphosate ($C_3H_8NO_5P$)
  - Heptachlor ($C_{10}H_5C_{17}$)
  - Heptachlor epoxide ($C_{10}H_5C_{17}O$)
  - Lindane (gamma-hexachlorocyclohexane; gamma-BHC; $C_6H_6Cl_6$)
  - Methoxychlor ($C_{16}H_{15}Cl_3O_2$)
  - Oxamyl (Vydate; $C_7H_{13}N_3O_3S$)
  - Picloram ($C_6H_3Cl_3N_2O_2$)
  - Simazine ($C_7H_{12}ClN_5$)
  - Toxaphene ($C_{10}H_8Cl_8$)
  - 2,4,5-TP (Silvex; 2-(2,4,5-trichlorophenoxy)propionic acid; $C_9H_7Cl_3O_3$)

f. ***Radionuclides:*** alpha/photon emitters, beta photon emitters, radium-226 (Ra-226) and radium 228 (Ra-228), and uranium (U).

2.1.3 Clean Water Act

CWA establishes the basic structure to regulate discharges of pollutants into the waters of the United States and develops quality standards for surface waters. It is to restore and maintain the chemical, physical, and biological integrity of the nation's water, whether on public or private land. The USEPA has also developed criteria for surface water quality that reflect the latest scientific knowledge on the impacts of pollutants on human health and the environment. USEPA's Effluent Guidelines are national regulatory standards for wastewater discharged to surface waters and to municipal sewage treatment plants. The USEPA issues these regulations for industrial categories, based on the performance of treatment and control technologies (USEPA, 2023b). USEPA's National Pollutant Discharge Elimination System (NPDES) permitting program controls discharges of any pollutants from point sources as well as from stormwater runoff (USEPA, 2023c).

USEPA's *Priority Pollutants* are a set of chemical pollutants that they regulate, and for which they have developed analytical test methods. The current list of 126 Priority Pollutants can be found in 40 CFR Part 423, Appendix A. These are not the only pollutants regulated by the CWA programs; however, the list is an important starting point for the USEPA to consider, for example, in developing national discharge standards (such as Effluent Guidelines) or in national permitting programs (such as NPDES) (USEPA, 2023c).

It should be noted that some emerging contaminants, such as per- and polyfluoroalkyl substances (PFAS), have not been added to the list yet. These 126 Priority Pollutants are tabulated below [Note: They are grouped by the author to facilitate later discussion]:

a. ***Disinfection byproducts:*** chloroform ($CHCl_3$), dichlorobromomethane ($CHCl_2Br$), chlorodibromomethane ($CHClBr_2$), and bromoform ($CHBr_3$)
b. ***Inorganic chemicals:***
   - antimony (Sb), arsenic (As), beryllium (Be), cadmium (Cd), chromium (Cr), copper (Cu), lead (Pb), mercury (Hg), nickel (Ni), selenium (Se), silver (Ag), thallium (Tl), and zinc (Zn)
   - cyanide (CN^-), total
   - asbestos
c. ***Organic chemicals:***

   [Note: (i) They are grouped by the author, mainly based on their functional groups, to facilitate later discussion. However, some of them have more than one functional group so that they can be put in groups different from what they are in. (ii) An underline indicates that the compound is also on the list of the SDWA's primary MCLs. (iii) Some chemical compounds may have more than one name (e.g., a common name, a trade name, and/or another one following the nomenclature of IUPAC.]
   - *Simple aromatic hydrocarbons*: benzene (C_6H_6), toluene ($C_6H_5CH_3$), and ethylbenzene ($C_6H_5C_2H_5$)
   - *Phenolic compounds*: phenol (C_6H_5OH), 2-chlorophenol (C_6H_4ClOH), 2,4-dichlorophenol ($C_6H_3Cl_2OH$), 2,4-dimethylphenol (2,4-xylenol, $C_8H_{10}O$), pentachlorophenol (C_6Cl_5OH), parachlorometa cresol (p-chloro-m-cresol; C_7H_7ClO), 4,6-dinitro-o-cresol ($C_7H_6N_2O_5$)
   - *Halogenated compounds:*
     - methyl chloride (CH_3Cl), methyl bromide (CH_3Br), dichloromethane (methylene chloride, CH_2Cl_2), carbon tetrachloride (CCl_4), chloroethane (C_2H_5Cl), 1,1-dichloroethane ($C_2H_4Cl_2$), 1,2-dichloroethane ($C_2H_4Cl_2$), 1,2-dichloropropane ($C_3H_6Cl_2$), 1,1,1-trichloroethane ($C_2H_3Cl_3$), 1,1,2-trichloroethane ($C_2H_3Cl_3$), 1,1,2,2-tetrachloroethane ($C_2H_2Cl_4$), hexachloroethane (C_2Cl_6), vinyl chloride (C_2H_3Cl), 1,1-dichloroethylene ($C_2H_2Cl_2$), trans-1,2-dichloroethylene ($C_2H_2Cl_2$), trichloroethylene (C_2HCl_3), tetrachloroethylene (C_2Cl_4), 1,3-dichloropropylene ($C_3H_4Cl_2$), hexachlorobutadiene (C_4Cl_6), and hexachlorocyclopentadiene (C_5Cl_6)
     - chlorobenzene (C_6H_5Cl), o-dichlorobenzene ($C_6H_4Cl_2$), m-dichlorobenzene ($C_6H_4Cl_2$), p-dichlorobenzene ($C_6H_4Cl_2$), 1,2,4-trichlorobenzene ($C_6H_3Cl_3$), hexachlorobenzene (C_6Cl_6)
     - bis(2-chloroethoxy) methane ($C_5H_{10}Cl_2O_2$)
   - *Alcohols:* (none)

- *Aldehydes and ketones*: acrolein (C_3H_4O), isophorone ($C_9H_{14}O$)
- *Carboxylic acids, anhydrides, esters, acetates, acrylates, and phthalates:*
  - dimethyl phthalate ($C_{10}H_{10}O_4$), diethyl phthalate ($C_{12}H_{14}O_4$), di-n-butyl phthalate ($C_{16}H_{20}O_4$), butyl benzyl phthalate ($C_{19}H_{20}O_4$), bis(2-ethylhexyl) phthalate (DEHP; $C_{24}H_{38}O_4$), and di-n-octyl phthalate ($C_{24}H_{38}O_4$)
- *Ethers and epoxides:* bis(2-chloroethyl) ether ($C_4H_8Cl_2O$), bis(2-chloroisopropyl) ether ($C_6H_{12}Cl_2O$), 2-chloroethyl vinyl ether (C_4H_7ClO), 4-chlorophenyl phenyl ether ($C_{12}H_9ClO$), and 4-bromophenyl phenyl ether ($C_{12}H_9BrO$)
- *Amines, amides, imines, and imides:* benzidine ($C_{12}H_{12}N_2$), 3,3-dichlorobenzidine ($C_{12}H_{10}Cl_2N_2$), N-nitrosodimethylamine (NDMA, $C_2H_6N_2O$), N-nitrosodi-n-propylamine ($C_6H_{14}N_2O$), and N-nitrosodiphenylamine ($C_{12}H_{10}N_2O$)
- *Nitriles, isocyanates, hydrazines, and pyridines:* acrylonitrile (C_3H_3N), 1,2-diphenylhydrazine ($C_{12}H_{12}N_2$)
- *Thiols, sulfides, and sulfonic acids:* (none)
- *Nitro and nitroso compounds:* nitrobenzene ($C_6H_5NO_2$), 2,4-dinitrotoluene ($C_7H_6N_2O_4$), 2,6-dinitrotoluene ($C_7H_6N_2O_4$), 2,4,6-trichlorophenol ($C_6H_2Cl_3OH$), 2-nitrophenol ($C_6H_5NO_3$), 4-nitrophenol ($C_6H_5NO_3$), and 2,4-dinitrophenol ($C_6H_4N_2O_5$),
- *PAHs:* naphthalene ($C_{10}H_8$), 2-chloronaphthalene ($C_{10}H_7Cl$), acenaphthylene ($C_{12}H_8$), acenaphthene ($C_{12}H_{10}$), fluorene ($C_{13}H_{10}$), anthracene ($C_{14}H_{10}$), phenanthrene ($C_{14}H_{10}$), fluoranthene ($C_{16}H_{10}$), pyrene ($C_{16}H_{10}$), benzo(a)anthracene ($C_{18}H_{12}$), chrysene ($C_{18}H_{12}$), benzo(a)pyrene ($C_{20}H_{12}$), benzo(b)fluoranthene ($C_{20}H_{12}$), benzo(k)fluoranthene ($C_{20}H_{12}$), benzo(ghi) perylene ($C_{22}H_{12}$), and dibenzo(a,h)anthracene ($C_{22}H_{14}$)
- *PCBs:* Arochlor 1016, Arochlor 1221, Arochlor 1232, Arochlor 1242, Arochlor 1248, Arochlor 1254, and Arochlor 1260
- *Dioxin and furans:* 2,3,7,8-TCDD (2,3,7,8-tetrachlorodibenzodioxin; $C_{12}H_4Cl_4O_2$)
- *Pesticides*:
  - Aldrin ($C_{12}H_8Cl_6$)
  - Alpha- and beta- endosulfan ($C_9H_6Cl_6O_3S$) and endosulfan sulfate ($C_9H_6Cl_6O_4S$)
  - Chlordane ($C_{10}H_6Cl_8$)
  - 4,4-DDD (dichlorodiphenyldichloroethane; $C_{14}H_{10}Cl_4$)
  - 4,4-DDE (dichlorodiphenyldichloroethylene; $C_{14}H_8Cl_4$)
  - 4,4-DDT (dichlorodiphenyltrichloroethane; $C_{14}H_9Cl_5$)
  - Dieldrin ($C_{12}H_8Cl_6O$)
  - Endrin ($C_{12}H_8Cl_6O$)
  - Endrin aldehyde ($C_{12}H_8Cl_6O$)
  - Heptachlor ($C_{10}H_5C_{17}$)
  - Heptachlor epoxide ($C_{10}H_5C_{17}O$)
  - Lindane (gamma-BHC; $C_6H_6Cl_6$), alpha-BHC, beta-BHC, and delta-BHC
  - Toxaphene ($C_{10}H_8Cl_8$)

2.1.4 Clean Air Act

Clean Air Act (CAA) is a comprehensive federal law that regulates air emissions from stationary and mobile sources. The CAA authorizes the USEPA to establish National Ambient Air Quality Standards (NAAQS) to protect public health and public welfare and to regulate emissions of hazardous air pollutants (USEPA, 2022).

The CAA, which was last amended in 1990, requires the USEPA to set NAAQS for six principal air pollutants (often referred to as the "criteria air pollutants" which can be harmful to public health and the environment. These six *criteria air pollutants* are carbon monoxide (CO), lead (Pb), nitrogen dioxide (NO_2), ozone (O_3), particulate matter ($PM_{2.5}$ & PM_{10}), and sulfur dioxide (SO_2). The standards are reviewed periodically and sometimes may be revised (USEPA, 2023d).

The CAA also requires the USEPA to regulate toxic air pollutants (also known as "*air toxics*" or "*hazardous air pollutants* (HAPs)) from large industrial facilities in two phases. The first phase is "technology-based", where the USEPA develops maximum achievable control technology (MACT) standards for controlling the emissions of HAPs from sources in an industry group (or "*source category*"). The MACTs are based on emissions levels that are already being achieved by the controlled and low-emitting sources in that industry. The second phase is a "risk-based" approach (i.e., residual risk) in which the USEPA determines whether more health-protective standards are necessary. Every 8 years, the USEPA would assess the remaining health risks from each source category to determine whether the standards are adequate to protect public health against adverse environmental effects (USEPA, 2023e). The original list of HAPs included 189 pollutants, and it has been modified to 188 HAPs. These 188 HAPs are tabulated below (they are grouped by the author to facilitate later discussion):

a. ***Disinfection byproducts:*** chloroform ($CHCl_3$), bromoform ($CHBr_3$)
b. ***Inorganic chemicals:***
   - antimony (Sb) compounds, arsenic (As) compounds, beryllium (Be) compounds, cadmium (Cd) compounds, chromium (Cr) compounds, cobalt (Co) compounds, lead (Pb) compounds, manganese (Mn) compounds, mercury (Hg) compounds, nickel (Ni) compounds, and selenium (Se) compounds
   - calcium cyanamide ($CaCN_2$), cyanide (CN^-) compounds
   - chlorine (Cl_2)
   - carbon disulfide (CS_2), carbonyl sulfide (OCS), and hydrogen sulfide (H_2S)
   - hydrazine (N_2H_4)
   - phosphine (PH_3), phosphorus (P)
   - hydrochloric acid (HCl), hydrofluoric acid (HF)
   - titanium tetrachloride ($TiCl_4$)
   - asbestos, fine mineral fibers, and coke oven emissions

c. ***Organic chemicals***

   [Note: (i) They are grouped by the author, mainly based on their functional groups, to facilitate later discussion. However, some of them have more than one functional group so that they can be put in groups different from what they are in. (ii) Some chemical compounds may have more than one name (e.g., a common name, a trade name, and/or another one following the nomenclature of IUPAC.]

- *Simple alkanes and alkenes:* 1,3-butadiene (C_4H_6), hexane (C_6H_{14}), 2,2,4-trimethylpentane (C_8H_{18})
- *Simple aromatic hydrocarbons*: benzene (C_6H_6), toluene ($C_6H_5CH_3$), ethylbenzene ($C_6H_5C_2H_5$), xylenes ($C_6H_3(CH_3)_2$), styrene ($C_6H_5CHCH_2$), and cumene (isopropylbenzene; $C_6H_5C_3H_7$)
- *Phenolic compounds*: phenol (C_6H_5OH), o-, m-, p-cresols ($C_6H_4CH_3OH$), hydroquinone ($C_6H_4(OH)_2$), 2,4,5-trichlorophenol ($C_6H_2Cl_3OH$), 2,4,6-trichlorophenol ($C_6H_2Cl_3OH$), 4-nitrophenol ($C_6H_5NO_3$), 2,4-dinitrophenol ($C_6H_4N_2O_5$), and pentachlorophenol (C_6Cl_5OH)
- *Halogenated compounds:*
  - Halogenated alkanes: methyl chloride (CH_3Cl), methyl bromide (CH_3Br), methyl iodide (CH_3I), dichloromethane (methylene chloride, CH_2Cl_2), ethylene dibromide (dibromoethane, $C_2H_4Br_2$), carbon tetrachloride (CCl_4), ethyl chloride (chloroethane, C_2H_5Cl), 1,1-dichloroethane ($C_2H_4Cl_2$), 1,2-dichloroethane (ethylene dichloride, $C_2H_4Cl_2$), 1,1,1-trichloroethane ($C_2H_3Cl_3$), 1,1,2-trichloroethane ($C_2H_3Cl_3$), 1,1,2,2-tetrachloroethane ($C_2H_2Cl_4$), hexachloroethane (C_2Cl_6), 1-bromopropane (C_3H_7Br), 1,2-dichloropropane ($C_3H_6Cl_2$), 1,2-dibromo-3-chloropropane ($C_3H_5Br_2Cl$), and epichlorohydrin (l-chloro-2,3-epoxypropane, C_3H_5ClO)
  - Halogenated alkenes: vinyl bromide (C_2H_3Br), vinyl chloride (C_2H_3Cl), allyl chloride (C_3H_5Cl), 1,1-dichloroethylene ($C_2H_2Cl_2$), trichloroethylene (C_2HCl_3), tetrachloroethylene (C_2Cl_4), 1,3-dichloropropylene ($C_3H_4Cl_2$), chloroprene (2-chlorobuta-1,3-diene, C_4H_5Cl), hexachlorobutadiene (C_4Cl_6), and hexachlorocyclopentadiene (C_5Cl_6)
  - Halogenated arenes: chlorobenzene (C_6H_5Cl), p-dichlorobenzene ($C_6H_4Cl_2$), 1,2,4-trichlorobenzene ($C_6H_3Cl_3$), hexachlorobenzene (C_6Cl_6), benzyl chloride ($C_6H_5CH_2Cl$), and benzotrichloride ($C_6H_5CCl_3$)
- *Alcohols:* catechol ($C_6H_4(OH)_2$), cresols/cresylic acid (isomers and mixture), 4,6-dinitro-o-cresol ($C_7H_6N_2O_5$) & salts, methanol (CH_3OH), and ethylene glycol $(CH_2OH)_2$
- *Aldehydes and ketones*:
  - formaldehyde (HCHO), acetaldehyde (CH_3CHO), propionaldehyde (C_2H_5CHO), and acrolein (C_3H_4O)
  - isophorone ($C_9H_{14}O$), methyl isobutyl ketone ($C_6H_{12}O$), beta-propiolactone ($C_3H_4O_2$), quinone, acetophenone ($C_6H_5COCH_3$), and 2-chloroacetophenone ($C_6H_5COCH_2Cl$)
- *Carboxylic acids, anhydrides, esters, acetates, acrylates, and phthalates:*
  - acrylic acid (C_2H_3COOH), chloroacetic acid ($ClCH_2COOH$)
  - vinyl acetate ($C_4H_6O_2$)

- ethyl acrylate ($C_5H_8O_2$), methyl methacrylate ($C_5H_8O_2$)
- dimethyl phthalate ($C_{10}H_{10}O_4$), di-n-butyl phthalate ($C_{16}H_{20}O_4$), and bis(2-ethylhexyl) phthalate (DEHP; $C_{24}H_{38}O_4$)
- phthalic anhydride ($C_8H_4O_3$), maleic anhydride ($C_4H_2O_3$)

• *Ethers and epoxides:*
  - bis(chloromethyl)ether ($C_2H_4Cl_2O$), chloromethyl methyl ether (C_2H_5ClO), bis(2-chloroethyl) ether ($C_4H_8Cl_2O$), methyl tert-butyl ether ($C_5H_{12}O$), and glycol ethers
  - ethylene oxide (C_2H_4O), propylene oxide (C_3H_6O), 1,2-epoxybutane (C_4H_8O), and styrene oxide ($C_6H_5CHCH_2O$)

• *Amines, amide, urea, azo compounds, imines, and imides:*
  - triethylamine ($(C_2H_5)_3N$), diethanolamine ($(CH_2CH_2OH)_2NH$), N-nitrosodimethylamine (NDMA, $C_2H_6N_2O$), p-phenylenediamine ($C_6H_4(NH_2)_2$), 1,2-propyleneimine ($CH_3CH(NH)CH_2$), 2,4-toluene diamine ($C_6H_3(NH_2)_2CH_3$), and ethyleneimine (aziridine; $(CH_2)_2NH$)
  - aniline ($C_6H_5NH_2$), N,N-dimethylaniline ($C_6H_5N(CH_3)_2$, o-anisidine (o-methoxyaniline, $CH_3OC_6H_4NH_2$), 4,4′-methylenedianiline ($CH_2(C_6H_4NH_2)_2$), 4,4′-methylenebis(2-chloroaniline) ($C_{13}H_{12}Cl_2N_2$), and o-toluidine ($C_6H_4CH_3NH_2$)
  - benzidine ($C_{12}H_{12}N_2$), 3,3'-dimethyl benzidine ($C_{14}H_{16}N_2$), 3,3′-dimethoxybenzidine ($C_{14}H_{16}N_2O_2$), and 3,3′-dichlorobenzidine ($C_{12}H_{10}Cl_2N_2$)
  - acetamide (CH_3CONH_2), acrylamide (C_3H_5NO), dimethyl formamide (C_3H_7NO), and hexamethylphosphoramide ($C_6H_{18}N_3OP$)
  - N-nitroso-N-methylurea ($C_2H_5N_3O_2$), ethylene thiourea
  - diazomethane (CH_2N_2), 4-(dimethylamino)azobenzene (methyl yellow; $C_6H_5N_2C_6H_4N(CH_3)_2$)

• *Nitriles, isocyanates, hydrazine, and pyridines:*
  - acetonitrile (CH_3N), acrylonitrile (C_3H_3N)
  - hexamethylene-1,6-diisocyanate ($C_8H_{12}N_2O_2$), methyl isocyanate (C_2H_3NO), methylene diphenyl diisocyanate (MDI, $C_{15}H_{10}N_2O_2$), and 2,4-toluene diisocyanate ($C_9H_6N_2O_2$)
  - methyl hydrazine (CH_3NHNH_2), 1,2-diphenylhydrazine ($C_{12}H_{12}N_2$)
  - quinoline (C_9H_7N)

• *Nitro and nitroso compounds:*
  - nitrobenzene ($C_6H_5NO_2$), 2,4-dinitrotoluene ($C_7H_6N_2O_4$), 2-nitropropane ($C_3H_7NO_2$), and pentachloronitrobenzene (quintobenzene, $C_6Cl_5NO_2$)
  - N-nitrosomorpholine ($C_4H_8N_2O_2$)

• *Thiols, sulfides, sulfates, and sulfonic acids:*
  - diethyl sulfate $(C_2H_5)_2SO_4$, dimethyl sulfate $(CH_3)_2SO_4$
  - 1,3-propane sultone ($C_3H_6O_3S$)

• *PAHs:* 2-acetylaminofluorene ($C_{15}H_{13}NO$), naphthalene ($C_{10}H_8$), polycyclic organic matter (including organic compounds with more than one benzene ring, which have a boiling point greater than or equal to 100°C).

- *Biphenyls and polychlorinated biphenyls (PCBs):* 4-aminobiphenyl ($C_{12}H_{11}N$), biphenyl ($C_{12}H_{10}$), 4-nitrobiphenyl ($C_{12}H_9NO_2$), and PCBs (Arochlors)
- *Dioxin and furans:* dibenzofurans ($C_{12}H_8O$), 2,3,7,8-TCDD (2,3,7,8-tetrachlorodibenzodioxin, $C_{12}H_4Cl_4O_2$)
- *Other industrial chemicals*: phosgene ($COCl_2$), 1,4-dioxane (1,4-diethyleneoxide, $C_4H_8O_2$), ethyl carbamate (urethane, $C_3H_7NO_2$), and dimethyl carbamoyl chloride (C_3H_6ClNO)
- *Pesticides*:
  - Captan ($C_9H_8Cl_3NO_2S$)
  - Carbaryl ($C_{12}H_{11}NO_2$)
  - Chloramben ($C_7H_5Cl_2NO_2$)
  - Chlordane ($C_{10}H_6Cl_8$)
  - Chlorobenzilate ($C_{16}H_{14}Cl_2O_3$)
  - 2,4-D (2,4-dichlorophenoxyacetic acid; $C_8H_6Cl_2O_3$), salts, and esters
  - DDE (dichlorodiphenyldichloroethylene, $C_{14}H_8Cl_4$)
  - Dichlorvos ($C_4H_7Cl_2O_4P$)
  - Heptachlor ($C_{10}H_5C_{17}$)
  - Lindane (gamma-hexachlorocyclohexane; gamma-BHC; $C_6H_6Cl_6$), alpha-BHC, beta-BHC, delta-BHC
  - Methoxychlor ($C_{16}H_{15}Cl_3O_2$)
  - Parathion ($C_{10}H_{14}NO_5PS$)
  - Propoxur (Baygon, $C_{11}H_{15}NO_3$)
  - Toxaphene ($C_{10}H_8Cl_8$)
  - Trifluralin ($C_{13}H_{16}F_3N_3O_4$)

d. ***Radionuclides***: Radionuclides (including radon (Rn))

2.2 INTRODUCTION TO ORGANIC CHEMISTRY

According to American Chemical Society (ACS), organic chemistry studies structure, properties, composition, reaction, and preparation of carbon-containing compounds. Most organic compounds contain carbon and hydrogen, but they may also include any number of other elements such as oxygen, nitrogen, sulfur, phosphorus, silicon, and halogens. Many organic compounds are being used in our lives. Some are naturally occurring compounds, while many synthetic compounds have also been made. It should be noted that some carbon compounds are not considered as organics such as carbon oxides (i.e., carbon monoxide (CO) and carbon dioxide (CO_2)), bicarbonate (HCO_3^-), carbonate ($CO_3^=$), cyanide (CN^-), cyanogen chloride (CNCl), cyanate (OCN^-), thiocyanate (SCN^-), carbide (i.e., a compound composed of carbon and a metal), and graphite (i.e., a crystalline form of the elemental carbon).

Carbon can form four covalent bonds, with each other and/or with other atoms, to form a wide variety of chemical compounds from simple to complex. Methane (CH_4) can be considered as the simplest organic compound, which contains one carbon atom; while a DNA molecule can contain carbon atoms in millions. Although there are numerous organic compounds, they are often classified into groups that have similar characteristics/properties.

General properties of organic compounds include the following:

- Most are insoluble (or slightly soluble) in water, but soluble in organic solvents.
- Most are flammable/combustible (i.e., having a large energy content).
- Most have relatively low melting points and low boiling points.
- Bonds are covalent bonds (versus ionic bonding for inorganics).
- Aqueous solutions are not conductive (i.e., when dissolved, not dissociated into ions).
- Their functional groups often decide the properties that they possess.

Several adjectives are commonly used in describing organic compounds, depending on their origins and degradability. They are:

- *Anthropogenic*: resulted from or related to human activities.
- *Biogenic*: produced or brought about by living organisms.
- *Synthetic*: made from chemical synthesis, especially to imitate a natural product.
- *Xenobiotic*: a synthetic compound that is foreign to an ecological system (i.e., organisms, plants, or human beings) and/or the environment.
- *Degradable*: capable of being chemically degraded/decomposed. The opposite of degradable is "*non-degradable*".
- *Biodegradable*: can be utilized/degraded by microorganisms within a reasonable length of time. The opposite of biodegradable is "*non-biodegradable*".
- *Incalcitrant/recalcitrant*: those are not biodegradable or are only biodegradable under certain conditions.
- *Persistent*: resistant (or slow to break down) to degradation through chemical, biological, and photolytic processes.

2.3 HYDROCARBONS

Hydrocarbons are organic compounds composed of carbon and hydrogen only. This section presents four subgroups of hydrocarbons: (i) saturated, (ii) cyclic, (iii) unsaturated, and (iv) aromatic hydrocarbons.

2.3.1 Saturated Hydrocarbons

Alkanes are the sub-group of hydrocarbons containing no multiple bonds (i.e., no double or triple bond); thus, they are also called *saturated* hydrocarbons. Names of alkanes end with "-ane".

Alkanes can be further divided into three subgroups: (i) straight-chain alkanes, (ii) branched alkanes, and (iii) cycloalkanes. The *straight-chain alkanes* (or called *normal alkanes*) start with methane (CH_4), followed by ethane (C_2H_6), propane (C_3H_8), butane (C_4H_{10}), pentane (C_5H_{12}), hexane (C_6H_{14}), heptane (C_7H_{16}), octane (C_8H_{18}), nonane (C_9H_{20}), decane ($C_{10}H_{22}$), etc. Normal alkanes have a general chemical formula of C_nH_{n+2}. These compounds are homologous. The prefix "*n*-" is sometimes used when all carbons in an organic compound form a continuous linear (unbranched) chain.

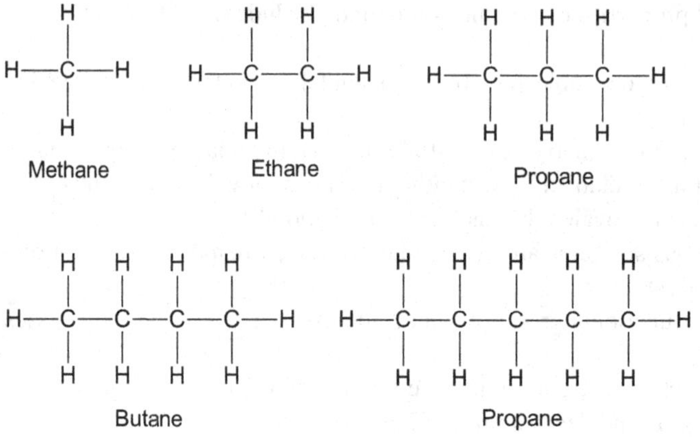

FIGURE 2.1 Chemical structural formula of five straight-chain alkanes.

A *homologous series* is a series of compounds with the same functional group and similar chemical properties (e.g., the normal alkanes). A *functional group* is a specific group of atoms that is part of a molecule, and it gives the molecule distinct properties and reactivity. The single C–C covalent bond is considered as a functional group by many. Figure 2.1 illustrates the chemical structural formula of five simplest straight-chain alkanes. It should be noted that they are in 3-D configurations, not planar as shown.

A *branched-chain* (or *branched*) *alkane* is an alkane that contains branches (i.e., alkyl groups). The smallest and common *alkyl groups* are methyl ($-CH_3$), ethyl ($-C_2H_5$), propyl ($-C_3H_7$), and butyl ($-C_4H_9$), with a general chemical formula of $-C_nH_{2n+1}$. There are millions of known organic compounds, many of them have several different names (e.g., common names and trade names) and often cause confusion. Consequently, a more consistent nomenclature is important and some of the basic guidelines are shown below (and more later) [Note: IUPAC has set up guidelines for chemical nomenclature and terminology, including naming of new elements in the periodic table].

- Find the longest carbon chain in the molecule, and it gives the base of its name.
- Indicate the position of the alkyl group and the functional group (i.e., the carbon number in the main chain that it is attached to); the names of the groups are given in alphabetical order.
- If there is more than one group of the same type, the group name is prefixed with a counting prefix (e.g., di-, tri-, tetra-, penta-, and hexa-) and preceded by the location numbers of the group.

Figure 2.2 illustrates the chemical structural formula of four branched alkanes: 2-methyl pentane, 2-methyl hexane, 2,3-dimethyl hexane, and 3-ethyl-2-methyl hexane.

Hexane (C_6H_{14}) and 2,2,4-trimethylpentane (C_8H_{18}) are on the list of HAPs.

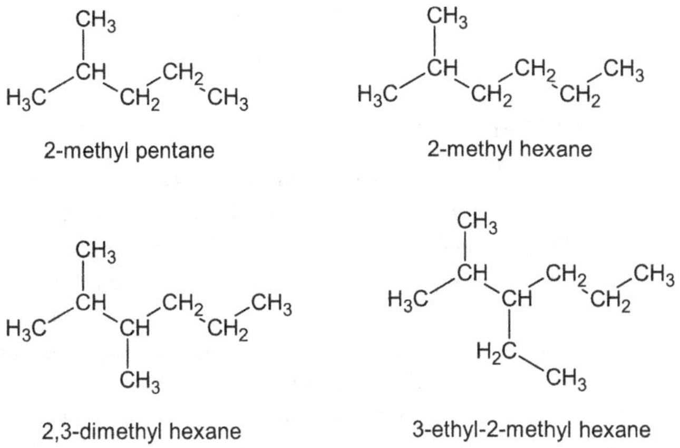

FIGURE 2.2 Chemical structural formula and names of four branched alkanes.

2.3.2 Cyclic Hydrocarbons

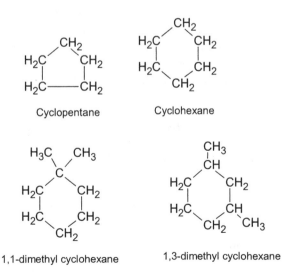

A *cycloalkane* still consists only of carbon and hydrogen atoms, but its chemical structure contains a single ring and all the C–C bonds are single. Cycloalkanes can have branches. Figure 2.3 illustrates the chemical structural formula of cyclopentane (C_5H_{10}), cyclohexane (C_6H_{12}), 1,1-dimethyl cyclohexane (C_8H_{16}), and 1,3-dimethyl cyclohexane (C_8H_{16}). These cycloalkanes have a general chemical formula of C_nH_{2n}, different from that of straight-chain alkanes (i.e., C_nH_{2n+2}). It should also be noted that 1,1-dimethyl cyclohexane and 1,3-dimethyl cyclohexane have the same chemical formula of C_8H_{16}, but different chemical structures. They are isomers. Chemical

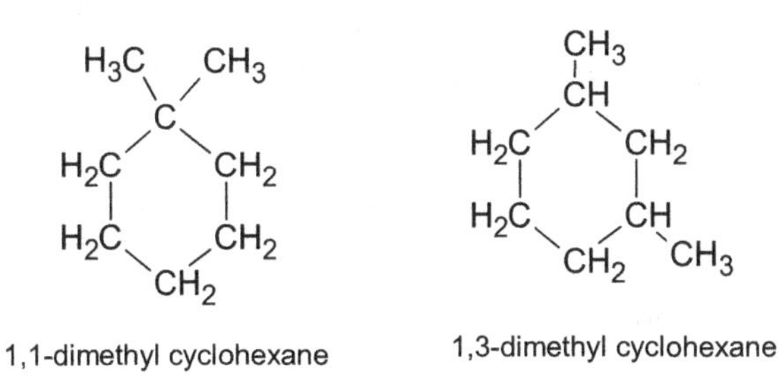

FIGURE 2.3 Chemical structural formula and names of four cycloalkanes.

compounds that have identical chemical formula but differ in arrangement of atoms (and in properties) are called *isomers*.

2.3.3 Unsaturated Hydrocarbons

An *unsaturated* compound is a chemical compound that contains carbon–carbon double bond(s) (i.e., C=C) or triple bond(s) (i.e., C≡C). The term "unsaturated" means that hydrogen atoms could be added to the hydrocarbon molecule to make it saturated (i.e., consisting of single C–C bonds only). The configuration of an unsaturated hydrocarbon can be straight chain, branched, or aromatic compounds (see next sub-section). Except for aromatic compounds, unsaturated hydrocarbons are relatively more reactive due to their multiple carbon–carbon bonds.

Alkenes are a class of "unsaturated" hydrocarbons that contain carbon-carbon double bond(s). Straight-chain alkenes have an empirical formula of C_nH_{2n}. Names of alkenes end with "-ene"; and the simplest alkene is ethene (C_2H_4), which also has a common name of ethylene. The homologous series of n-alkenes are ethene, propene or propylene (C_3H_6), butene or butylene (C_4H_8), pentene or pentylene (C_5H_{10}), etc.

Alkynes are another class of unsaturated hydrocarbon with carbon-carbon triple bond(s). They have an empirical formula of C_nH_{2n-2}. Names of alkynes end with a suffix of "-yne"; and the simplest alkyne is ethyne (C_2H_2), which also has a common name of acetylene. The first four in the homologous series of n-alkynes are ethyne, propyne (C_3H_4), butyne (C_4H_6), and pentyne (C_5H_8).

Figure 2.4 illustrates the chemical structural formula and names of four unsaturated hydrocarbons: ethene or ethylene, 2-heptene (C_7H_{14}), ethyne or acetylene, and 3-heptyne (C_7H_{12}). Ethene is the simplest alkene and ethyne is the simplest alkyne.

Vinyl group. In organic chemistry, a *vinyl group* is a functional group with the formula of "$H_2C{=}CH{-}$". It is essentially an ethylene molecule ($H_2C{=}CH_2$) with one fewer hydrogen atom. Vinyl chloride ($H_2C{=}CHCl$) is very toxic, and it is on the lists of USEPA's MCL, Priority Pollutants, and HAPs. Vinyl chloride is used primarily to make polyvinyl chloride (PVC), one of the commonly used plastics in our daily lives; and it can also be a product from anaerobic biodegradation of common chlorinated solvents.

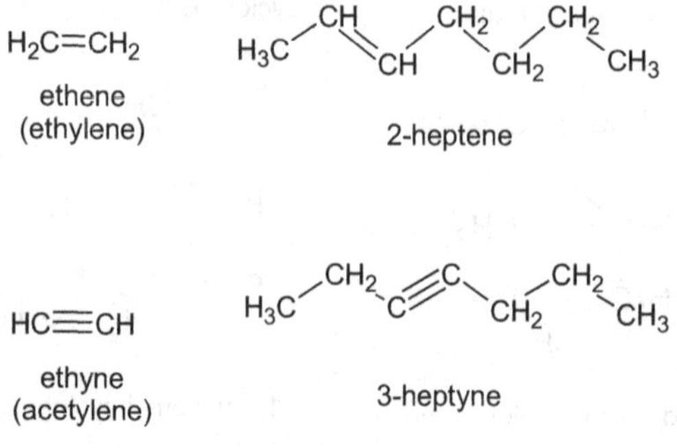

FIGURE 2.4 Chemical structural formula and names of four unsaturated hydrocarbons.

Allyl group. An *allyl group* is similar to a vinyl group, but with a small variation. Both groups own a double bond between two carbon atoms. An allyl group is a functional group with a chemical formula of "$H_2C{=}CH{-}CH_2{-}$". It is essentially the propylene molecule ($H_2C{=}C_2H_4$) with one fewer hydrogen atom from the third carbon atom. Allyl chloride ($H_2C{=}C_2H_3Cl$) is on the list of HAPs.

Dienes. *Diene* is an unsaturated organic compound containing two double bonds between carbon atoms; dienes are used in industries as monomers for the synthesis of polymer. A good example of dienes is 1,3-butadiene ($CH_2{=}CH{-}CH{=}CH_2$), which is on the list of HAPs.

2.3.4 Aromatic Hydrocarbons

In organic chemistry, hydrocarbons can also be classified into two groups: aliphatic compounds and aromatic compounds. *Aromatic compounds* (sometimes called *arenes* or *aryl hydrocarbons*) are hydrocarbons containing one or more aromatic rings, whereas *aliphatic compounds* are hydrocarbons connected by single, double, or triple carbon–carbon bonds to form non-aromatic structures (in other words, they are non-aromatic hydrocarbons).

Benzene (C_6H_6) is the simplest aromatic hydrocarbon and the six carbon atoms connected as a hexagon (see Figure 2.5). As mentioned, a carbon atom can form four covalent bonds (i.e., having four valence electrons to share). For each of the six carbon atoms in benzene, one of the four valence electrons shares with a hydrogen atom, each of the next two shares with each of the two neighboring carbon atoms, and the fourth one shares with one of these two neighboring carbon atoms (often shown as a double bond). Figure 2.5 also illustrates two other representations of a *benzene ring*. Although double bonds appear in the representation of the benzene ring, the benzene ring is relatively stable and does not break into smaller compounds readily.

Crude oil and coal contain many compounds having benzene rings. There are several groups of aromatic compounds that are especially of environmental professionals' concern (see below).

Benzene/toluene/ethylbenzene/xylenes (B/T/E/X). Benzene (C_6H_6), toluene (methyl benzene, $C_6H_5CH_3$), ethylbenzene ($C_6H_5C_2H_5$), and xylenes (dimethyl benzene, $C_6H_4(CH_3)_2$) occur naturally in crude oil and also

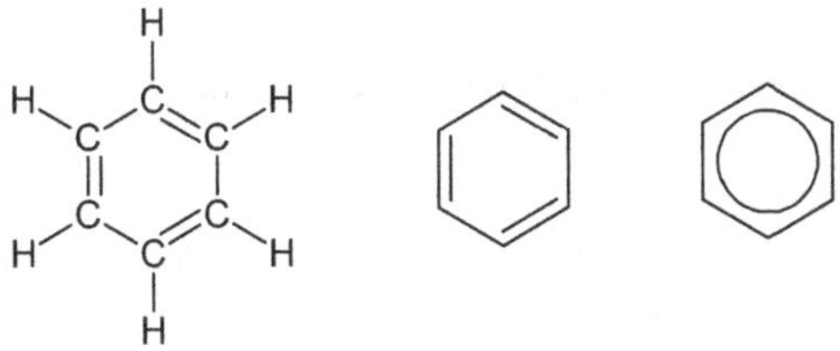

FIGURE 2.5 Chemical structural formula of benzene and two common representations of the benzene ring.

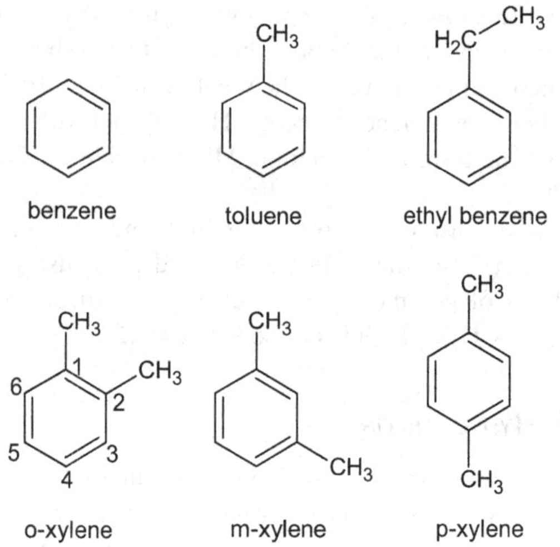

FIGURE 2.6 Chemical structural formula of benzene, toluene, ethyl benzene, and xylenes.

commonly present in gasoline. They are widely used in many industrial applications, as starting materials for a wide range of consumer products. Consequently, they are commonly found in water bodies and in ambient air as pollutants. Figure 2.6 illustrates the chemical structural formula of B/T/E/X. As shown, xylene has three isomers: 1,2-dimethyl benzene (ortho xylene; o-xylene), 1,3-dimethyl benzene (meta xylene; m-xylene), and 1,3-dimethyl benzene (para xylene; p-xylene). Environmental professionals often come across the term "o-, m-, p-xylenes"; "xylenes" is in a plural sense. B/T/E/X are on the lists of MCL, Priority Pollutants, and HAPs [Note: Xylenes are not on the list of Priority Pollutants].

Aryl group. In organic chemistry, an *aryl group* is a functional group, or a substituent derived from an aromatic ring. The simplest aryl group is the "*phenyl*" group (C_6H_5–), which is derived from benzene (C_6H_6). A "*benzyl*" group consists of a phenyl group (i.e., a benzene ring) attached to a methyl group with a chemical formula of "$C_6H_5CH_2$–". [Note: A benzene ring is a phenyl group, not a benzyl group.]

Styrene (C_8H_8, or $H_2C{=}CHC_6H_5$) is on the list of HAPs; it is a vinyl group (i.e., $H_2C{=}CH$–) with a phenyl group (i.e., C_6H_5–).

Biphenyl. Biphenyl ($C_{12}H_{10}$ or $C_6H_5C_6H_5$) is also known as phenylbenzene or diphenyl; it consists of two benzene rings connected by a single carbon–carbon covalent bond. Biphenyl is on the list of HAPs; and it is naturally present in crude oil, coal tar, and natural gas.

Polycyclic aromatic hydrocarbons (PAHs). *PAHs* are hydrocarbons that have two or more "fused" benzene rings. PAHs are a class of chemicals that occur naturally in coal, crude oil, and gasoline. They are also emitted from burning fossil fuels, wood, garbage, and tobacco. PAHs can bind to or form small particulates in the air, and they are toxic and persistent. Naphthalene

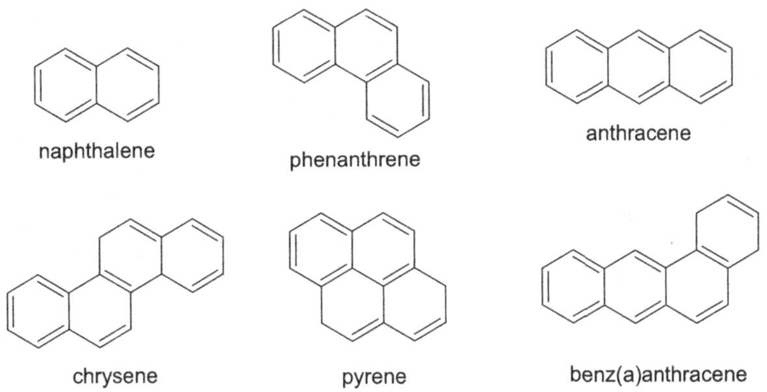

FIGURE 2.7 Chemical structural formula of some 2-, 3-, and 4-aromatic ring PAHs.

($C_{10}H_8$) forms the smallest PAH ring structure with two fused benzene rings, and it is a manmade PAH used to make other chemicals and mothballs. Figure 2.7 illustrates the chemical structural formula of naphthalene, two PAHs of three fused benzene rings [Note: phenanthrene and anthracene have the same chemical formula of $C_{14}H_{10}$], and three PAHs with four fused benzene rings (i.e., chrysene, pyrene, and benz(a)anthracene having the same chemical formula of $C_{18}H_{12}$).

PAHs can have six or even more fused benzene rings. They can also contain atoms other than carbon and hydrogen, such as O, N, and S, within the structure of their aromatic rings (and they are often referred to as *heteroatom PAHs*). With regard to PAHs, the MCL list only contains benzo(a)pyrene ($C_{20}H_{12}$); the Priority Pollutants list contains 16 specific PAH molecules, including benzo(a)pyrene; while the HAPs list specifies the regulated HAPs as 2-acetylaminofluorene ($C_{15}H_{13}NO$), naphthalene ($C_{10}H_8$), and polycyclic organic matter (including organic compounds with more than one benzene ring, and which have a boiling point ≥100°C).

2.4 FUNCTIONAL GROUPS OF ORGANIC COMPOUNDS

Organic compounds are often classified according to the type(s) of functional group(s) that they have. A *functional group* refers to an atom or a special grouping of atoms within a molecule that exhibits characteristic properties, independent of the "remaining" atoms present in that molecule. The previously mentioned alkanes, alkenes, alkynes, and aromatics are considered having their own functional groups (i.e., single, double, and triple carbon–carbon covalent bonds, and the benzene ring, respectively) by some. A molecule may contain more than one functional group.

More on nomenclature of organic compounds. Below are more basic guidelines for naming "simple" organic compounds:

1. Find the longest carbon chain in the molecule and that will give us the base of the name (e.g., pentane (C_5H_{12})).

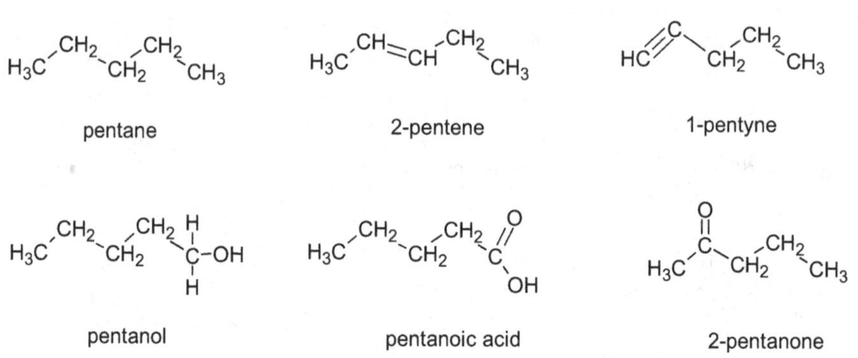

FIGURE 2.8 Pentane and some pentane-based organic compounds with different functional groups.

2. Determine the principal functional group and its position (e.g., 2-pentene (C_5H_{10}), 1-pentyne (C_5H_8), pentanol ($CH_3(CH_2)_4OH$), pentanoic acid ($CH_3(CH_2)_3COOH$), and 2-pentanone (or methyl propyl ketone, $CH_3C(=O)CH_2CH_2CH_3$) – see Figure 2.8 (and more later). Position is indicated by numbering the carbons in the main chain and positioning numbers are flanked by the dashed signs. Multiple positions for a given functional group are separated by commas, and indicated by the prefixes (i.e., di-, tri-, tetra-, penta-, hexa-, etc.).
3. Ancillary functional groups are given in the alphabetical order, with their positions at the beginning of the name. The common *ancillary functional groups* include (i) alkyl groups (e.g., methyl ($-CH_3$), ethyl ($-C_2H_5$), propyl ($-C_3H_7$), butyl ($-C_4H_9$)), (ii) halogen groups (e.g., fluoro (−F), chloro (−Cl), bromo (−Br), and iodo (−F)), (iii) amino ($-NH_2$), nitro ($-NO_2$), nitroso (−NO), cyano (−CN), (iv) phenyl ($-C_6H_5$) and benzyl ($-CH_2C_6H_5$), (v) hydroxyl (−OH), and (vi) methoxy ($-O-CH_3$). An *organyl group* (usually denoted as −R, −R′, or −R″) is any organic substitute group, regardless of functional type, with one free valence electron at a carbon atom (i.e., it can form a covalent bond with another atom).
4. The term "Bis" is a prefix used in naming chemical compounds, and it is used to denote the presence of two identical but separated complex groups in one molecule. On the other hand, the term "di-" is also used to say "two" or "twice". Prefixes "Bis" and "di-" are the same in meaning, but different in usage. The main difference is that "Bis" is often used to denote the presence of two identical, but "separate and more complex" groups.

In the following subsections, we will discuss functional groups of organics that we may come across as environmental professionals. These functional groups are (i) alcohols and phenolic compounds, (ii) aldehydes and ketones, (iii) carboxylic acids, anhydrides, esters, acetates, acrylates, and phthalates, (iv) ethers and epoxides, (v) amines, amides, urea, azo, imines, and imides, (vi) nitriles, isocyanates, hydrazine, and pyridines, (vii) nitro and nitroso compounds, (viii) thiols, sulfides, and sulfonic acid, and (ix) halogenated hydrocarbons.

2.4.1 Alcohols and Phenolic Compounds

Alcohols. If a hydrogen atom in a hydrocarbon molecule is substituted by a hydroxyl group (i.e., –OH), the compound is an alcohol. The general chemical formula of alcohols is R–OH; and the simplest two are methanol (or methyl alcohol, CH_3OH) and ethanol (or ethyl alcohol, C_2H_5OH). The "O–H" bond is highly polarized and participates in hydrogen bonding; and the hydroxyl group (–OH) also increases the water solubility (i.e., alcohols are very soluble in water).

Alcohol molecules are polar because the oxygen atom is much more negative than carbon and hydrogen; consequently, smaller alcohols (with six or fewer carbons) are miscible with water. Like water, alcohols can be weak acids and/or bases.

Alcohol can also be grouped into three classes. A *primary (1°) alcohol* is one in which the carbon with the –OH group is attached to zero or to only one carbon atom, and its general formula is RCH_2OH. If that carbon is attached to two carbon atoms, it is a *secondary (2°) alcohol*; and its general chemical formula is RR′CHOH. If that carbon has covalent bonds with three carbon atoms, it is a *tertiary (3°) alcohol*; and its general formula is RR′R″COH. Figure 2.9 illustrates four isomers of butyl alcohols in three different classes and their properties are different [Note: Please also see them in four different butyl substitute groups (i.e., butyl, isobutyl, sec-butyl, and tert-butyl)]. The prefix "iso-" is commonly given to 2-methyl alkanes, in which there is a methyl group located on the second carbon of a carbon chain [Note: Both butyl alcohol and isobutyl alcohol are primary alcohols]. Methanol (CH_3OH) is on the list of HAPs.

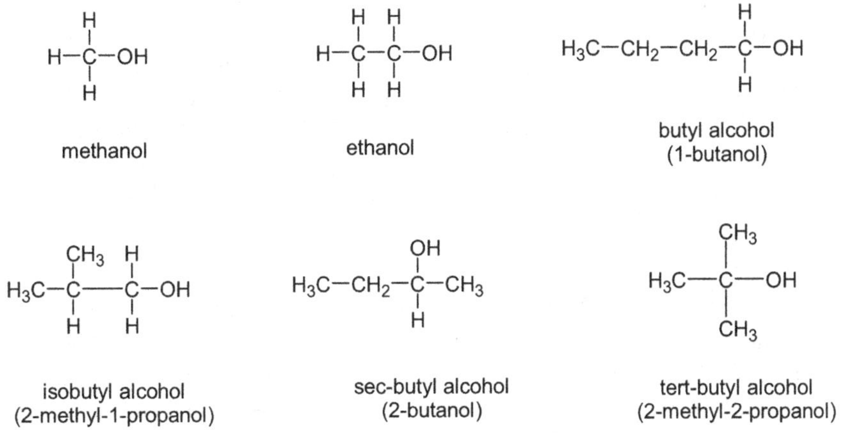

FIGURE 2.9 Chemical structural formula of some alcohols.

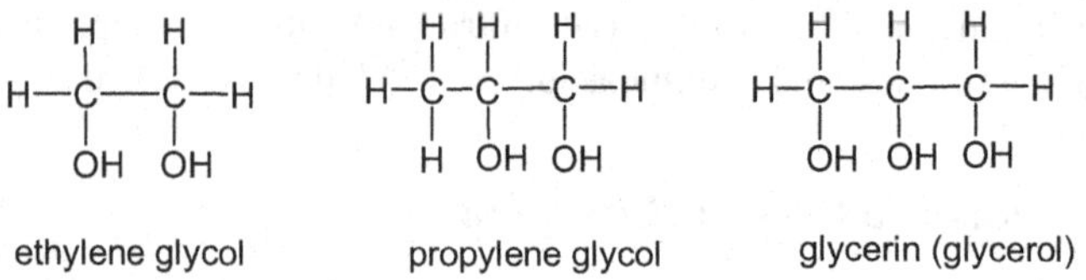

FIGURE 2.10 Structural formula of two glycols and glycerin.

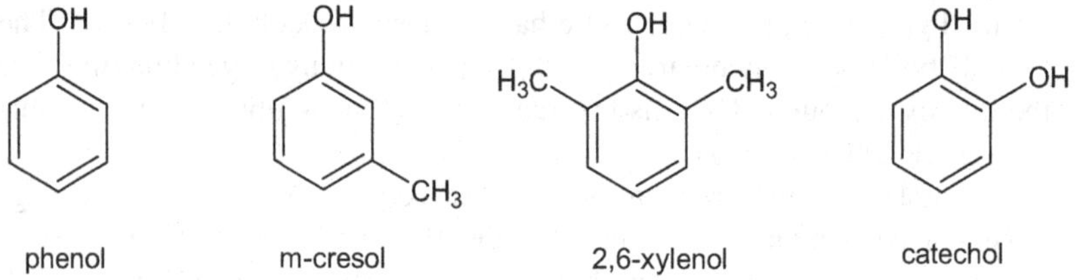

FIGURE 2.11 Chemical structural formula of phenol, o-cresol, 2,6-xylenol, and catechol.

Glycols and glycerin. *Glycols* are alcohols with two "–OH" groups, and each on two adjacent carbon atoms. The most common glycols are 1,2-ethanediol (or ethylene glycol) and 1,2-propanediol (or propylene glycol). 1,2,3-propanetriol (or glycerol, glycerin) is the most important trihydroxy alcohol. The chemical structural formula of these three compounds is shown in Figure 2.10. Ethylene glycol $(CH_2OH)_2$ is on the list of HAPs.

Phenols and phenolic compounds. Phenol (C_6H_5OH) is commonly used in household products and in industry as a starting material for many chemical products. It is a combination of the aromatic ring (the phenyl group) and the hydroxyl (–OH) group (see Figure 2.11); and it may work as a weak acid by releasing a proton (H^+) to become phenolate ($C_6H_5O^-$). Catecol (1,2-benzenediol, 2-hydroxyphenol, $C_6H_4(OH)_2$) comprises a benzene ring and two hydroxyl groups "ortho" to each other.

Cresols ($C_6H_4CH_3OH$), also known as hydroxytoluene, toluenol, or cresylic acid, are mono-methyl phenol; and all three isomers of cresols (i.e., o-cresol (2-methyl phenol), m-cresol (3-methyl phenol), and p-cresol (4-methyl phenol)), along with phenol are on the list of HAPs. Xylenols are dimethyl phenols with the chemical formula of $(CH_3)_2C_6H_3OH$; there are six isomers of xylenols. Figure 2.11 illustrates the chemical structural formula of phenol, m-cresol, 2,6-xylenol, and catechol. Catechol and o-, m-, and p- cresols are on the list of HAPs.

2.4.2 Aldehydes and Ketones

Carbonyl group. The *carbonyl group* (C=O) is simply a carbon double-bonded to an oxygen atom; the other two covalent bonds on the carbon can be anything, from hydrogen, oxygen, and carbon, to a complicated carbon chain (R). It should be noted that the carbonyl group is just "C=O" (i.e., it does not include any of the other two connections on the carbon).

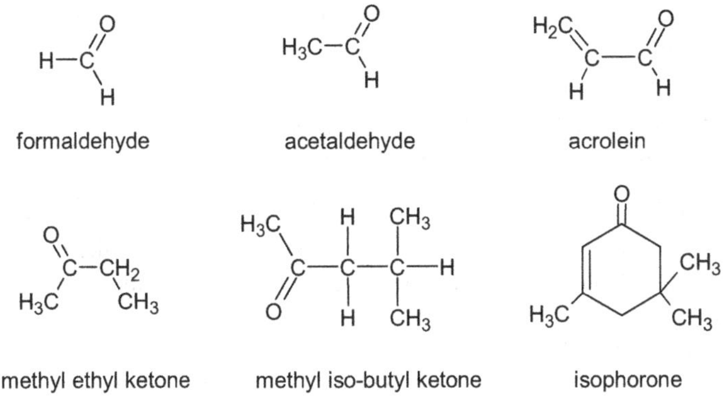

FIGURE 2.12 Chemical structural formula of some aldehydes and ketones.

Aldehydes and ketones are two classes of organic compounds containing a carbonyl group. If at least one of the two remaining carbon covalent bonds is with hydrogen, the compound is an *aldehyde*; if neither is with hydrogen, the compound is a *ketone.*

Aldehydes. The general chemical formula of aldehydes is R–CHO, and two simplest ones are formaldehyde (HCHO) and acetaldehyde (CH_3CHO) [Note: In organic chemistry, "*acetyl*" is a functional group with a chemical formula of $-C(=O)CH_3$].

Acrolein (C_3H_4O or $H_2C{=}CHCHO$) is the simplest unsaturated aldehyde. Aldehydes have polar covalent bonding, but they are not hydrogen bond donors. Formaldehyde, acetaldehyde, propionaldehyde (C_2H_5CHO), and acrolein are on the list of HAPs; Figure 2.12 shows the chemical structural formula of three of them.

Ketones. The general chemical formula of ketones is R–C(=O)–R′ and two simplest ketones are acetone (CH_3COCH_3), which is commonly used as a nail polisher, and methyl ethyl ketone ($CH_3COC_2H_5$). Acetophenone (C_8H_8O), isophorone ($C_9H_{14}O$), methyl ethyl ketone (MEK, 2-butanone, C_4H_8O), methyl isobutyl ketone (MIBK, $C_6H_{12}O$), β-propiolactone ($C_3H_4O_2$), and quinones are on the list of HAPs. Acetophenone ($C_6H_5COCH_3$) is the simplest aromatic ketone. Please be noted that the names of some ketones do not provide straight information about their chemical structures. Isophorone is an α, β-unsaturated cyclic ketone; the α-carbon refers to the first carbon atom that attaches to a functional group (such as a carbonyl group) and β-carbon is the second carbon atom. The carbon of the carbonyl group of β-propiolactone ($C_3H_4O_2$) is part of the four-member ring (three carbons and one oxygen). *Quinones* are a special class of ketones in which carbonyl groups are a part of aromatic ring of benzene, anthracene, or naphthalene (e.g., 1,4-benzoquinone, $C_6H_4O_2$). The chemical structural formula of MEK, MIBK, and isophorone is shown in Figure 2.12. The name of hydroquinone ($C_6H_4(OH)_2$) implies that it is in the ketone family, but it is in the phenol family. Its chemical structure is two hydroxyl groups (–OH) bonded to a benzene ring in a para position (it is also known as benzene-1,4-diol). Hydroquinone is on the list of HAPs.

2.4.3 Carboxylic Acids, Anhydrides, Esters, Acetates, Acrylates, and Phthalates

Acyl group. The *acyl group* can be considered as a specific type of a carbonyl group (C=O), in which one of the R groups connected to the carbonyl, as "R–C=O" (Note: The fourth carbon covalent bond is not part of the acyl group).

Carboxylic acids. The *carboxyl group* has a carbon which is double-bonded to an oxygen atom, and the carbon atom is also bonded to a hydroxyl group (–OH). Compounds with a carboxyl functional group (–COOH) are called *carboxylic acids*. The general chemical formula of carboxylic acids is R–COOH. Three simplest carboxylic acids are formic acid (HCOOH), acetic acid (CH_3COOH) which is the main ingredient of vinegar, and propanoic acid (C_2H_5COOH). Carboxylic acids have higher boiling points because the hydroxyl group participates in hydrogen bonding. Carboxylic acids are relatively weak acids and not undergoing full dissociation (i.e., complete release of protons) in water when compared to strong inorganic acids such as hydrochloric acid (HCl) and sulfuric acid (H_2SO_4). Chloroacetic acid ($ClCH_2COOH$) and acrylic acid ($H_2C{=}CHCOOH$), an unsaturated carboxylic acid, are on the list of HAPs.

Anhydrides. An *anhydride* is a compound that has two acyl groups bonded to the same oxygen atom; the general chemical formula of anhydrides is $(R{-}C{=}O)_2O$. The chemical formula of acetic anhydride is $(CH_3C{=}O)_2O$ and that of acetic formic anhydride is $CH_3C({=}O){-}O{-}CHO$. Maleic anhydride ($CH_2(C{=}O)_2O$) is the acid hydride of maleic acid ($HO_2CCH{=}CHCO_2H$). Maleic acid and phthalic anhydride ($C_6H_4(C{=}O)_2O$) are on the list of HAPs (see below for phthalic acid).

Esters. *Esters* are similar to carboxylic acids, except the hydroxyl group is replaced by an *alkoxy* group (–OR′). The general chemical formula of esters is R–C(=O)–OR′. Methyl formate ($HC({=}O){-}O{-}CH_3$) is the methyl ester of formic acid (HCOOH), and it is the simplest carboxylate ester.

Acetates and acrylates. An *acetate* is a salt or ester of acetic acid, containing a functional group of "$CH_3C({=}O)O{-}$". Vinyl acetate ($CH_3C({=}O){-}O{-}CHCH_2$) is unsaturated (with a C=C bond), and it is primarily used as a monomer in the production of polyvinyl acetate (PVA), a water-soluble synthetic polymer. An *acrylate* is a salt or ester of acrylic acid ($H_2C{=}CHCOOH$), containing a functional group of "$H_2C{=}CHCOO{-}$". Ethyl acrylate ($H_2C{=}CHCOOC_2H_5$), methyl methacrylate ($CH_2{=}C(CH_3)COOCH_3$), and vinyl acetate are on the list of HAPs. [Note: both ethyl acrylate and methyl methacrylate have the same chemical formula of $C_5H_8O_2$.]

Figure 2.13 illustrates the chemical structural formula of acetic acid, acrylic acid, acetic anhydride, acetic formic anhydride, vinyl acetate, and ethyl acrylate.

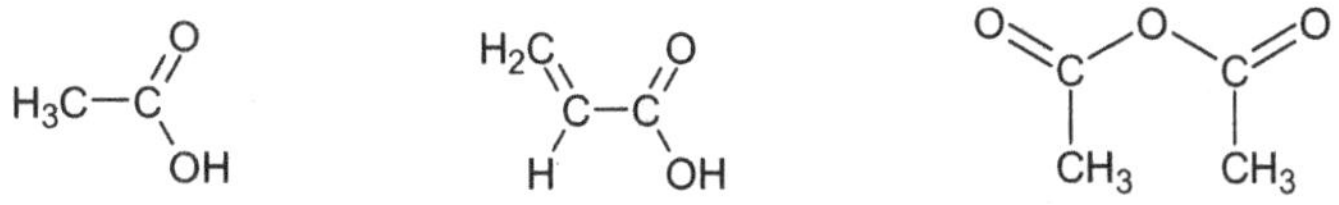

acetic acid acrylic acid acetic anhydride

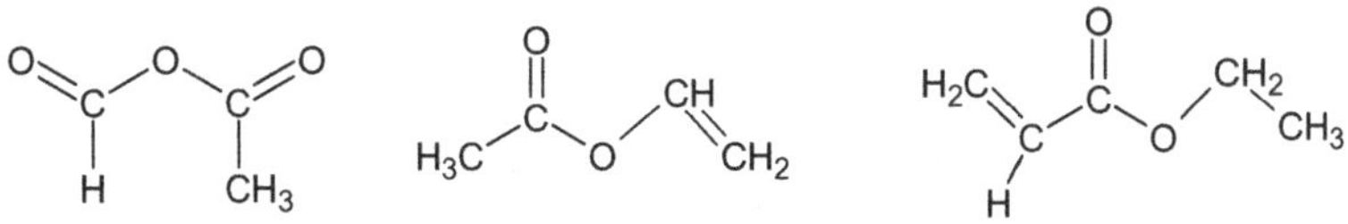

acetic formic anhydride vinyl acetate ethyl acrylate

FIGURE 2.13 Chemical structural formula of two carboxylic acids, two anhydrides, one acetate, and one acrylate.

Phthalates. *Phthalates* (or *phthalate esters*) are esters of phthalic acid, which is an aromatic dicarboxylic acid with the chemical formula of $C_6H_4(COOH)_2$. Phthalates are a group of chemicals used to make plastics more durable, and they are often called *plasticizers*. They are in hundreds of products such as vinyl flooring, lubricating oils, and personal care products such as soaps, shampoos, and hair sprays. Many of them are considered as endocrine disruptors.

Dimethyl phthalate ($C_{10}H_{10}O_4$), diethyl phthalate ($C_{12}H_{14}O_4$), di-n-butyl phthalate ($C_{16}H_{20}O_4$), butyl benzyl phthalate ($C_{19}H_{20}O_4$), bis(2-ethylhexyl) phthalate (DEHP; $C_{24}H_{38}O_4$), di-n-octyl phthalate ($C_{24}H_{38}O_4$) are on the list of Priority Pollutants. Dimethyl phthalate and di-n-butyl phthalate are also on the list of HAPs; while DEHP is also on the list of MCLs and HAPs. Phthalic anhydride ($C_6H_4(C{=}O)_2O$) is the anhydride of the phthalic acid, and it is on the list of HAPs. Figure 2.14 illustrates the chemical structural formula of phthalic acid, diethyl phthalate, and DEHP.

2.4.4 Ethers and Epoxides

Ethers. *Ethers* are a class of organic compounds that contain an "ether" group, in which an oxygen atom is connected to two alkyl groups. The general chemical formula of ethers is R–O–R′; and three simplest ones are dimethyl ether (CH_3OCH_3), methyl ethyl ether ($CH_3OC_2H_5$), and diethyl ether ($C_2H_5OC_2H_5$). Ethers cannot serve as hydrogen bond donors, so that their boiling points are lower than those of alcohols with similar molecular weight. Bis(chloromethyl) ether ($C_2H_4Cl_2O$), chloromethyl methyl ether (C_2H_5ClO), bis(2-chloroethyl) ether ($C_4H_8Cl_2O$), methyl tert-butyl ether ($C_5H_{12}O$) and *glycol ethers* (i.e., a large group of organic compounds consist of alkyl ethers which are based on glycols such as ethylene glycol or propylene glycol) are on the list of HAPs.

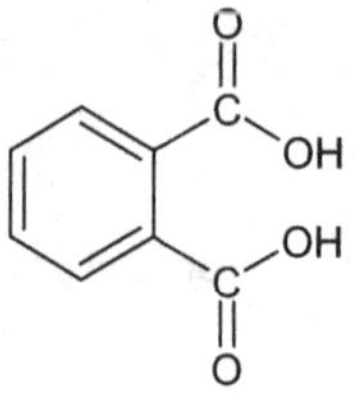

phthalic acid

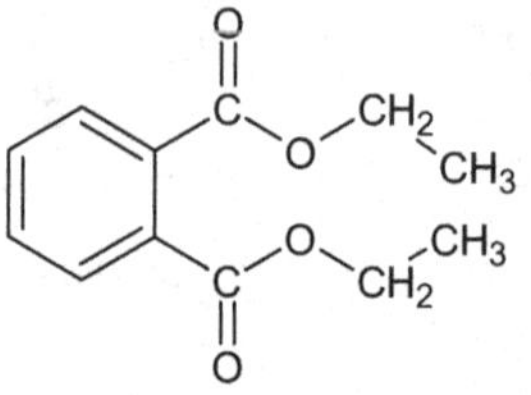

diethyl phthalate

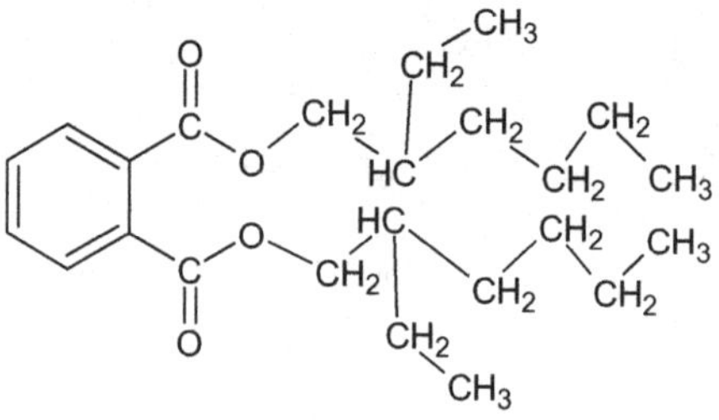

Bis(2-ethylhexyl) phthalate

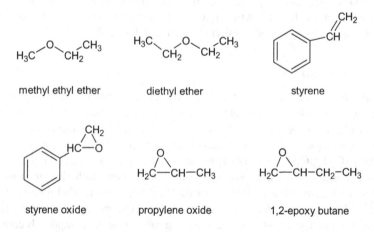

FIGURE 2.14 Chemical structural formula of three phthalates.

Epoxides. An *epoxide* is a cyclic ether, in which the ether forms a three-atom ring, with one atom of oxygen and two atoms of carbon. This triangular structure makes epoxides more reactive than other ethers.

Figure 2.15 shows the chemical structural formula of methyl ethyl ether, diethyl ether, styrene, styrene oxide ($C_6H_5CHCH_2O$), propylene oxide (C_3H_6O), 1,2-epoxybutane (C_4H_8O); the last three and ethylene oxide (C_2H_4O) are on the list of HAPs, and styrene is also on the list of MCLs.

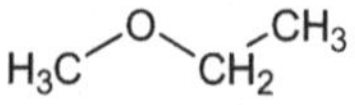

methyl ethyl ether

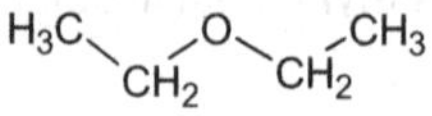

diethyl ether

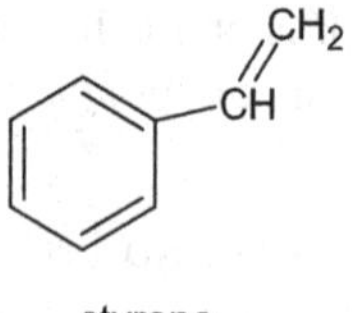

styrene

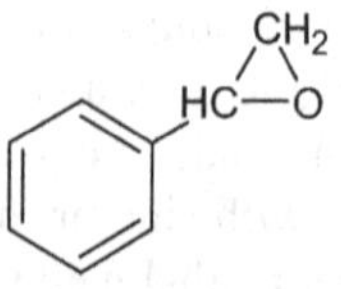

styrene oxide

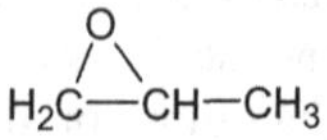

propylene oxide

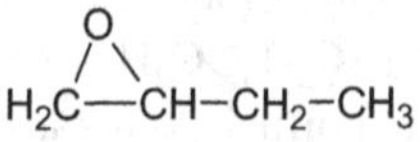

1,2-epoxy butane

FIGURE 2.15 Chemical structural formula of methyl ethyl ether, diethyl ether, styrene, styrene oxide, propylene oxide, and 1,2-epoxy butane.

2.4.5 Amines, Amides, Urea, Azo Compounds, Imines, and Imides

Amines. *Amines* are organic derivatives of ammonia (NH_3), in which one, two, or all three hydrogen atoms of ammonia are replaced by organic groups. Compounds R–NH_2 are called *primary amines*, RR'NH are *secondary amines*, and RR'R"N are *tertiary amines*. The simplest primary amine is methylamine (CH_3NH_2); dimethylamine ($(CH_3)_2NH$) is the simplest secondary amine; and trimethylamine ($(CH_3)_3N$) is the simplest tertiary amine. Since nitrogen is less electronegative than oxygen, the N–H bond is less polar than the O–H bonds; consequently, hydrogen bondings from the N–H bonds are less strong than those resulting from the O–H bonds. Amines of small molecular weights are generally water-soluble; and they tend to have odors similar to ammonia.

The name of the "–NH_2" substituent is *amino*. The letter "N" in the name of N-methylethanamine ($H_3C–NHC_2H_5$) stands for "nitrogen", indicating that the nitrogen atom in the molecule is part of a nitrogen-carbon bond. "N,N" in the name of N,N-dimethylethylamine (DMEA, $(CH_3)_2NC_2H_5$) means that two methyl groups attached to the same N atom and indicates that the subsequently-named substituent is attached to the nitrogen and not elsewhere.

Triethylamine, diethanolamine ($(CH_2CH_2OH)_2NH$), N-nitrosodimethylamine (NDMA, $(CH_3)_2NNO$), p-phenylenediamine ($C_6H_4(NH_2)_2$), 1,2-propyleneimine ($CH_3CH(NH)CH_2$), and 2,4-toluene diamine ($C_6H_3(NH_2)_2CH_3$) are on the list of HAPs.

Aniline ($C_6H_5–NH_2$) is the simplest aromatic amine, consisting of a phenyl group attached to an amine. N,N-dimethylaniline ($C_6H_5N(CH_3)_2$, o-anisidine (o-methoxyaniline, $CH_3OC_6H_4NH_2$), 4,4′-methylenedianiline ($CH_2(C_6H_4NH_2)_2$), 4,4′-methylenebis(2-chloroaniline) ($C_{13}H_{12}Cl_2N_2$), and o-toluidine ($C_6H_4CH_3NH_2$) are on the list of HAPs.

Benzidine is also an aromatic amine; it is also called 1,1′-biphenyl-4,4′-diamine with a chemical formula of $(C_6H_4NH_2)_2$. It has a biphenyl structure with an –NH_2 functional group at the end of each benzene ring. Benzidine, 3,3′-dimethyl benzidine ($C_{14}H_{16}N_2$), 3,3′-dimethoxybenzidine ($C_{14}H_{16}N_2O_2$), and 3,3′-dichlorobenzidine ($C_{12}H_{10}Cl_2N_2$) are on the list of HAPs. Figure 2.16 illustrates the structural formula of several amines.

Amides. An *amide* contains a nitrogen which is directly attached to a carbon in a carbonyl (C=O) group, and its general chemical formula is "RC(=O)NH_2". Amides are named by changing the ending of corresponding carboxylic acid to "-amide". The *carboxamides* (RC(=O)NR'R"), which are derived from carboxylic acids (R–COOH) are the most important group of amides.

Acetamide (CH_3CONH_2), acrylamide (C_3H_5NO), dimethyl formamide (C_3H_7NO), hexamethylphosphoramide ($C_6H_{18}N_3OP$) are on the list of HAPs. Figure 2.17 illustrates the chemical structural formula of acetamide, acrylamide, and dimethyl formamide.

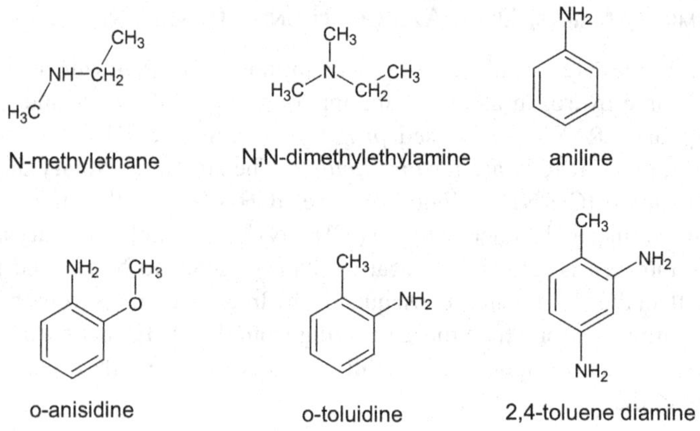

FIGURE 2.16 Chemical structural formula of some organic amines.

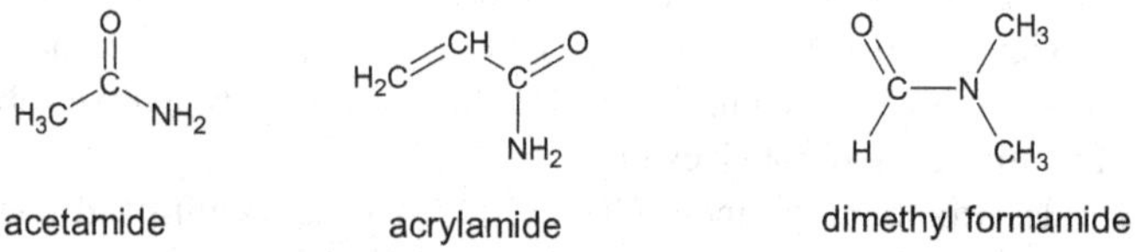

FIGURE 2.17 Chemical structural formula of some amide compounds.

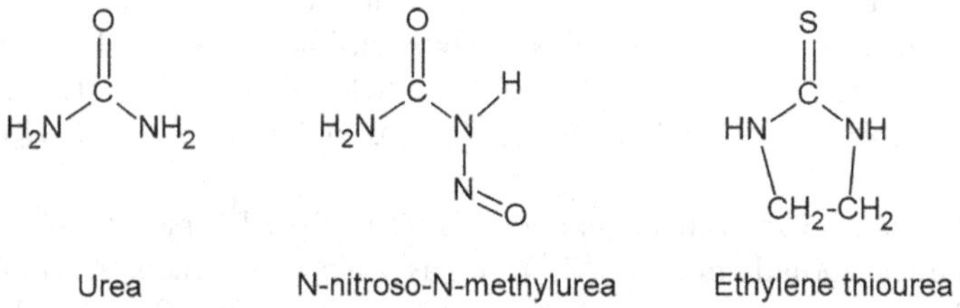

FIGURE 2.18 Chemical structural formula of some urea compounds.

Urea. Urea ($CO(NH_2)_2$) is also called carbamide because it is a diamide of carbonic acid. N-nitroso-N-methylurea ($C_2H_5N_3O_2$) and ethylene thiourea ($C_3H_6N_2S$) are on the list of HAPs (see Figure 2.18 for their chemical structural formula). The prefix "*thio*-" in chemistry means that an oxygen atom in a compound is replaced by a sulfur atom.

Azo compounds. Azo compounds are organic compounds bearing the functional *azo* (or called *diazinyl*) group (R–N=N–R′), in which R and R′ can be either aryl or alkyl groups. Diazomethane (CH_2N_2) is the simplest diazo compound. Azo compounds with alkyl groups are unstable and can be explosive. Azo compounds with aryl groups have vivid colors, especially red, orange, and yellow; for example, methyl yellow (4-(dimethylamino)

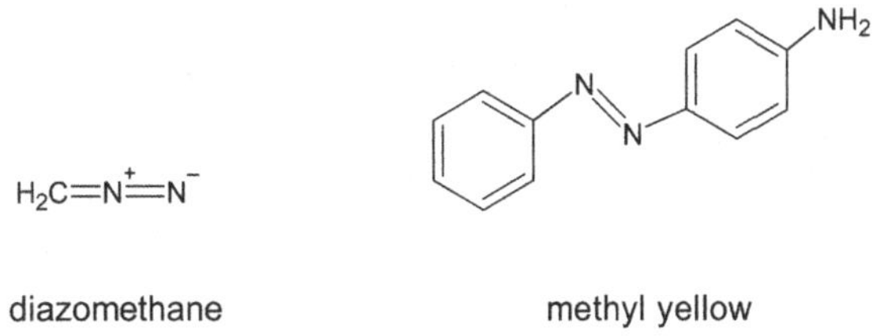

FIGURE 2.19 Chemical structural formula of diazo methane and methyl yellow.

azobenzene; $C_6H_5N_2C_6H_4N(CH_3)_2$). Diazomethane and methyl yellow are on the list of HAPs. Figure 2.19 illustrates the chemical structural formula of diazomethane and methyl yellow.

Imines. An *imine* is a chemical compound containing a carbon-nitrogen double bond (C=N), with a general chemical formula of RR'C=NR" [Note: An amine is a compound containing a carbon-nitrogen single bond]. Ethyleneimine (Aziridine; $(CH_2)_2NH$) is on the list of HAPs; its name ends with "imine" seems confusing because it does not contain a C=N bond; instead, it has a three-member carbon ring with two carbon and one nitrogen atoms.

Imides. An *imide* is a functional group consisting of two acyl groups (R–C=O) bonded to nitrogen, with a general chemical formula of $(RC(=O))_2NR'$; the compounds are structurally related to acid anhydrides.

2.4.6 Nitriles, Isocyanates, Hydrazine, and Pyridines

Nitriles. A *nitrile* is an organic compound having a "–C≡N" functional group, with a general chemical formula of R–C≡N. The simplest organic nitrile is acetonitrile (ethanenitrile; $H_3C–C≡N$), which is a common solvent. Acetonitrile and acrylonitrile ($H_2C=CHC≡N$) are on the list of HAPs.

Isocyanates. The cyanate ion is an anion with the chemical formula of OCN^-. *Isocyanate* is a functional group with a chemical formula of "–N=C=O". Organic compounds that contain an isocyanate group are referred to as *isocyanates*, which are used in many consumer products such as polyurethane foams and automotive paints. Hexamethylene-1,6-diisocyanate ($C_8H_{12}N_2O_2$), methyl isocyanate ($CH_3N=C=O$), methylene diphenyl diisocyanate (MDI, $C_{15}H_{10}N_2O_2$) are on the list of HAPs.

Hydrazine compounds. Hydrazine ($H_2N–NH_2$) is a toxic inorganic compound, which is a colorless liquid with ammonia-like odor. *Hydrazines* refer to a group of organic compounds derived by replacing one or more hydrogen atoms in hydrazine with organyl group(s). Methyl hydrazine (CH_3NHNH_2) and 1,2-diphenylhydrazine $(C_6H_5NH)_2$ are on the list of HAPs.

Pyridines. *Pyridines* are a class of organic compounds having a six-member ring structure composed of five carbon atoms and one nitrogen atom. The simplest member of the pyridines is pyridine (C_6H_5N). Quinoline (C_9H_7N)

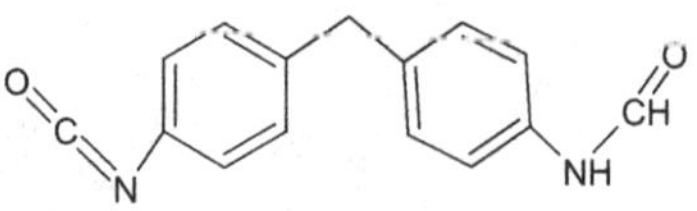

MDI

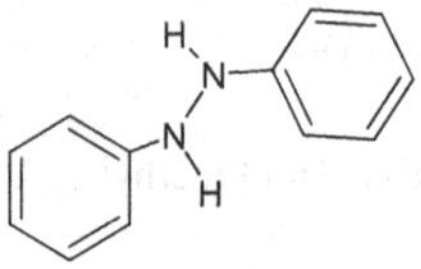

1,2-diphenylhydrazine

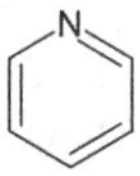

pyridine

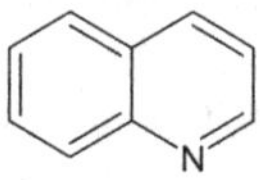

quinoline

FIGURE 2.20 Chemical structural formula of isocyanate, hydrazine, and pyridine compounds.

is on the list of HAPs. Figure 2.20 illustrates the chemical structural formula of MDI, 1,2-diphenylhydrazine, pyridine, and quinoline.

2.4.7 Nitro and Nitroso Compounds

Nitro compounds. *Nitro organic compounds* are those that contain one or more *nitro* functional group ($-NO_2$). The nitro compounds on the lists of Priority Pollutants and/or HAPs include: 2-nitropropane ($C_3H_7NO_2$), nitrobenzene ($C_6H_5NO_2$), 2,4-dinitrotoluene ($C_7H_6N_2O_4$), 2,6-dinitrotoluene ($C_7H_6N_2O_4$), pentachloronitrobenzene (quintobenzene, $C_6Cl_5NO_2$), 4-nitrophenol ($C_6H_5NO_3$), 2,4-dinitrophenol ($C_6H_4N_2O_5$).

Nitroso compounds. In organic chemistry, *nitroso* refers to a functional group of nitric oxide ($-NO$). The nitroso compounds on the lists of Priority Pollutants and/or HAPs include N-nitrosodimethylamine (NDMA, $C_2H_6N_2O$), N-nitrosodi-n-propylamine ($C_6H_{14}N_2O$), N-nitrosodiphenylamine ($C_{12}H_{10}N_2O$). Figure 2.21 illustrates the chemical structural formula of some nitro and nitroso compounds.

2.4.8 Thiols, Sulfides, Sulfates, and Sulfonic Acid

Thiols. In organic chemistry, *thiols* (also called mercaptan) are a family of organosulfur compounds that contains a "$-SH$" (*thiol, mercapto, sulfanyl,* or *sulfhydryl*) functional group. It is similar to alcohols and phenols, but a sulfur atom relaces the oxygen atom in the hydroxyl ($-OH$) group. They are odorous; for example, methyl mercaptan (CH_3SH) is a colorless flammable gas with unpleasant odor as rotten cabbage.

Thioethers (sulfides). *Thioethers* (*sulfides*) are the sulfur equivalent of ethers (R–O–R′). with a general chemical formula of R–S–R′. It should be noted that the term "thio" is also used in inorganic chemistry; for example, $S_2O_3^{-2}$

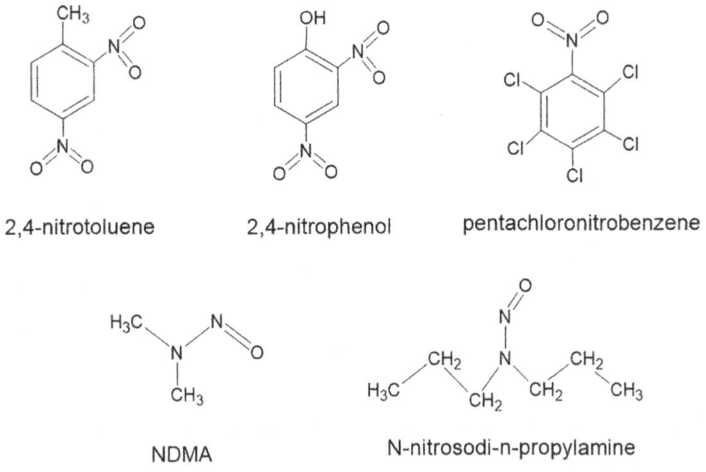

FIGURE 2.21 Chemical structural formula of some nitro- and nitroso- organic compounds.

is called thiosulfate because one of the oxygen atoms in sulfate ion $\left(SO_4^{-2}\right)$ is replaced by a sulfur atom. Dimethyl sulfide (DMS, methylthiomethane, CH_3SCH_3)) is the simplest thioether; it is a flammable liquid with a strong odor.

Sulfates. Dimethyl sulfate $(CH_3O)_2SO_2$ is the diester of methanol and sulfuric acid; and diethyl sulfate $(C_2H_5O)_2SO_2$ is the diester of ethanol and sulfuric acid. Both are on the list of HAPs [Note: Be aware of the difference between "sulfates" here and the sulfate ion $(SO_4^{=})$].

Sulfonic acid. In organic chemistry, *sulfonic acids* are a family of organosulfur compounds with a general chemical formula of $R{-}S(=O)_2{-}OH$. Perfluorooctane sulfonic acid (PFOS) is one of "forever chemicals". Figure 2.22 illustrates the chemical structural formula of methyl mercaptan, dimethyl sulfide, and PFOS.

2.4.9 Halogenated Hydrocarbons

Halogenated organic compounds are organic compounds in which one or more hydrogen atoms are replaced by halogens (i.e., chlorine, bromine, fluorine, and iodine). Some of the general characteristics of halogenated hydrocarbons include:

- Many of them are toxic.
- The toxicity increases with the number of halogens present in the molecule.
- Many of them are used as industrial solvents (e.g., chlorinated solvents, especially).
- They are formed as reaction products when chlorine is used as the oxidizing agent (e.g., pulp bleaching and chlorine disinfection of water and wastewater).

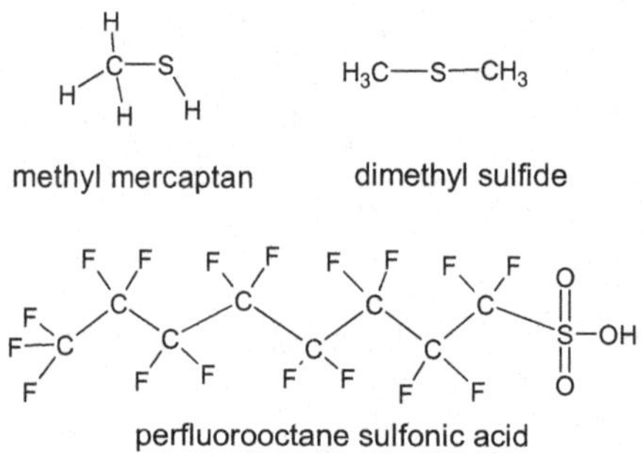

FIGURE 2.22 Chemical structural formula of a thiol, a sulfide, and a sulfonic compound.

- Halogenated organic compounds could also be generated as byproducts during incineration of organic wastes.
- Those of smaller molecular weights are often volatile.
- Many of them are not readily biodegradable, especially under aerobic conditions. They become more persistent with the number of halogens present in the molecule.

Below are the halogenated compounds on the lists of MCLs, Priority Pollutants, and HAPs (they are grouped by the author).

a. ***"Simple" halogenated hydrocarbons.***
   - Halogenated alkanes: methyl chloride (CH_3Cl), methyl bromide (CH_3Br), methyl iodide (CH_3I), dichloromethane (methylene chloride, CH_2Cl_2), ethylene dibromide (dibromoethane, $C_2H_4Br_2$), carbon tetrachloride (CCl_4), ethyl chloride (chloroethane, C_2H_5Cl), 1,1-dichloroethane ($C_2H_4Cl_2$), 1,2-dichloroethane (ethylene dichloride, $C_2H_4Cl_2$), 1,1,1-trichloroethane ($C_2H_3Cl_3$), 1,1,2-trichloroethane ($C_2H_3Cl_3$), 1,1,2,2-tetrachloroethane ($C_2H_2Cl_4$), hexachloroethane (C_2Cl_6), 1-bromopropane (C_3H_7Br), 1,2-dichloropropane ($C_3H_6Cl_2$), 1,2-dibromo-3-chloropropane ($C_3H_5Br_2Cl$), 1,2-dichloropropane ($C_3H_6Cl_2$), epichlorohydrin (l-chloro-2,3-epoxypropane, C_3H_5ClO), and bis(2-chloroethoxy) methane ($C_5H_{10}Cl_2O_2$)
   - Halogenated alkenes: vinyl bromide (C_2H_3Br), vinyl chloride (C_2H_3Cl), allyl chloride (C_3H_5Cl), 1,1-dichloroethylene ($C_2H_2Cl_2$), cis-1,2-dichloroethylene ($C_2H_2Cl_2$), trans-1,2-dichloroethylene ($C_2H_2Cl_2$), trichloroethylene (C_2HCl_3), tetrachloroethylene (C_2Cl_4), 1,3-dichloropropylene ($C_3H_4Cl_2$), chloroprene (2-chlorobuta-1,3-diene, C_4H_5Cl), hexachlorobutadiene (C_4Cl_6), and hexachlorocyclopentadiene (C_5Cl_6)
   - Halogenated arenes: chlorobenzene (C_6H_5Cl), o-dichlorobenzene ($C_6H_4Cl_2$), m-dichlorobenzene ($C_6H_4Cl_2$), p-dichlorobenzene ($C_6H_4Cl_2$), 1,2,4-trichlorobenzene ($C_6H_3Cl_3$), hexachlorobenzene (C_6Cl_6), benzyl chloride ($C_6H_5CH_2Cl$), and benzotrichloride ($C_6H_5CCl_3$)

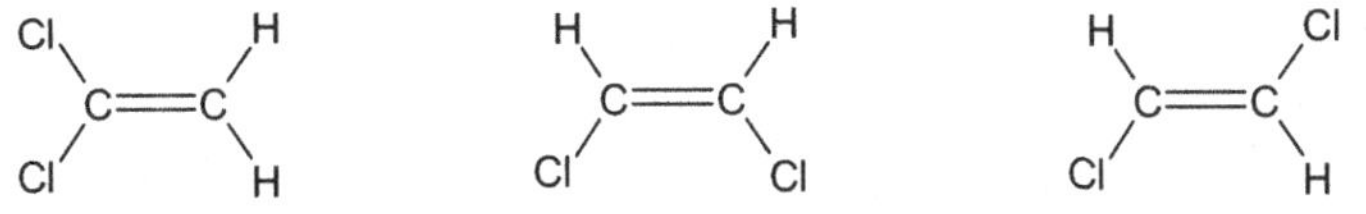

FIGURE 2.23 Chemical structural formula of DCE isomers.

As mentioned, disinfection byproducts (DBPs) may be formed during chlorine disinfection because chlorine would react with natural organic matters (NOMs) present in water. These DBPs include (i) trihalomethanes (i.e., chloroform, bromoform, bromodichloromethane, and dibromochloromethane, and haloacetic acids (those commonly found in drinking water include monochloroacetic acid, dichloroacetic acid, trichloroacetic acid, monobromoacetic, and dibromoacetic acid).

Chlorinated solvents are industrial chemicals that are widely used for metal cleaning and in the production of many consumer products such as plastics, lacquers, and perfumes. Common chlorinated solvents include dichloroethane (DCA), dichloroethylene (DCE), trichloroethane (TCA), trichloroethylene (TCE), and tetrachloroethylene (PCE). DCA has two isomers: 1,1-DCA and 1,2-DCA; while DCE has three isomers: 1,1-DCE, cis-1,2-DCE, and trans-1,2-DCE (see Figure 2.23). The reason, that 1,2-DCE has cis- and trans- isomers, is because the C=C double bond is rigid. It will not allow free rotation about the carbon–carbon double bond. The cis-1,2-DCE (both chlorines are on the same side of the C=C bond) is more stable than trans-1,2-DCE.

b. ***Polychlorinated biphenyls (PCBs).***

Polychlorinated biphenyls (PCBs) are a class of chemical compounds in which 2–10 chlorine atoms are attached to the biphenyl molecule, with a general chemical formula of $C_{12}H_{10-n}Cl_n$. PCBs were domestically manufactured from 1929 until manufacturing was banned in 1979 by the Toxic Substances Control Act (TSCA). They have a range of toxicity, depending mainly on the number of chlorine atoms and their locations in a PCB molecule. Due to their non-flammability, chemical stability, high boiling point, and electrical insulating properties, PCBs were used in hundreds of industrial and commercial applications (USEPA, 2023f). PCBs are on the lists of Priority Pollutants and HAPs. The general chemical structural formula of PCBs and 3,3′,4,4,′,5,5′-hexachlorobiphenyl are shown in Figure 2.24.

A *PCB congener* is any single and uniquely defined chemical compound in the PCB family. The name of a congener specifies the total number of chlorine constituents and the position of each chlorine (e.g., 3,3′,4,4′,5,5′-hexachlorobiphenyl, a PCB congener, in Figure 2.24). *Homologs* are subcategories of PCB congeners that have equal numbers of chorine

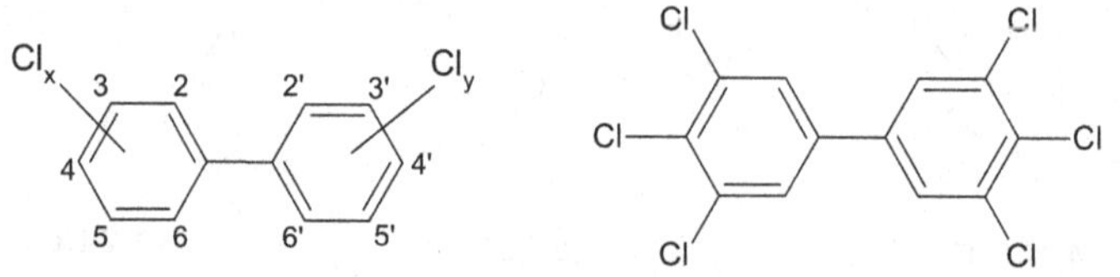

FIGURE 2.24 Chemical structural formula of PCBs and 3,3′,4,4′,5,5′-hexachlorobiphenyl.

substituents. For example, the hexachlorobiphenyls are all PCB congeners with exactly six chlorine substituents that can be in any arrangement (e.g., 3,3′,4,4′,5,5′-hexachlorobiphenyl is in the homolog of hexachlorobiphenyls). Arochlor is the most known trade name for PCB mixtures. There are many types of Aroclor and each has a distinguishing suffix number (see the List of Priority Pollutants) that indicates the degree chlorination. The first two digits usually refer to the number of carbon atoms in the phenyl rings (i.e., 12 for PCBs) and the second (and the last) two numbers indicate the percentage of chlorine by weight in the mixture. For example, the name Aroclor 1254 means that this PCB mixture contains approximately 54% chlorine by weight (USEPA, 2023f).

c. ***Dioxins and furans and dioxin-like biphenyls (PCBs).***
Dioxins refer to a group of toxic compounds that share certain chemical and biological characteristics. Several hundreds of these chemicals exist and they are members of three closely-related families:
   - polychlorinated dibenzo-p-dioxins (PCDDs)
   - polychlorinated dibenzofurans (PCDFs)
   - certain polychlorinated PCBs (e.g., 3,3′,4,4′,5,5′-hexachlorobiphenyl)

The family of dioxins is characterized by the presence of two benzene rings connected by a pair of oxygen atoms; while *furan* is a heterocyclic organic compound, consisting of a five-membered ring with four carbon atoms and one oxygen atom. Each of the eight carbons on the benzene rings that are not part of the dioxin ring or the furan ring can bind with hydrogen or other atoms; these positions are assigned the numbers from 1 to 4 and from 6 to 9. The more toxic dioxins and furans carry chlorine atoms at these positions; in addition, those with chlorines at positions 2, 3, 7, and 8 are more toxic (2,3,7,8-tetrachloro-p-dioxin is considered as the most toxic) (USEPA, 2023g). Figure 2.25 illustrates the chemical structural formula of 2,3,7,8-tetrachloro-p-dioxin [Note: The prefix "p" indicates the two oxygen atoms are in the "para" positions] and 2,3,7,8-tetrachlorodibenzofuran.

PCDDs and PCDFs are not created intentionally, but they are produced as a result of human activities (e.g., fossil fuel combustion, waste incineration, cigarette smoking, chlorine bleaching of pulp, and production of some chlorinated compounds); and they also come from natural processes like forest fires.

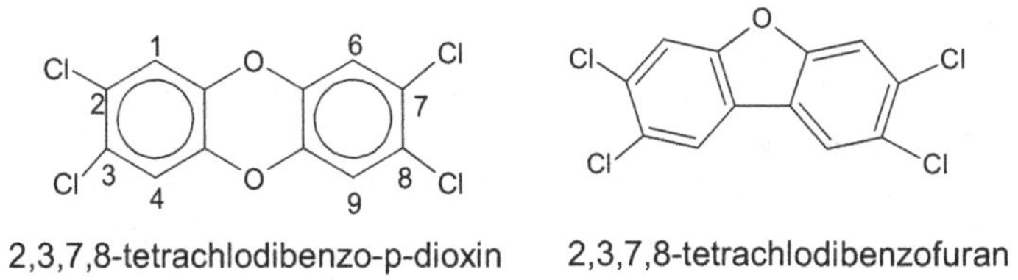

FIGURE 2.25 Chemical structural formula of dioxin and furan.

2.5 ORGANIC COMPOUNDS RELATED TO BIOCHEMISTRY

Biological processes are commonly applied in wastewater treatment and remediation of contaminated soil and groundwater. To gain the insights into biological treatment processes (as well as water and wastewater disinfection), environmental engineers may want to have in-depth knowledge in biochemistry. *Biochemistry* is the study of chemistry related to biomolecules and living organisms as well as biological processes at the cellular and molecular levels [Note: Biochemistry is out of the scope of this book]. Here, we just have a brief coverage on the groups of organic compounds relevant to biochemistry.

2.5.1 Carbohydrates

Carbohydrates, along with lipids, proteins, nucleic acids, and other compounds, are known as biomolecules because they are closely associated with living organisms. Most of the oxygen atoms in carbohydrates are found in the hydroxyl (–OH) groups, but one of them is part of the carbonyl (C=O) group. Carbohydrates (or saccharides), with an empirical chemical formula of $C_m(H_2O)_n$, are made of carbon, hydrogen, and oxygen in a ratio of roughly one carbon atom to one water molecule.

Monosaccharides are simple sugar (e.g., glucose, $C_6H_{12}O_6$). Disaccharides consist of two monosaccharide units linked together by a covalent bond (e.g., sucrose, $C_{12}H_{22}O_{11}$). Polysaccharides contain very long chains of monosaccharide units, in hundreds/thousands (e.g., cellulose, starch).

2.5.2 Proteins

Proteins are one of the most abundant organic molecules in living organisms, and they have the most diverse range of functions of all macromolecules. The major building blocks of protein are amino acids. The general chemical formula of *amino acids* is R–CH(NH_2)–COOH, which consists of a basic amino group (–NH_2) and a carboxylic group (–COOH) with a unique organyl group as the side chain. In other words, in each molecule of amino acid, there is a central carbon atom (often called α-carbon), connected by an amino group on one side and a carboxylic acid group on the other side. A hydrogen atom and the R-group then satisfy the remaining two covalent bonds of the α-carbon.

2.5.3 Lipids

Lipids are the polymers of fatty acids, which have hydrocarbon chains of differing lengths, with various degrees of saturation (i.e., some come with carbon–carbon double bonds), that end with carboxylic acid (–COOH). Lipids include fats and oils (triglycerides), phospholipids, waxes, and steroids. Lipids contribute to some of the vital processes of living organisms, and they are an essential component of the cell membrane.

2.5.4 Nucleic Acids

Nucleic acids are complex organic substances present in living cells, and they are the main information-carrying molecules that make up genetic materials. The two main classes of nucleic acids are deoxyribonucleic acid (DNA) and ribonucleic acid (RNA). A *nucleotide* is a basic structural unit and building block for nucleic acids; a nucleotide is composed of three parts: five-sided sugar, phosphate group, and nitrogen-containing base.

There are four types of nitrogen-containing bases in the nucleic acid of DNA: adenine (A), guanine (G), thymine (T), and cytosine (C). UV disinfection is becoming popular in water and wastewater disinfection. Thymine of microorganisms undergoes more readily chemical changes, than the other bases, under UV light irradiation to produce a thymine dimer from two adjacent thymine molecules. If the thymine dimers are in the vital areas of the DNA, the microorganisms cannot reproduce. It is the main mechanism for effective UV disinfection.

2.6 OTHER ORGANIC COCs

2.6.1 Per- and Polyfluoroalkyl Substances

Hundreds of products used in our daily lives are made with highly toxic fluorinated chemicals in the family of PFAS, having carbon atoms linked to each other and bonded to fluorine atoms at most or all of the available carbon bonding sites [Note: Prefix "per-" in chemistry means (i) a molecule containing an oxygen-oxygen single bond (e.g., peracetic acid ($CH_3C(=O)-O-OH$)) or (ii) a molecule in which all hydrogens bonded to carbon have been replaced by an atom of the same element such as fluorine or chlorine (e.g., tetrachloroethylene (C_2Cl_4) is often named as perchloroethylene (PCE)).

PFAS are widely used, long-lasting chemicals; they break down very slowly over time. They are commonly called "forever chemicals". Many PFAS are found in the blood of people and animals all over the world and are present at low levels in a variety of food products and in the environment. PFAS are found in all environmental media (i.e., water, air, and soil) as well as fish at locations across the nation and the globe. Thousands of PFAS have been manufactured, used, and found in many different consumer, commercial, and industrial products. On March 14, 2023 USEPA proposed NPDWR for six PFAS including perfluorooctanoic acid (PFOA, $C_7F_{15}COOH$), perfluorooctane sulfonic acid (PFOS, $C_8F_{17}S(=O)_2(OH)$ - Figure 2.22), perfluorononanoic acid (PFNA, $C_8F_{17}COOH$), hexafluoropropylene oxide dimer acid (HFPO-DA, commonly known as GenX Chemicals, 2,3,3,3-tetrafluoro-2-(heptafluoropropoxy) propanoic acid, $C_6HF_{11}O_3$), perfluorohexane sulfonic acid (PFHxS, $C_6F_{13}S(=O)_2(OH)$), and perfluorobutane sulfonic acid (PFBS, $C_4F_9S(=O)_2(OH)$) (USEPA, 2023h).

2.6.2 Pesticides

The Federal Insecticides, Fungicides, and Rodenticides Act (FIFRA) is the U.S. federal law that provides the USEPA with the authority to oversee registration, distribution, sale and use of pesticides. A *pesticide* is any substance or mixture of substances intended for (i) preventing, destroying, repelling, or mitigating, any pest, (ii) use as a plant regulator, defoliant, or desiccant, or (iii) use as a nitrogen stabilizer. Pesticide products contain both "active" and "inert" ingredients. Active ingredients are the chemicals in the product that act to prevent, destroy, repel, or mitigate pests; all the other ingredients are called inert ingredients, which may also be toxic (USEPA, 2023i). From this definition, "pesticides" is the more general term which includes insecticides, fungicides, herbicides, termiticides, rodenticides, etc.

The pesticides on the lists of MCLs, Priority Pollutants, and HAPs include:

- Alachlor ($C_{14}H_{20}ClNO_2$)
- Aldrin ($C_{12}H_8Cl_6$)
- Atrazine ($C_8H_{14}ClN_5$)
- Captan ($C_9H_8Cl_3NO_2S$)
- Carbaryl ($C_{12}H_{11}NO_2$)
- Carbofuran ($C_{12}H_{15}NO_3$)
- Chloramben ($C_7H_5Cl_2NO_2$)
- Chlordane ($C_{10}H_6Cl_8$)
- Chlorobenzilate ($C_{16}H_{14}Cl_2O_3$)
- 2,4-D ($C_8H_6Cl_2O_3$)
- Dalapon ($C_3H_4Cl_2O_2$)
- DBCP ($C_3H_5Br_2Cl$)
- 4,4-DDD ($C_{14}H_{10}Cl_4$)
- 4,4-DDE ($C_{14}H_8Cl_4$)
- 4,4-DDT ($C_{14}H_9Cl_5$)
- Dichlorvos ($C_4H_7Cl_2O_4P$)
- Dieldrin ($C_{12}H_8Cl_6O$)
- Dinoseb ($C_{10}H_{12}N_2O_5$)
- Diquat ($C_{12}H_{12}Br_2N_2$)
- Endosulfan sulfate ($C_9H_6Cl_6O_4S$)
- Endosulfans ($C_9H_6Cl_6O_3S$)
- Endothall ($C_8H_{10}O_5$)
- Endrin ($C_{12}H_8Cl_6O$)
- Glyphosate ($C_3H_8NO_5P$)
- Heptachlor ($C_{10}H_5C_{17}$)
- Heptachlor epoxide ($C_{10}H_5C_{17}O$)

- Lindane ($C_6H_6Cl_6$)
- Methoxychlor ($C_{16}H_{15}Cl_3O_2$)
- Oxamyl (Vydate; $C_7H_{13}N_3O_3S$)
- Parathion ($C_{10}H_{14}NO_5PS$)
- Picloram ($C_6H_3Cl_3N_2O_2$)
- Propoxur (Baygon, $C_{11}H_{15}NO_3$)
- Simazine ($C_7H_{12}ClN_5$)
- 2,4,5-TP ($C_9H_7Cl_3O_3$)
- Toxaphene ($C_{10}H_8Cl_8$)
- Trifluralin ($C_{13}H_{16}F_3N_3O_4$)

2.6.3 Fats, Oils, and Grease

Animal and vegetable fats and oils are just large and complex esters. The main difference between a fat (e.g., butter) and an oil (e.g., peanut oil) is simply in the melting points of esters they contain. Grease is a general term used to describe melted animal fat or a lubricant.

The presence of *fats, oils, and grease* (FOG) in wastewater presents a challenge for Publicly Owned Treatment Works (POTWs). The USEPA's National Pretreatment Program implements the CWA's requirements to control pollutants that are introduced to POTWs. Meeting these requirements may require the elimination of discharge of FOG from food service establishments to POTWs. The regulations at 40 CFR 403.5(b)(3) specifically prohibit "solid or viscous pollutants in amounts which will cause obstruction" in a POTW and its collection system (USEPA, 2012).

2.6.4 Surfactants

Surfactants (or called *surface active agents*) are used to reduce the surface tension of liquids or the interfacial tensions of two-phase systems due to their adsorption at the surface or the interface. Most surfactants have a common chemical structure with at least one polar region and one non-polar region. The non-polar region normally consists of a long-chain alkyl group and the polar region is a polar functional group. They are often referred to a hydrophobic tail and a hydrophilic head, respectively. Surfactants are widely used as part of detergents and personal care products as well as used in industrial products as the wetting agent or the emulsifier.

Surfactants may not be readily biodegradable. Since they are not fully removed by POTWs, they may enter ambient water bodies through wastewater discharges. These surfactants may cause foam and froth and pose toxicity to living organisms in the impacted water bodies. The USEPA published Safer Choice Criteria for Surfactants and it requires that surfactants with higher aquatic toxicity demonstrate a faster rate of biodegradation without generating degradation products of concern (USEPA, 2023j).

2.6.5 Volatile Organic Compounds

Volatile organic compounds (VOCs) are a group of chemicals that can readily transform into vapor at lower temperatures. Semi-volatile organic compounds (SVOCs) are more likely to be liquids or solids at lower temperatures. Examples of SVOCs are pesticides, oil-based products, and fire retardants.

There is a regulatory definition of VOC, which can be found in 40 CFR 51.100. VOC means any compound of carbon (excluding carbon monoxide, carbon dioxide, carbonic acid, metallic carbides or carbonates, and ammonium carbonate) that participates in atmospheric photochemical reactions [Note: Many organic compounds (e.g., methane, ethane, methylene chloride, and 1,1,1-trichloroethane) are not considered as VOCs because of their negligible photochemical reactivity].

The general definition of VOCs is organic vapor compounds whose compositions make them possible to evaporate under normal atmospheric conditions. The European Union uses the boiling point, instead of volatility, to define VOCs as "any organic compound has an initial boiling point ≤250°C measured at standard pressure of 101.3 kPa". VOCs are of concern as potential indoor and outdoor air pollutants (USEPA, 2023k).

REFERENCES

USEPA (2012). "National Pretreatment Program (40 CFR 403) - Controlling Fats, Oils, and Grease Discharges from Food Service Establishments", EPA-833-F-12-003, US Environmental Protection Agency, https://www3.epa.gov/npdes/pubs/pretreatment_foodservice_fs.pdf.

USEPA (2022). "Summary of the Clean Air Act", US Environmental Protection Agency, https://www.epa.gov/laws-regulations/summary-clean-air-act, last updated September 12, 2022.

USEPA (2023a). "National Primary Drinking Water Regulations", US Environmental Protection Agency, https://www.epa.gov/ground-water-and-drinking-water/national-primary-drinking-water-regulations, last updated January 9, 2023.

USEPA (2023b). "Effluent Guidelines", US Environmental Protection Agency, https://www.epa.gov/eg, last updated June 21, 2023.

USEPA (2023c). "Summary of the Clean Water Act", US Environmental Protection Agency, https://www.epa.gov/laws-regulations/summary-clean-water-act, last updated June 22, 2023.

USEPA (2023d). "NAAQS Table", US Environmental Protection Agency, https://www.epa.gov/criteria-air-pollutants/naaqs-table, last updated March 15, 2023.

USEPA (2023e). "Controlling Hazardous Air Pollutants", US Environmental Protection Agency, https://www.epa.gov/haps/controlling-hazardous-air-pollutants, last updated May 3, 2023.

USEPA (2023f). "Learn about Polychlorinated Biphenyls", US Environmental Protection Agency, https://www.epa.gov/pcbs/learn-about-polychlorinated-biphenyls, last updated April 12, 2023.

USEPA (2023g). "Learn about Dioxin", US Environmental Protection Agency, https://www.epa.gov/dioxin/learn-about-dioxin, last updated June 1, 2023.

USEPA (2023h). "Per- and Polyfluoroalkyl Substances (PFAS)", US Environmental Protection Agency, https://www.epa.gov/sdwa/and-polyfluoroalkyl-substances-pfas, last updated June 6, 2023.

USEPA (2023i). "Pesticides", https://www.epa.gov/pesticides, US Environmental Protection, last updated August 28, 2023.

USEPA (2023j). "Safer Choice Criteria for Surfactants", https://www.epa.gov/saferchoice/safer-choice-criteria-surfactants, US Environmental Protection, last updated August 24, 2023.

USEPA (2023k). "Technical Overview of Volatile Organic Compounds", US Environmental Protection, https://www.epa.gov/indoor-air-quality-iaq/technical-overview-volatile-organic-compounds, last updated March 14, 2023.

EXERCISE QUESTIONS

[*Note: The exercise questions below are designed to help the readers of this book to be more comfortable when they come across the compounds on the lists of MCLs, Priority Pollutants, and HAPs. They are placed in a sequence following the discussion in the main text of this chapter. Going over all these questions is highly recommended. The answers to all the questions can be found readily from the main text of this chapter with some quick Internet searches, if necessary.*]

1. Draw the chemical structural formula of hexane (C_6H_{14}) and 2,2,4-trimethylpentane (C_8H_{18}) which are on the list of HAPs.
2. Vinyl bromide, vinyl chloride, and allyl chloride are on the list of HAPs. What are their chemical formula and chemical structural formula?
3. a. What are the differences among an acryl group, a phenyl group, a benzyl group, and a biphenyl group?
   b. Is "C_6H_5–" a phenyl group or a benzyl group?
4. 1,3-Butadiene, hexachlorobutadiene, and hexachlorocyclopentadiene are on the list of HAPs. What are their chemical formula and chemical structural formula?
5. Cumene (isopropylbenzene) is on the list of HAPs. (a) What are its chemical formula and chemical structural formula? (b) Why a prefix "iso" is added to the substitute "propyl"?
6. Naphthalene is a polyaromatic hydrocarbon (PAH) with two fused aromatic rings. What are its chemical formula and chemical structural formula?
7. Phenanthrene and anthracene are polyaromatic hydrocarbons (PAHs) with three fused aromatic rings. What are their chemical formula and chemical structural formula?
8. Chrysene, pyrene, and benz(a)anthracene are polyaromatic hydrocarbons (PAHs) with four fused aromatic rings. What are their chemical formula and chemical structural formula?
9. 2-Acetylaminofluorene is a polyaromatic hydrocarbon (PAH) and it is on the list of HAPs. What are its chemical formula and chemical structural formula?
10. (a) What are the chemical and structural formula of propylene glycol and glycerin? (b) How many hydroxyl groups does each propylene glycol molecule have? (c) How many hydroxyl groups does each glycerin molecule have?
11. "Cresols/cresylic acid (isomers and mixtures)" are on the list of HAPs. The three isomers of cresols (i.e., o-, m-, p-cresol) have different industrial uses; and mixed cresols are used as disinfectants, preservatives, and wood preservatives. What are the chemical formula of cresols and the chemical structural formula of these three isomers?

12. Phenol, o-, m-, and p-cresols, and catechol are on the list of HAPs. Xylenols are dimethyl phenols. (a) What are their chemical formula and chemical structural formula? (b) How many hydroxyl (–OH) group(s) does each molecule have? (c) How many methyl (–CH_3) group(s) does each molecule have?
13. What are the functional groups of ketones and aldehydes?
14. Both quinone and hydroquinone are on the list of HAPs. Quinones are ketones in which carbonyl groups are a part of aromatic ring of benzene, anthracene, or naphthalene. Hydroquinone is in the phenol family. What are the chemical formula and chemical structural formula of 1,4-benzoquinone and hydroquinone?
15. Both quinone and quinoline are on the list of HAPs. Quinones are ketones in which carbonyl groups are a part of aromatic ring of benzene, anthracene, or naphthalene. Quinoline is in the amine family. What are the chemical formula and chemical structural formula of 1,2-benzoquinone and quinoline?
16. Both acetophenone ($C_6H_5COCH_3$) and 2-chloroacetophenone ($C_6H_5COCH_2Cl$) are in the ketone family and they are on the list of HAPs. What are their chemical structural formula?
17. What are the functional groups of carboxylic acids and anhydrides?
18. What are the functional groups of acetates and acrylates?
19. Both ethyl acrylate and methyl methacrylate are on the list of HAPs. They have the same chemical formula of $C_5H_8O_2$. What are their chemical structural formula?
20. What are the functional groups of esters and ethers?
21. Bis(2-ethylhexyl) phthalate (DEHP; $C_{24}H_{38}O_4$) is on the lists of MCLs, Priority Pollutants, and HAPs. What is its chemical structural formula?
22. Phthalic anhydride ($C_6H_4(C{=}O)_2O$) is the anhydride of the phthalic acid, and it is on the list of HAPs. What is its chemical structural formula?
23. Bis(chloromethyl) ether ($C_2H_4Cl_2O$) and chloromethyl methyl ether (C_2H_5ClO) are on the list of HAPs. What are their chemical structural formula?
24. Diethanolamine is on the list of HAPs. What are its chemical formula and chemical structural formula? How many functional groups does it have and what are their types?
25. Triethylamine ($(C_2H_5)_3N$), diethanolamine ($(CH_2CH_2OH)_2NH$), N-nitrosodimethylamine (NDMA, $C_2H_6N_2O$), p-phenylenediamine ($C_6H_4(NH_2)_2$), 1,2-propylenimine ($CH_3CH(NH)CH_2$), and 2,4-toluene diamine ($C_6H_3(NH_2)_2CH_3$) are on the list of HAPs. What are their chemical structural formula?
26. N,N-dimethylaniline ($C_6H_5N(CH_3)_2$, o-anisidine (o-methoxyaniline, $CH_3OC_6H_4NH_2$), 4,4′-methylenedianiline ($CH_2(C_6H_4NH_2)_2$), 4,4-methylene bis(2-chloroaniline) ($C_{13}H_{12}Cl_2N_2$), and o-toluidine ($C_6H_4CH_3NH_2$) are on the list of HAPs. What are their chemical structural formula?
27. Benzidine ($C_{12}H_{12}N_2$), 3,3′-dimethyl benzidine ($C_{14}H_{16}N_2$), 3,3′-dimethoxybenzidine ($C_{14}H_{16}N_2O_2$), and 3,3′-dichlorobenzidine ($C_{12}H_{10}Cl_2N_2$) are on the list of HAPs. What are their chemical structural formula?

28. Hexamethylphosphoramide ($C_6H_{18}N_3OP$) is a phsophoramide and it is on the list of HAPs. What is its chemical structural formula?
29. Aziridine (ethylene imine, $(CH_2)_2NH)$) is on the list of HAPs. What is its chemical structural formula?
30. Acetonitrile and acrylonitrile (C_3H_3N) are on the list of HAPs. What are their chemical structural formula?
31. Hydrazine, an inorganic compound, and methyl hydrazine, an organic compound, are on the list of HAPs. What are their chemical formula and chemical structural formula?
32. What are the functional groups of nitro- and nitroso- compounds?
33. Both quinone and quinoline are on the list of HAPs. (a) What are the chemical formula and chemical structural formula of quinones and quinoline? (b) Which one is in the pyridine family?
34. Both phosgene and phosphine are on the list of HAPs. (a) What are their chemical formula and chemical structural formula? (b) Are they organic or inorganic?
35. Dimethyl sulfate and diethyl sulfate are on the list of HAPs. What are their chemical formula and chemical structural formula?
36. 1,3-Butadiene and chloroprene are on the list of HAPs. What is the IUPAC name of chloroprene? What are the chemical formula and chemical structural formula of these two compounds?
37. What are the "methoxy", "ethoxy", and "phenoxy" functional groups?
38. Phosgene ($COCl_2$), 1,4-dioxane (1,4-diethyleneoxide, $C_4H_8O_2$), ethyl carbamate (urethane, $C_3H_7NO_2$), and dimethyl carbamoyl chloride (C_3H_6ClNO) are on the list of HAPs [Note: They were not discussed in the main text of this Chapter]. What are their chemical structural formula?

3 Inorganic Chemistry for Environmental Engineers

3.1 INTRODUCTION

According to American Chemical Society (ACS), inorganic chemistry is concerned with the properties of inorganic compounds, which include metals, minerals, and organometallic compounds. While organic chemistry is the study of carbon-containing compounds (with some exceptions), inorganic chemistry is the study of the remaining subset of compounds.

3.1.1 Types of Inorganic Compounds

In general, two types of inorganic compounds can be formed; they are (i) ionic compounds that are formed between metals and nonmetals (e.g., NaCl, $Ca(OH)_2$, $CaCl_2$) and (ii) molecular compounds that are formed between nonmetals (e.g., NH_3, H_2O, CO_2).

3.1.2 Properties of Inorganic Compounds

General properties of inorganic compounds include the following:

- Most of them are not flammable/combustible (i.e., having a low energy content).
- Most of them have relatively high melting points and high boiling points.
- Bonds are ionic bonds for inorganic ionic compounds and covalent bonds for inorganic molecular compounds.
- Inorganic ionic compounds are soluble in water and will dissociate into ions readily.
- Aqueous solutions of inorganic ionic compounds are conductive.
- Inorganic compounds do not have isomers.

3.1.3 Classification of Inorganic Compounds

Inorganic compounds are also often classified into the following groups: (i) oxides, (ii) hydroxides, (iii) hydrides, (iv) inorganic acids, (v) inorganic bases, (vi) salts, (vii) carbonates, and (viii) silicates and aluminosilicates.

1. ***Oxides.*** Oxides are binary compounds of oxygen atom(s) with another element. Oxides can be further grouped into (i) acidic, (ii) basic, (iii) amphoteric, and (iv) neutral oxides.

DOI: 10.1201/9781003502661-3

Acidic oxides are oxides of nonmetals (e.g., CO_2), and they are *acid hydrides* because they form acids (e.g., carbonic acid (H_2CO_3)) with water as:

$$CO_{2(g)} + H_2O_{(l)} \rightarrow H_2CO_{3(aq)} \tag{3.1}$$

- *Basic oxides* are oxides of metals, especially Group 1 (i.e., alkali metals) and Group 2 metals (i.e., alkaline earth metals), for example, sodium oxide (Na_2O) and calcium oxide (CaO). Basic oxides are *base hydrides* because they form bases (e.g., sodium hydroxide (NaOH) and calcium hydroxide ($Ca(OH)_2$) with water as:

$$Na_2O_{(s)} + H_2O_{(l)} \rightarrow 2\ NaOH_{(aq)} \tag{3.2}$$

$$CaO_{(s)} + H_2O_{(l)} \rightarrow Ca(OH)_{2(s)} \tag{3.3}$$

- Oxides of metals are often referred to as *metallic oxides*; oxides of non-metals are referred to as *non-metallic oxides*.
- *Amphoteric oxides* are oxides that can react with a base as an acid, and they can also react with an acid as a base. Zinc oxide (ZnO) and aluminum oxide (Al_2O_3) are good examples of amphoteric oxides:

$$ZnO_{(s)} + 2HCl_{(l)} \rightarrow ZnCl_{2(s)} + H_2O_{(l)} \tag{3.4}$$

$$ZnO_{(s)} + 2OH^-_{(aq)} \rightarrow ZnO_{2\ (aq)}^{-2} + H_2O_{(l)} \tag{3.5}$$

- Carbon monoxide (CO), nitric oxide (NO), and nitrous oxide (N_2O) are *neutral oxides* because they show neither acidic nor basic properties. It should be noted that (i) CO_2 is acidic, (ii) nitrous oxide is a major greenhouse gas (GHG), and (iii) nitrogen dioxide (NO_2) and sulfur oxides (i.e., sulfur dioxide (SO_2) and sulfur trioxide (SO_3)) are the main gaseous pollutants in the air that are responsible for the acid rain problems.

2. ***Hydroxides.*** In Chapter 2, we discussed the fact that the hydroxyl (–OH) group is the functional group of alcohols and phenolic compounds. A *hydroxide* (OH^-) is an anion (also called a *hydroxyl ion*) consisting of one oxygen atom and one hydrogen atom bonded by a covalent bond, and it functions as a base. The inorganic hydroxides consist of Group 1 and Group 2 metals, and these hydroxides are important industrial alkali (see Eqs. 3.2 and 3.3).
3. ***Hydrides.*** *Hydrides* are binary compounds that contain hydrogen atom(s) with other elements in the periodic table. The *covalent hydrides* consist of hydrogen and other nonmetals, held together by covalent bonds; for example, NH_3, HF, and HCl. The *ionic hydrides* are often referred to as compounds consisting of hydrogen and the alkali metals or the alkaline earth metals (e.g., NaH and MgH_2), in which hydrogen acts as a hydride ion (H^-) with a negative charge. The third group of hydrides is *metallic hydrides* consisting of hydrogen and transition metals(e.g., GeH_4 and AsH_3).

4. ***Inorganic acids.*** One definition of acids is "compounds that can generate hydrogen ions (H^+) when dissolved in water". Acetic acid (CH_3COOH) is an organic acid, while HCl, HNO_3, H_2SO_4, and H_2CO_3 are inorganic acids.
5. ***Inorganic bases.*** One definition of bases is "compounds that can generate hydroxyl ions (OH^-) when dissolved in water". Methylamine (CH_3NH_2) is an organic base, whereas KOH, NaOH, and $Ca(OH)_2$ are inorganic bases.
6. ***Salts.*** A salt consists of the cation of a base and the anion of an acid. For example, the table salt consists of Na^+ (a cation) of NaOH (a base) and Cl^- (an anion) of HCl (an acid).
7. ***Carbonates.*** Carbonate ion is a polyatomic ion with a chemical formula of CO_3^{-2}. A *carbonate* is a salt of carbonic acid (H_2CO_3) or carbonate minerals; the commonly known carbonates are sodium carbonate (Na_2CO_3) and calcium carbonate ($CaCO_3$) which is the main constituent of limestone. *Calcination* is to decompose limestone (i.e., a sedimentary rock composed mainly of $CaCO_3$) in an oven into calcium oxide (CaO) and carbon dioxide; and *slaking* is to add water to quick lime (i.e., burnt lime, CaO) to form hydrated lime (i.e., slaked lime, $Ca(OH)_2$):

$$\text{Calcination}: CaCO_{3(s)} \xrightarrow{\text{Heat}} CaO_{(s)} + CO_{2(g)} \tag{3.6}$$

$$\text{Slaking}: CaO_{(s)} + H_2O_{(l)} \rightarrow Ca(OH)_{2(s)} \tag{3.7}$$

Carbonation is to convert $Ca(OH)_2$ in a slurry form to the $CaCO_{3(s)}$ precipitate:

$$\text{Carbonation}: Ca(OH)_{2(s)} + CO_{2(g)} \rightarrow CaCO_{3(s)} + H_2O_{(l)} \tag{3.8}$$

These processes are commonly practiced in water treatment and in air pollution control.
8. ***Silicates and aluminosilicates.*** *Silicates* are a large class of chemical compounds composed of silicon (Si), oxygen, and at least one metal element. The basic structural unit of all silicate minerals is the silicon tetrahedron $\left(SiO_4^{-4}\right)$ in which a silicon atom is covalent-bonded to four oxygen atoms, each at the corners of the tetrahedron. The silicates make up ~97% of the Earth's crust. Silica sand is made of silicon dioxide (SiO_2). Asbestos is a naturally occurring fibrous silicate mineral, and it is on the lists of MCLs, Priority Pollutants, and HAPs (see Chapter 2).

Aluminosilicates are compounds containing oxides of both silicon and aluminum. Clay minerals are a class of aluminosilicates with a layer structure from weathering of other silicate minerals. The presence of clayey materials complicates the fate and transport of the contaminants in soil and groundwater and the effectiveness of many groundwater and soil remediation processes. *Zeolites* are aluminum silicates with porous structures and commonly used as adsorbents for removal of contaminants and as catalysts for some chemical reactions.

3.2 CHEMISTRY OF METALS

Metals are natural elements of Earth's crust, in which they are generally found in the form of metal ores (i.e., mineral-bearing substances), bonding with each other or with many other elements. They are also naturally present in rocks and in atmospheric dusts (many metal compounds are on the list of HAPs – see Chapter 2). The metals in the rocks or in the groundwater matrices could be leached out by surface water or groundwater through weathering and/or dissolution processes.

Approximately three-quarters of all known chemical elements are metals. They are present on the left-hand side of the periodic table (see Figure 3.1). Metals are often classified based on their positions in the periodic table. One way is to group them into (i) main-group metals (i.e., elements in Groups 1, 2, 13, and 14 of the periodic table, and bismuth (Bi) in Group 15 by some) and (ii) transition metals (elements in Groups 3–12 of the periodic table).

3.2.1 MAIN-GROUP METALS

Group 1 metals – alkali metals. Alkali metals include lithium (Li), sodium (Na), potassium (K), rubidium (Rb), cesium (Cs), and francium (Fr). Their general properties include:

- Their valence shells contain only one electron; consequently, they tend to lose that valence electron to become single-valent cations. They are the most electropositive elements (in other words, the least electronegative elements). The ionization energies are very low.
- They are widely used as reducing agents; the oxidation number would change from zero to +1 in the final product (see more later for the oxidation state/number). They are extremely reactive when exposed to oxidizing agents such as oxygen and halogens.
- They have low melting points, boiling points, and densities.
- They are "soft" metals.

	1	2	3	4	5	6	7	8	9	10	11	12	13	14	15	16	17	18
1	1																1	2
2	3 Li	4 Be											5	6	7	8	9	10
3	11 Na	12 Mg											13 Al	14	15	16	17	18
4	19 K	20 Ca	21 Sc	22 Ti	23 V	24 Cr	25 Mn	26 Fe	27 Co	28 Ni	29 Cu	30 Zn	31 Ga	32	33	34	35	36
5	37 Rb	38 Sr	39 Y	40 Zr	41 Nb	42 Mo	43 Tc	44 Ru	45 Rh	46 Pd	47 Ag	48 Cd	49 In	50 Sn	51	52	53	54
6	55 Cs	56 Ba	57-71 La	72 Hf	73 Ta	74 W	75 Re	76 Os	77 Ir	78 Pt	79 Au	80 Hg	81 Tl	82 Pb	83 Bi	84	85	86
7	87 Fr	88 Ra	89-103 Ac	104 Rf	105 Db	106 Sg	107 Bh	108 Hs	109 Mt	110 Ds	111 Rg	112 Cn	113 Nh	114 Fl	115 Mc	116 Lv	117 Ts	118

FIGURE 3.1 Metals on the periodic table.

- Oxides of Group 1 metals are basic oxides (e.g., Na_2O, K_2O), and they react with water to form hydroxide (e.g., NaOH and KOH).

Group 2 metals – alkaline earth metals. Alkaline earth metals include beryllium (Be), magnesium (Mg), calcium (Ca), strontium (Sr), barium (Ba), and radium (Ra). Their general properties include:

- Their valence shells contain two electrons. Consequently, they tend to lose those two valence electrons to become di-valent cations.
- Their oxidation number/state in chemical compounds is typically +2.
- They have higher melting points, boiling points, and densities than alkali metals.
- Calcium ions (Ca^{+2}) and magnesium ions (Mg^{+2}) are the main species responsible for water hardness [Note: The hardness of water is a measure of multi-valent cation concentrations in water].
- Oxides of Group 2 metals are basic oxides (e.g., MgO, CaO) and they react with water to form hydroxide (e.g., $Mg(OH)_2$ and $Ca(OH)_2$).

Group 13 metals. Aluminum (Al), gallium (Ga), indium (In), and thallium (TI) are the four metals in Group 13 [Note: Properties of nihonium (Nh) have not been well determined/tested]. Their valence shells contain three electrons. Aluminum is abundant in soil and aluminum ions (Al^{+3}) are commonly present in water bodies. Alum (aluminum sulfate) is an inorganic compound containing aluminum, and it is commonly used as a coagulant in water treatment for the removal of fine particulates.

Group 14 metals. Tin (Sn) and lead (Pb) are the two metal elements in Group 14. They have two oxidation states: +2 and +4. The presence of lead in environmental media, especially air and water, is of great concern because of its toxicity [Note: Pb is not a transition metal].

3.2.2 Transition Metals

Transition metals are the elements in Groups 3–12 of the periodic table (see Figure 3.2). The general characteristics of the transition metals include:

- They have two valence electron shells, instead of one.
- They can have several different oxidation states.
- They are more electronegative than the main-group metals so that they are more likely to form covalent compounds (versus ionic compounds).
- Most of them are hard and strong and have high melting and boiling points.
- They are good conductors of heat and electricity.
- Many of them, at elevated concentrations, are considered toxic to human and aquatic species.

The lanthanides and the actinides at the bottom of the periodic table are sometimes known as the *inner transition metals* because their atomic numbers fall between the first and the second elements in the last two rows of the transition metals.

Some scientists believe that Group 12 elements (i.e., zinc (Zn), cadmium (Cd), and mercury (Hg)) should be classified as the main group elements/metals because

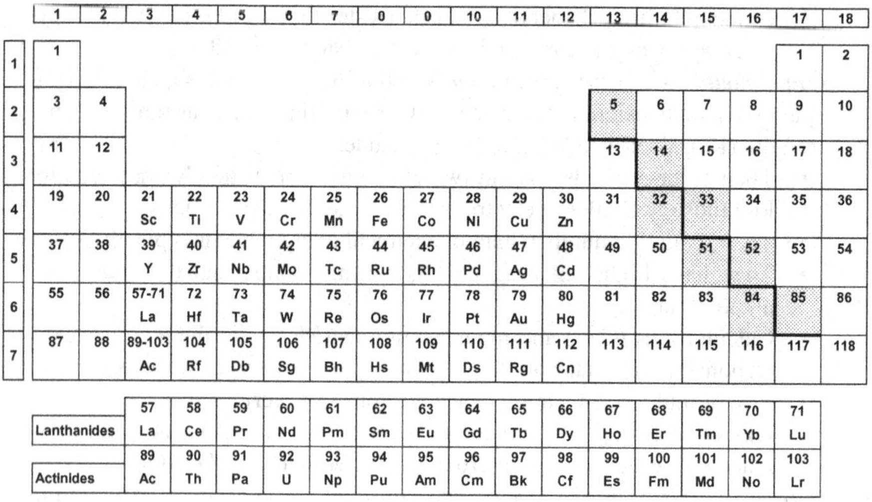
1 2 3 4 5 6 7 8 9 10 11 12 13 14 15 16 17 18
1 1 1 2
2 3 4 5 6 7 8 9 10
3 11 12 13 14 15 16 17 18
4 19 20 21 Sc 22 Ti 23 V 24 Cr 25 Mn 26 Fe 27 Co 28 Ni 29 Cu 30 Zn 31 32 33 34 35 36
5 37 38 39 Y 40 Zr 41 Nb 42 Mo 43 Tc 44 Ru 45 Rh 46 Pd 47 Ag 48 Cd 49 50 51 52 53 54
6 55 56 57-71 La 72 Hf 73 Ta 74 W 75 Re 76 Os 77 Ir 78 Pt 79 Au 80 Hg 81 82 83 84 85 86
7 87 88 89-103 Ac 104 Rf 105 Db 106 Sg 107 Bh 108 Hs 109 Mt 110 Ds 111 Rg 112 Cn 113 114 115 116 117 118
Lanthanides 57 La 58 Ce 59 Pr 60 Nd 61 Pm 62 Sm 63 Eu 64 Gd 65 Tb 66 Dy 67 Ho 68 Er 69 Tm 70 Yb 71 Lu
Actinides 89 Ac 90 Th 91 Pa 92 U 93 Np 94 Pu 95 Am 96 Cm 97 Bk 98 Cf 99 Es 100 Fm 101 Md 102 No 103 Lr

FIGURE 3.2 Transition metals on the periodic table.

they have some properties similar to those of the main-group metals. They are soft, diamagnetic, and divalent metals (mercury can form both mono-valent and di-valent ions). They are the transition metals with the lowest melting points (mercury is the only "liquid" metal under ambient conditions).

3.2.3 Other Classifications of Metals

Heavy metals. The term "heavy metals" is commonly used, but many different definitions exist (based on density, atomic number, or toxicity). Heavy metals are generally referred to as any metallic element that has a relatively high density (>5 g/cm^3) with an atomic mass greater than 20 [Note: Scandium(Sc) is the transition metal with the smallest atomic mass of 21] and is toxic or poisonous at low concentrations. Heavy metals are dangerous because they tend to accumulate in living organisms and human beings (i.e., bioaccumulation). Examples of heavy metals include mercury (Hg), cadmium (Cd), arsenic (As), chromium (Cr), nickel (Ni), copper (Cu), zinc (Zn), and lead (Pb). Elevated concentrations of heavy metals in environmental media (i.e., water, air, and soil) are of concern.

Metals of environmental concern. Metals (and metalloids) are electropositive elements that occur in all ecosystems. Natural concentrations of metals in water bodies vary according to local geology. All the environmental systems are interconnected. Using water bodies as an example, metals can reach water bodies when they are released into air, water, and soil. While some metals are essential as nutrients for living organisms and humans, all metals can be toxic at some level. Some metals are even toxic in minute amounts (USEPA, 2023a).

Many metals are on the USEPA's lists of MCLs, Priority Pollutants, and HAPs (see Chapter 2). There are 13 elements on EPA's Priority Pollutant List; they are antimony (Sb), arsenic (As), beryllium (Be), cadmium (Cd), chromium (Cr), copper (Cu), lead (Pb), mercury (Hg), nickel (Ni), selenium (Se), silver (Ag), thallium (Tl), and zinc (Zn). They are often referred to as "*13 Priority Pollutant Metals*"; however, it should be noted that antimony (Sb) and arsenic (As) are semi-metal, and selenium (Se) is considered as a nonmetal or semi-metal by some. Eight metals (i.e., arsenic (As), barium (Ba), cadmium (Cd), chromium (Cr), lead (Pb), mercury (Hg), selenium (Se), and silver (Ag)) are also regulated by USEPA's Resource Conservation and Recovery Act (RCRA); they are often referred to as "*RCRA 8 metals*". RCRA is a federal law that provides a structure for proper management of hazardous and non-hazardous wastes with regards to their generation, transportation, storage, treatment, and final disposal (USEPA, 2023b).

Table 3.1 tabulates the names and their group numbers in the periodic table of metals that are on the lists of MCL, Pollutant Metals, HAPs, and RCRA. The regulatory limits for metals on the lists of MCLs and Pollutant Pollutants are set on their dissolved concentrations in water. Those for RCRA 8 metals are the leachate concentrations extracted from solid waste samples using the Toxicity Characteristic Leaching Procedure (TCLP) test (USEPA, 2023c). The metals on the HAPs list are more inclusive; they are not the metal elements, but the metal compounds).

TABLE 3.1
Metals on the Lists of MCLs, Priority Pollutants, HAPs, and RCRA

Metal	Group	MCLs	Priority Pollutants	HAPs	RCRA
Barium (Ba)	2	√			√
Beryllium (Be)	2	√	√	√	
Chromium (Cr)	6	√	√	√	√
Manganese (Mn)	7			√	
Cobalt (Co)	9			√	
Nickel (Ni)	10		√	√	
Copper (Cu)	11	√	√		
Mercury (Hg)	11	√	√	√	√
Silver (Ag)	11		√		√
Cadmium (Cd)	12	√	√	√	√
Zinc (Zn)	12		√		
Thallium (Tl)	13	√	√		
Lead (Pb)	14	√	√	√	√
Arsenic (As)	15	√	√	√	√
Antimony (Sb)	16	√	√	√	
Selenium (Se)	16	√	√	√	√

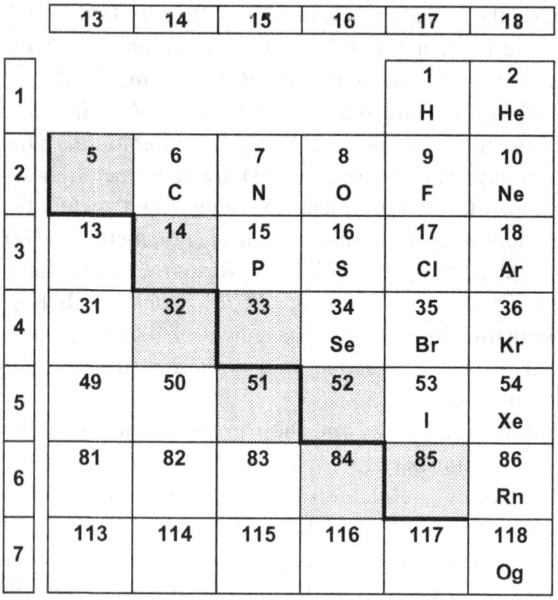
13 14 15 16 17 18
1 H 2 He
5 6 C 7 N 8 O 9 F 10 Ne
13 14 15 P 16 S 17 Cl 18 Ar
31 32 33 34 Se 35 Br 36 Kr
49 50 51 52 53 I 54 Xe
81 82 83 84 85 86 Rn
113 114 115 116 117 118 Og
1
2
3
4
5
6
7

FIGURE 3.3 Nonmetals on the periodic table.

3.3 CHEMISTRY OF NONMETALS

The non-metal elements are located in the upper-right corner of the periodic table, and they include the halogens, the noble gases, and several other elements (i.e., H, C, N, P, O, S, and Se) – see Figure 3.3.

3.3.1 General Properties of Nonmetals

General properties of nonmetals include the following:

- Poor conductivity of heat and electricity
- High electronegativity and ionization energies
- Lower melting points and boiling points (when compared to metals).

3.3.2 The Noble Gases

The noble gases are in Group 18 (the last column) of the periodic table, and they contain helium (He), neon (Ne), argon (Ar), krypton (Kr), xenon (Xe), radon (Rn), and oganesson (Og) [Note: Og is a synthetic chemical element and its properties have not been well determined/tested].

Since their shell of valance electrons is "full", the noble gases are typically unreactive except under extreme conditions. However, it should be noted that radon is radioactive and its presence in ambient soil and water is of environmental concern.

3.3.3 The Halogens

The halogens are in Group 17 (the second from the last column) of the periodic table; and they contain fluorine (F), chlorine (Cl), bromine (Br), iodine (I), astatine (At), and tennessine (Ts) [Note: At is often omitted from discussions of the halogens; and the properties of Ts are not well determined/tested].

The halogens have seven valence electrons, and they are strong oxidizing agents to react with other substances to achieve its "–1" oxidation state. The halogens can combine with other elements to form compounds such as halides, such as fluorides, chlorides, bromides, and iodides (e.g., hydrogen fluoride (HF) and sodium chloride (NaCl)). The halogenated hydrocarbons (i.e., hydrocarbons contain one or more halogen element) are of environmental concern because they are often toxic and pose risks to human and the environment. Many chlorinated organic compounds are toxic, and they are on the lists of MCLs, Priority Pollutants, and HAPs.

Chlorine gas (Cl_2) is a very strong oxidizing agent. Chlorine, chloramines (i.e., monochloramine (NH_2Cl) and dichloramine ($NHCl_2$)), and chlorine dioxide (ClO_2) are disinfectants used in water treatment. Bromate $\left(BrO_3^-\right)$, chlorite (ClO^-), haloacetic acid (HAAs) and total trihalomethanes (THMs) are common disinfection byproducts (DBPs). They are all on the list of MCLs. Hydrochloric acid (HCl) and HF are strong acids, and they are on the list of HAPs [Note: Chlorine gas is also on the list].

3.3.4 Other Nonmetal Elements

Other nonmetal elements of concern include hydrogen (H), carbon (C), nitrogen (N), oxygen (O), phosphorus (P), sulfur (S), and selenium (Se).

1. ***Hydrogen.*** With an atomic number of 1, hydrogen (H) is the lightest atomic element. Hydrogen is a nonmetal, and it is placed on the tops of Group 1 (alkali metals) and Group 17 (halogens) on most periodic tables. The reason for being placed on the top of Group 1 is mainly because it has one valence electron, same as the alkali metals. Under ambient conditions, hydrogen is a diatomic gas (H_2), which is highly combustible (often considered as a clean energy source). Hydrogen is non-metallic (except it becomes metallic at extremely high pressures and that is another reason why it shows up on both sides of the periodic table in Figure 1.1). Hydrogen readily forms a single covalent bond with most non-metallic elements to form inorganic molecules such as water (H_2O) and organic molecules (e.g., methanol CH_3OH). It can form inorganic acids such as hydrochloric acid (HCl), sulfuric acid (H_2SO_4), nitric acid (HNO_3), and phosphoric acid (H_3PO_4) as well as organic acids such as acetic acid (CH_3COOH). Consequently, it plays an important role in acid-base reactions in the form of proton (i.e., hydrogen ion, H^+). Hydrogen can also have an oxidation state of "–1" in metal hydrides (e.g., calcium hydride CaH_2, aluminum hydride AlH_3).
2. ***Carbon.*** Carbon (C) has an atomic number of 6 and belongs to Group 14 of the periodic table. Carbon can form many *allotropes* (i.e., structurally different forms of the same element), including graphite diamond, amorphous

carbon, graphene, fullerene, and nanotubes. Carbon atoms are the back bones of the organic compounds.

The largest sources of inorganic carbon are limestones (i.e., calcium carbonate $CaCO_3$), dolomite (calcium magnesium carbonate $CaMg(CO_3)_2$), and carbon dioxide (CO_2). A *carbide* is a compound composed of carbon and a metal (e.g., sodium carbide Na_2C_2 and calcium carbide CaC_2).

3. ***Nitrogen.*** Nitrogen (N) has an atomic number of 7 and belongs to group 15 of the periodic table. Nitrogen is a constituent of all living matter; nitrogen along with phosphorus are considered as essential nutrients when environmental engineers deal with biological processes and environmental pollution issues related to biological activities.

   Nitrogen gas (N_2) is the most abundant element in our atmosphere (~79% by volume). With regards to air pollution, nitrogen dioxide (NO_2) is one of the six ambient air criteria pollutants regulated by CAA's National Ambient Air Quality Standards (NAAQS). Nitrous oxide (N_2O), along with CO_2 and CH_4, are the top three greenhouse gases (GHGs) of concern. Nitrogen dioxide along with sulfur oxides (i.e., sulfur dioxide (SO_2) and sulfur trioxide (SO_3)) are considered as the main precursors for the acid rain problems. Hydrazine (N_2H_4), calcium cyanamide ($CaCN_2$), and cyanide (CN^-) compounds are on the list of HAPs.

   With regard to water quality, nitrite $\left(NO_2^-\right)$ and nitrate $\left(NO_3^-\right)$ are on the list of MCLs because of their toxicity to human. Cyanide is on the list of MCLs; and total cyanide (i.e., the sum of all cyanide species that are converted to hydrogen cyanide following the laboratory reflux procedure) is on the list of Priority Pollutants because of its toxicity to human and aquatic species. Ammonium (NH_4^+) and nitrate are also regulated in wastewater discharge because of their aquatic toxicity and toxicity to human, respectively. Nitric acid (HNO_3) is a strong acid.

4. ***Phosphorus.*** Phosphorus (P) has an atomic number of 15 and belongs to Group 15 of the periodic table. Phosphorus is a constituent of all living matter; phosphorus along with nitrogen are considered as essential nutrients when environmental engineers deal with biological processes and environmental pollution issues related to biological activities.

   Phosphoric acid (H_3PO_4) is a weak acid. *Orthophosphates* are derivatives of phosphoric acid, including dihydrogen phosphate $\left(H_2PO_4^-\right)$, hydrogen phosphate $\left(HPO_4^{-2}\right)$, and phosphate $\left(PO_4^{-3}\right)$. *Polyphosphate* is the polymer of orthophosphates (e.g., sodium triphosphate $Na_3P_3O_{10}$) and can be found in all living species. Phosphine (PH_3) and phosphorus (P) are on the list of HAPs.

5. ***Oxygen.*** Oxygen (O) has an atomic number of 8 and belongs to Group 16 of the periodic table. Oxygen is the most abundant element in the Earth's crust (~46% of the mass) and in the seawater (~90% of the mass). Oxygen gas (O_2) occupies ~21% by volume of our atmosphere. In soil/rocks, oxygen is combined with metals and metalloids in the form of oxides such as metal oxides, silicates, and aluminosilicates.

Oxygen is essential for aerobic species and human (e.g., oxygen content in the air and dissolved oxygen (DO) in aqueous systems are critical in supporting lives).

Oxygen has two allotropes: oxygen gas (O_2) and ozone gas (O_3). Both O_2 and O_3 are oxidizing agents, and O_3 is a stronger one. Ozone has been used as a water disinfectant. Ozone in the stratosphere protects us from harmful UV irradiation, while ozone in the troposphere poses hazard to human beings and the environment.

The typical oxidation state of oxygen in oxides is −2; while those in peroxides (e.g., hydrogen peroxide, H_2O_2) and in superoxide (e.g., potassium superoxide, KO_2) are −1 and −1/2, respectively. Hydrogen peroxide and potassium peroxide (K_2O_2) are strong oxidizing agents.

6. ***Sulfur.*** Sulfur (S) has an atomic number of 16 and belongs to Group 16 of the periodic table (same as oxygen). Sulfur is reactive, and it can react with most metals to form sulfides as well as forms compounds with several non-metallic elements.

   Many fossil fuels (i.e., coal, petroleum, and natural gas) contain elemental sulfur and sulfur compounds. Combustion of these fossil fuels could release sulfur dioxide (SO_2) and sulfur trioxide (SO_3) into the atmosphere; sulfur oxides along with nitrogen dioxide are precursors of acid rain problems.

   Reduced sulfur compounds (RSCs) are formed from anaerobic degradation of organic materials. The RSCs of environmental concern include hydrogen sulfide (H_2S), methyl mercaptan (CH_3SH), dimethyl sulfide ($(CH_3)_2S$), dimethyl disulfide ($(CH_3S)_2$), and other volatile compounds containing reduced sulfur. Hydrogen sulfide, carbon disulfide (CS_2), and carbonyl sulfide (O=C=S) gases are on the list of HAPs. Sulfuric acid (H_2SO_4) is a strong acid, and sulfate ion $\left(SO_4^{-2}\right)$ is a water quality parameter.
7. ***Selenium.*** Selenium is a chemical element with the symbol Se and an atomic number of 34. It is in Group 16 of the periodic table and placed below sulfur (S). It is interesting to note that it is considered as nonmetal by some, while considered as metalloids by others. It is on the lists of MCLs, Priority Pollutants, and HAPs. It is often included as one of the "13 Priority Pollutant Metals".

3.4 CHEMISTRY OF SEMI-METALS

Semi-metals or *metalloids* are chemical elements that display properties of both metals and nonmetals. For example, some metalloids can have a high melting point (like a metal), but a low density (like a nonmetal). In general, metalloids have the physical properties of metals, but the chemical properties of nonmetals.

The metalloids are boron (B), silicon (Si), germanium (Ge), arsenic (As), antimony (Sb), tellurium (Te), and polonium (Po). These elements are in Groups of 13 to 16 of the periodic table, along a zig-zag line between boron and aluminum down to polonium and astatine (At) – see Figure 3.4. They have lower heat and electrical conductivity than metals. They tend to make good semiconductors.

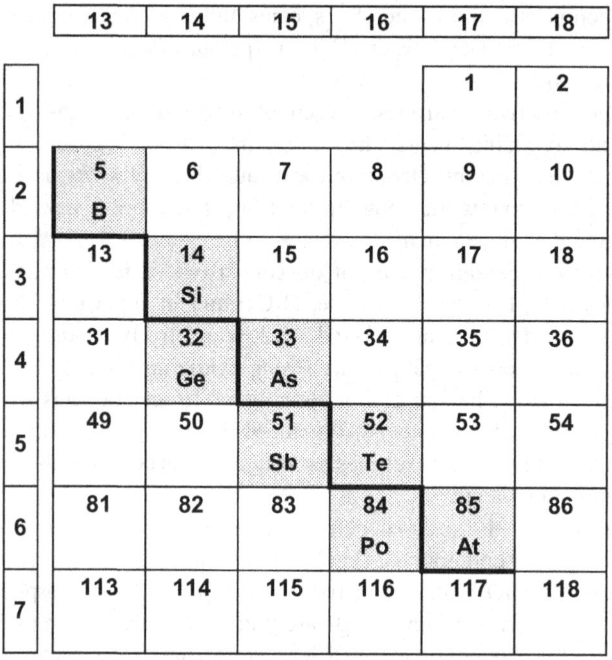
13 14 15 16 17 18
1
1 2
2
5 B 6 7 8 9 10
3
13 14 Si 15 16 17 18
4
31 32 Ge 33 As 34 35 36
5
49 50 51 Sb 52 Te 53 54
6
81 82 83 84 Po 85 At 86
7
113 114 115 116 117 118

FIGURE 3.4 Metalloids on the periodic table.

3.5 ORGANOMETALLIC COMPOUNDS

Organometallic compounds are chemical compounds that contain at least one bond between a metallic (or semi-metallic) element and a carbon atom belonging to an organic molecule. This bond is generally covalent in nature. The organometallic compounds are used in some industrial chemical reactions, often serving as catalysts. Many of these organometallic compounds have been found toxic to humans, especially those are volatile in nature.

Chlorophyll is the green pigment in algae, plants, and cyanobacteria that is essential for photosynthesis. It is an organometallic compound. Its central structure is a complex aromatic macrocyclic ring with a sequestered magnesium atom. Chlorophyll is not a single molecule; there are varieties that have various side groups on the rings. The bonding in many metal-organic coordination compounds allows them to absorb certain wavelengths of visible light to make them green. The empirical formula of chlorophyll is $C_{55}H_{72}MgN_4O_5$ (ACS, 2019).

3.6 NAMING OF INORGANIC COMPOUNDS

In general, there are two types of inorganic compounds, and they can be grouped into (i) ionic compounds between metals and nonmetals, and (ii) molecular compounds between nonmetals and nonmetals.

3.6.1 Compounds between Metals and Nonmetals

Inorganic compounds made of a metal and a nonmetal are ionic compounds and their names have an ending of "-ide". The guidelines for naming this type of compounds are:

1. The cation (metal) is always named first with its name unchanged (e.g., sodium (Na^+) and magnesium (Mg^{+2})); while the anion (nonmetal) is written after the cation, modified to end in "-ide" (see below):
   - hydride (H^-)
   - Group 14: carbide (C^{-4})
   - Group 15: nitride (N^{-3}), phosphide (P^{-3})
   - Group 16: oxide (O^{-2}), sulfide (S^{-2})
   - Group 17: fluoride (F^-), chloride (Cl^-), bromide (Br^-), iodide (I^-)

   Examples: lithium hydride (LiH), beryllium carbide (Be_2C), sodium nitride (Na_3N), potassium phosphide (K_3P), calcium oxide (CaO), barium sulfide (BaS), sodium chloride (NaCl), and calcium bromide ($CaBr_2$).
2. As mentioned, many transition metals have more than one oxidation state (i.e., they may form more than one type of ion). For those metals with multiple oxidation states, a Roman numeral is assigned after the name of the metal; and the Roman numeral denotes its charge and the oxidation state. Some of these metals also have common names, for example:
   - copper (I), Cu^+, cuprous
   - iron (II), Fe^{+2}, ferrous
   - mercury (I), Hg_2^{+2}, mercurous
   - copper (II), Cu^{+2}, cupric
   - iron (III), Fe^{+3}, ferric
   - mercury (II), Hg^{+2}, mercuric

   As shown in the above list, the ions of transition metals with a smaller charge (or the oxidation state) have common names ending with "-ous", while those with a larger charge (or the oxidation state) ending with "-ic". It should be noted that mercury (I) ion is diatomic as Hg_2^{+2} because it is not stable on its own [Note: This is not a common case for other metal ions].

   Examples: copper (I) chloride (cuprous chloride, CuCl) and copper (II) chloride (cupric chloride, $CuCl_2$); ferrous oxide (FeO) and ferric oxide (Fe_2O_3); and mercury (I) chloride (Hg_2Cl_2) and mercury (II) chloride ($HgCl_2$).

3.6.2 Compounds between Nonmetals and Nonmetals

Inorganic compounds that are composed of nonmetals or semi-metals only will form covalent bonds (instead of ionic bonds) and they are classified as molecular compounds. For example, carbon monoxide (CO) is composed of two non-metallic elements and silicon dioxide (SiO_2) is composed of a metalloid and a non-metallic element. Both are inorganic compounds and are covalent-bonded molecules.

A *polyatomic ion* is an ion composed of two or more elements. Common examples of polyatomic ions include hydroxide (OH^-), cyanide (CN^-), ammonium $\left(NH_4^+\right)$, nitrate $\left(NO_3^-\right)$, carbonate $\left(CO_3^{-2}\right)$, sulfate $\left(SO_4^{-2}\right)$, and phosphate $\left(PO_4^{-3}\right)$ ions – see a more complete list below. The atoms in a polyatomic ion are tightly bonded together so that the entire ion behaves as a single unit. Non-metallic atoms in the polyatomic ions are joined by covalent bonds, but the ions as a whole can form ionic bonds with other elements. For example, ammonium ion (a polyatomic ion) can join with chloride ion (Cl^-) or nitrate ion (another polyatomic ion) by covalent bonds to form ammonium chloride (NH_4Cl) and ammonium nitrate (NH_4NO_3), respectively. Within magnesium hydroxide ($Mg(OH)_2$), magnesium ion has ionic bonding with hydroxyl ions, but the bonding within the polyatomic ions (OH^-) is covalent. There are more anionic polyatomic ions than cationic polyatomic ions.

Naming binary (two-element) inorganic molecular compounds is similar to naming ionic compounds. The guidelines include:

1. The name of the element with the positive oxidation state (which is usually the first element in the chemical formula) is used first, followed by the stem of the name of the second element and the suffix "-ide".
2. Numeral prefixes (i.e., mono-, di-, tri-, tetra-, … etc.) are used to specify the number of atoms in a molecule. Normally, "mono-" is not added to the name of the first element if there is only one atom of the first element.
3. If the second element is oxygen, the trailing vowel is usually omitted from the end of the poly-syllabic prefix, but not a mono-syllabic one (i.e., monoxide, dioxide, trioxide, tetroxide, pentoxide, etc.).

3.6.3 Common Mono-Element and Poly-Element Ions

Environmental professionals need to be familiar with the names and forms of common ions that they may come across. The list below shows the common monoatomic cations and anions as well as polyatomic ions:

1. <u>Common monoatomic cations</u>: [Note: They are listed according to the sequence of the groups in the periodic table, when applicable; and many transition metals have more than one valence charges (e.g., Fe^{+2} and Fe^{+3})]:
   - *Mono-valent cations*: hydrogen (H^+), lithium (Li^+), sodium (Na^+), potassium (K^+), Cu(I) or cuprous (Cu^+), and silver (Ag^+)
   - *Di-valent cations*: magnesium (Mg^{+2}), calcium (Ca^{+2}), barium (Ba^{+2}), manganese (II) (Mn^{+2}), Fe(II) or ferrous (Fe^{+2}), Cu(II) or cupric (Cu^{+2}), zinc (Zn^{+2}), mercury (I) (Hg_2^{+2}) and mercury (II) (Hg^{+2}), tin (II) or stannous (Sn^{+2}), and lead(II) or plumbous (Pb^{+2})

- *Tri-valent cations*: chromium (III) (Cr^{+3}), Fe(III) or ferric (Fe^{+3}), and aluminum (Al^{+3})
- *Tetra-valent cations*: tin (IV) or stannic (Sn^{+4}), and Pb (IV) or plumbic (Pb^{+4})
- *Hexavalent cations*: chromium (VI) Cr^{+6}

2. Common monoatomic anions:
   - *Mono-valent anions*: hydride (H^-), fluoride (F^-), chloride (Cl^-), bromide (Br^-), and iodide(I^-)
   - *Di-valent anions*: oxide (O^{-2}) and sulfide (S^{-2})
   - *Tri-valent anions*: nitride (N^{-3})
3. Common polyatomic ions:
   - hydronium (H_3O^+) and hydroxide (OH^-)
   - peroxide $\left(O_2^{-2}\right)$
   - bicarbonate $\left(HCO_3^-\right)$ and carbonate $\left(CO_3^{-2}\right)$
   - ammonium $\left(NH_4^+\right)$, nitrite $\left(NO_2^-\right)$, and nitrate $\left(NO_3^-\right)$
   - bisulfate $\left(HSO_4^-\right)$, sulfate $\left(SO_4^{-2}\right)$, bisulfite $\left(HSO_3^-\right)$, sulfite $\left(SO_3^{-2}\right)$, and thiosulfate $\left(S_2O_3^{-2}\right)$
   - hypochlorite (OCl^-), chlorite $\left(ClO_2^-\right)$, chlorate $\left(ClO_3^-\right)$, and perchlorate $\left(ClO_4^-\right)$
   - dihydrogen phosphate $\left(H_2PO_4^-\right)$, hydrogen phosphate $\left(HPO_4^{-2}\right)$, and phosphate $\left(PO_4^{-3}\right)$
   - bromate $\left(BrO_3^-\right)$ and iodate $\left(IO_3^-\right)$
   - chromate $\left(CrO_4^{-2}\right)$ and dichromate $\left(Cr_2O_7^{-2}\right)$
   - cyanide (CN^-), cyanate (OCN^-), and thiocyanate (SCN^-)
   - permanganate $\left(MnO_4^-\right)$
   - acetate (CH_3COO^-) and oxalate $\left(C_2O_4^{-2}\right)$

Example 3.1: Bonding of Inorganic Compounds

Are the following compounds ionic or molecular? Are the bonds within these compounds ionic, covalent, or both?

a. Calcium chloride ($CaCl_2$)
b. Calcium oxide (CaO)
c. Carbon dioxide (CO_2)
d. Calcium hydroxide ($Ca(OH)_2$)
e. Sodium carbonate (Na_2CO_3)

Solution:

a. $CaCl_2$ is an ionic compound, and the two Ca–Cl bonds are ionic.
b. CaO is an ionic compound, and the Ca=O bond is ionic.

c. CO_2 is a molecular compound, and the two C=O bonds are covalent.
d. $Ca(OH)_2$ is an ionic compound. The two Ca–OH bonds are ionic, but the two O–H bonds are covalent.
e. Na_2CO_3 is an ionic compound. The bond between Na and the polyatomic ion $\left(CO_3^{-2}\right)$ is ionic, but the bonds within the polyatomic ion are covalent.

Example 3.2: Naming Ionic Compounds

Write the name for each of the following inorganic compounds:

a. NaH, NaF, and NaOH
b. $MgBr_2$ and MgO
c. CO and CO_2
d. NO, NO_2 and N_2O_5
e. $FeSO_4$ and Fe_2O_3
f. HCO_3^- and CO_3^{-2}

Solution:

a. Sodium hydride, sodium fluoride, and sodium hydroxide
b. Magnesium bromide and magnesium oxide
c. Carbon monoxide and carbon dioxide
d. Nitrogen monoxide (or nitric oxide), nitrogen dioxide, and dinitrogen pentoxide
e. Ferrous sulfate and ferric oxide (or iron (III) oxide)
f. Hydrogen carbonate ion (or bicarbonate ion) and carbonate ion.

Discussion:

1. $MgCl_2$ is an ionic compound, so the prefix "di-" is not needed before chloride.
2. NO is nitrogen monoxide, and nitric oxide is its common name.
3. The prefix "bi-" in bicarbonate ion indicates the presence of a single hydrogen. Although the IUPAC name of HCO_3^- is "hydrogen carbonate ion", "bicarbonate" is still commonly used.

Example 3.3: Chemical Formula of Inorganic Compounds

Write the formula for each of the following compounds:

a. Potassium dichromate
b. Mercury (I) chloride
c. Barium cyanide
d. Ferrous chloride
e. Ammonium carbonate
f. Sodium nitride
g. Hydrogen peroxide

h. Ammonium nitrate
i. Sulfur trioxide
j. Sodium bisulfate

Solution:

a. $K_2Cr_2O_7$
b. Hg_2Cl_2
c. $Ba(CN)_2$
d. $FeCl_2$
e. $(NH_4)_2CO_3$
f. Na_3N
g. H_2O_2
h. NH_4NO_3
i. SO_3
j. $NaHSO_4$

3.7 NUCLEAR CHEMISTRY AND RADIONUCLIDES

Nuclear chemistry deals with nuclear reactions (i.e., reactions that happen inside the nucleus of atoms). In chemical processes that we have talked about, atoms share their electrons with (or transfer their electrons to) other atoms, while their nuclei stay unaffected (i.e., identity of the element stays the same). In a *nuclear reaction*, the identity of the element would change (i.e., changes of number of protons and/or neutrons) and it often comes with a release of an enormous amount of energy.

3.7.1 Nuclides and Radionuclides

As mentioned in Chapter 1, an atom consists of a nucleus (where protons and neutrons stay) and electrons that orbit the nucleus. A *nuclide* is a specific type of atom characterized by the number of protons and neutrons in the nucleus. For example, a hydrogen atom (hydrogen-1,^{1}H) is a nuclide of hydrogen with one proton and no neutrons, while tritium (hydrogen-3,^{3}H) is another nuclide of hydrogen with one proton and two neutrons. They are isotopes of hydrogen, having the same mass (proton) number, but different neutron numbers [Note: There is another isotope of hydrogen; that is deuterium (hydrogen-2,^{2}H)].

When there is an imbalance between protons and neutrons, an isotope has the potential to transform itself into a different but more stable atom. When the transformation occurs, the atom decreases its mass through radioactive decay and emitting ionization radiation (i.e., alpha particles, beta particles, positrons, or gamma rays). The *radioactive decay* process is spontaneous. A *radionuclide* is an atom (element) with an unstable nucleus that can go through radioactive decay.

Elements in the periodic table can take on several different forms. Some of these forms are stable, while other forms are unstable. Although all elements may have different forms, the most stable form is the most common one in nature. In the case of hydrogen, hydrogen-1 is the most common in nature, tritium is radioactive while deuterium is not radioactive. There are some elements with no stable form, and they are always radioactive. The *radioactive elements* are technetium (Tc, atomic mass = 43), promethium (Pm, atomic mass = 61) and all the elements with atomic mass from 84 (i.e., polonium (Po)) to the last element oganesson (Og, atomic mass = 118) (EPA, 2023d). The other way to say, the radioactive elements include Tc (transition metal and in Row 14 of the periodic table); Pm (in the Lanthanide series), Po (Group 16), astatine (At, Group 17), and radon (Rn, Group 18, a noble gas), and all the elements in Row 6 including the Actinide series.

Every radionuclide has a specific decay rate, measured in terms of *half-life* (i.e., the time required for half of the radioactive atoms present to decay). Some radionuclides have half-lives of seconds, but others have half-lives of hundreds, thousands, millions, or billions of years. For example, the half-life of tritium is about 12 years, while carbon-14 undergoes radioactive decay with a half-life of about 5,730 years. The amount of carbon-14 is often used to show the object's age in "carbon dating".

When a radionuclide decays, it transforms into a different atom - a decay product. The atoms will keep transforming into new decay products until they reach a stable state and become non-radioactive. Most radionuclides only decay once before becoming stable, but some decay in more than one step. For example, unstable uranium-238 (U-238) will decay through many steps (i.e., a decay chain) until it becomes stable lead-206 (Pb-206).

3.7.2 Major Forms of Radioactive Decay

Major forms of radioactive decay include alpha decay, beta decay, and gamma decay/radiation.

1. ***Alpha decay***. *Alpha decay* occurs when an atom ejects an *alpha particle* (*α-particle*), consisting of two protons and two neutrons (identical to the nucleus of a helium atom), from its nucleus. Consequently, the atom transforms into a new element having its atomic number decreased by two and its atomic mass decreased by four. Alpha particles are positively charged.

   Alpha particles come from decay of heavy radionuclides, such as polonium (Po, atomic number = 84), radium (Ra, atomic number = 88), and uranium (U, atomic number = 92). Even though they are very energetic, they cannot travel very far once emitted because they use up their energy over short distances due to the relatively large mass. They cannot penetrate human's skin; however, once enter the body, they can be very harmful (USEPA, 2023d).
2. ***Beta decay***. *Beta decay* occurs when a neutron is turned into a proton (or a proton is turned into a neutron), and a beta particle is emitted from its nucleus. *Beta particles* are high-energy and high-speed electrons (β^-) or positrons (β^+). They are emitted by certain unstable nuclides such as hydrogen-3, carbon-14, and strontium-90. Beta particles are more penetrating

than alpha particles, but less damaging to living tissues and DNA. Some can penetrate the skin and cause skin burns. They are most hazardous once inhaled or swallowed (USEPA, 2023d).

3. ***Gamma decay***. *Gamma decay* occurs when there is residual energy in the nucleus following either alpha or beta decay or after neutron capture in a nuclear reaction. This residual energy is released as gamma rays which are weight-less packets of energy, called *photons*. Gamma rays can penetrate completely through the human body, and they are a radiation hazard for the entire body.

3.7.3 Radiation Sources and Units of Radioactivity

Sources of radiation are around us all the time; some are natural, and some are man-made. Background radiation is always present on Earth, and most of it occurs naturally from minerals. Naturally occurring radioactive minerals in soil and water produce background radiation. Uranium and thorium (Th, atomic number = 90) are naturally found in soil and they are the sources of terrestrial radiation. The radioactive gas radon (Rn, atomic number = 86) is created when other naturally occurring elements, mainly radium (Ra, atomic number = 88), undergo radioactive decay.

Radioactivity is a measure of the ionizing radiation released by a radioactive material. The radioactivity is measured in becquerels (Bq, international unit) and in curies (Ci, U.S. unit). Because a curie is a large unit, radioactivity values of environmental samples are usually shown in picocuries (pCi); 1 pCi = 10^{-12} Ci. The natural radium-226 level of surface water generally ranges from 0.0037 to 0.0185 Bq/L, or 0.1 to 0.5 pCi/L (USEPA, 2023d).

3.7.4 Radionuclides of Environmental Concern

The USEPA has "alpha/photon emitters, beta photon emitters, radium-226 (Ra-226) and radium 228 (Ra-228), and uranium (U)" on the list of MCLs. For example, the radium limit in drinking water for daily consumption is 0.185 Bq/L, or 5.0 pCi/L (USEPA, 2023d). The USEPA also has "radionuclides (including radon (Rn))" on the list of HAPs.

REFERENCES

ACS (2019). "Chlorophyll", American Chemical Society (ACS), https://www.acs.org/molecule-of-the-week/archive/c/chlorophyll.html, last updated January 21, 2019.

USEPA (2023a). "Metals", US Environmental Protection Agency, https://www.epa.gov/caddis-vol2/metals, last updated March 20, 2023.

USEPA (2023b). "Resource Conservation and Recovery Act (RCRA) Laws and Regulations", US Environmental Protection Agency, https://www.epa.gov/rcra, last updated August 25, 2023.

USEPA (2023c). "SW-846 Test Method 1311: Toxicity Characteristic Leaching Procedure", US Environmental Protection Agency, https://www.epa.gov/hw-sw846/sw-846-test-method-1311-toxicity-characteristic-leaching-procedure, last updated August 17, 2023.

USEPA (2023d). "Radiation Protection", US Environmental Protection Agency, https://www.epa.gov/radiation, last updated September 19, 2023.

EXERCISE QUESTIONS

1. Are the following compounds ionic or molecular? Are the bonds within these compounds ionic, covalent, or both?
   a. Magnesium fluoride (MgF_2)
   b. Magnesium oxide (MgO)
   c. Methane (CH_4)
   d. Sodium hydroxide (NaOH)
   e. Potassium bisulfite ($KHSO_3$)
2. Write the name for each of the following inorganic compounds:
   a. KH, KCl, and KOH
   b. $CaCl_2$ and CaO
   c. SO_2 and SO_3
   d. NH_4^+, NO_2^- and NO_3^-
   e. $FeCl_2$, $FeCl_3$, and Al_2O_3
   f. $S^=$, $S_2O_3^=$, $SO_3^=$, $SO_4^=$, HSO_4^-
3. Write the formula for each of the following compounds:
   a. Bromine dioxide
   b. Mercury (II) chloride
   c. Sodium cyanide
   d. Aluminum chloride
   e. Calcium carbonate
   f. Potassium nitride
   g. Potassium peroxide
   h. Ammonium chloride
   i. Sulfuric acid
   j. Sodium biphosphate

4 Fundamentals of Chemical Thermodynamics

4.1 INTRODUCTION

Thermodynamics deals with heat, work, and temperature as well as their relations to energy changes. It does not tell the route and the rate of an energy transformation, but the initial and the final states of the system undergoing the change. *Chemical thermodynamics* deals with the relationships between heat and work with chemical reactions or with physical changes of substances. From chemical thermodynamics, one can tell if a reaction/process could occur and how far it could proceed. However, it cannot tell how fast the reaction/process would be if it occurs, and that would be addressed by chemical kinetics (see Chapter 10).

4.1.1 Processes and Reactions

The term "process" is widely used in our daily lives. A *process* can be defined as an activity, or a set of activities, to produce a result. In environmental engineering, the main objective of a treatment process is to reduce the concentration/toxicity of contaminants in an environmental medium (e.g., water, air, or soil). A treatment process can be physical, chemical, biological, or thermal. A *physical treatment process* (e.g., filtration, volatilization, and settling) will just change the physical form of a substance. A *chemical treatment process* has chemical reaction(s) taking place. A *chemical reaction* (e.g., oxidation, reduction, and precipitation) is a process that involves rearrangement of the molecular or ionic structure of a substance. A *biological treatment process* (e.g., activated sludge in wastewater treatment, aerobic/anaerobic digestion of wastewater sludge) primarily employs bacteria, protozoa, and other microbial species to break down organic pollutants. The degradation of these pollutants is also a chemical reaction, but it is resulted from microbial activities. A *thermal treatment process* (e.g., incineration, combustion, and pyrolysis) applies heat to break down organic contaminants at elevated temperatures. It is essentially a chemical reaction process occurring at elevated temperatures.

It is often necessary to assemble a treatment process train, which consists of several unit processes in series, to meet the treatment goal. For example, treatment processes typically employed in municipal wastewater treatment are shown below:

- ***The Liquid Treatment Process Train***. Bar Screen (physical) → Aerated Grit Chamber (physical) → Primary Sedimentation Tank (physical) → Aeration Tank (biological) → Secondary Clarifier (physical) → Tertiary

DOI: 10.1201/9781003502661-4

Filter (physical) → Disinfection (chemical) → Dichlorination (chemical) → Discharge to water bodies (physical).

- ***The Sludge Treatment Process Train***: Sludge Thickening (physical) → Aerobic/Anaerobic Digestion (biological) → Dewatering (physical) → Composting (biological) → Land Applications (physical + chemical + biological); Incineration (thermal); Landfill (physical + chemical + biological).
- ***Control of Odor and Volatile Organic Compounds (VOCs) Emissions***: Biofilter (biological); Bio-trickling Filter (physical + biological); Scrubber (physical); Scrubber with Chlorination (physical + chemical), Granular Activated Carbon (GAC) Adsorption (physical/chemical); Advanced Oxidation Process (chemical); and Incineration (thermal).

4.1.2 System, Boundary, and Surroundings

To evaluate the effectiveness of a process/reaction, we need to define the system in which the process/reaction occurs. A *system* is a specific set of substances within a definite boundary (Note: The boundary can be real or imaginary); while its *surroundings* are everything outside the system (i.e., the rest of the universe). Depending on if there is mass or energy exchange between the system and its surroundings, a system would be called as:

- *Isolated*: with no mass and energy exchanges with its surroundings.
- *Closed*: with only energy exchange with its surroundings.
- *Open*: with both mass and energy exchanges with its surroundings.

There are also some specific terms used to describe a process, for example:

- *Adiabatic*: with no exchange of heat between a system and its surroundings.
- *Isothermal*: the process occurs at a constant temperature.
- *Isobaric*: the process occurs at a constant pressure.

4.1.3 State Functions

State functions describe the state of a system. The most common state functions are temperature (T), pressure (P), volume (V), and mass (m) of the system. A state function of a system depends only on its present state and is independent of the path taken to reach this state. In other words, a *state function* in thermodynamics is defined by the current/final state of the system, rather than the changes that have occurred between the initial and the final states.

4.2 ENERGY

4.2.1 Types of Energy

There are many different forms of energy; however, they all fall into two categories: potential energy and kinetic energy (US EIA, 2023).

When an object is subject to a force coming from a force field such as gravitational, magnetic, and electrostatic fields, *potential energy* is the energy that it has due to its location. Forms of potential energy include the following:

- *Chemical energy* is the energy stored in the chemical bonds of atoms, ions, and molecules.
- *Mechanical energy* is the energy stored in an object by tension (e.g., a stressed spring).
- *Nuclear energy* is the energy stored in the nucleus of an atom to hold the nucleus together.
- *Gravitational energy* is the energy stored in an object at a higher elevation, related to a reference elevation.

Kinetic energy is from the motion of waves, electrons, atoms, molecules, substances, or objects.

- *Radiation energy* is the electromagnetic energy that travels in transverse waves (e.g., light, X-rays, microwaves, and radio waves).
- *Thermal energy* (or *heat*) is the energy resulting from the movement of atoms and molecules within a substance.
- *Motion energy* is the energy stored in moving objects.
- *Sound energy* is produced when a force causes a substance or an object to vibrate and the energy moves through it in longitudinal waves.
- *Electrical energy* is delivered by electrons through a conductor (US EIA, 2023).

4.2.2 Macroscope and Microscope Kinetic and Potential Energy

For a macroscopic moving object (e.g., a ball), its kinetic energy (KE) is:

$$\text{KE} = \frac{1}{2} m v^2 \tag{4.1}$$

where m = its mass and v = its moving velocity. For a macroscopic object, its potential energy (PE) in a gravitational field is:

$$\text{PE} = mg\Delta h \tag{4.2}$$

where g = the gravitational constant (= 9.81 m/s^2 = 32.2 ft/s^2) and Δh = the elevation relative to a reference point.

In a microscopic scale, chemical compounds contain various forms of PE and KE because they consist of positively-charged nuclei surrounded by negatively-charged electrons. The total PE of a molecule is the sum of the attraction between nuclei and electrons and the repulsion between nuclei and nuclei and between electrons and electrons as:

$$(\text{PE})_{\text{total}} = (\text{PE})_{\text{nuclei-electron}} + (\text{PE})_{\text{nuclei-nuclei}} + (\text{PE})_{\text{electron-electron}} \tag{4.3}$$

In addition, molecules also possess PE in the form of chemical bonds (i.e., bonding energy).

A molecule can have translational motion (e.g., water vapor would have more translational motions than liquid water). A molecule can also have rotational motions. In addition, atoms of a molecule can have vibrational motions on each side of a chemical bond. Thus, the total KE of a molecule:

$$(\text{KE})_{\text{total}} = (\text{KE})_{\text{translation}} + (\text{KE})_{\text{rotation}} + (\text{KE})_{\text{vibration}} \tag{4.4}$$

4.2.3 Units of Energy

The SI unit for energy (and work) is Joule (J), and one Joule is equal to one Newton-meter (N·m) as:

$$1\ J = 1\ N \cdot m = 1\ \text{kg} \cdot m^2/\text{s}^2 \tag{4.5}$$

Please note that Newton (N) is the SI unit for force; and 1 N = 1 kg·m/s^2 [Note: Pound (lb_f) is the US customary unit for force; and 1 lb_f = 4.448 N].

Equation (4.5) shows that energy can have a unit of "force" times "distance" [Note: It can also have a unit of "pressure times volume" (i.e., $P \times V$)]. One Joule is the work done on an object when a force of 1 N acts on an object in the direction of its motion for 1 m. Other energy units used in engineering practices include calories, British thermal unit (BTU), ft-lb_f, and therm (a common energy unit for natural gas). One calorie is the energy needed to raise the temperature of 1 g of water by 1°C; while one BTU is the energy needed to raise the temperature of one pound (lb_m) of water by 1°F at the temperature when water has its largest density (~4°C, or ~39°F). Conversions among these common energy units are:

$$1\ \text{BTU} = 1{,}055\ \text{J}$$

$$1\ \text{calorie} = 4.184\ \text{J}$$

$$1\ \text{ft-lb}_f = 1.356\ \text{N·m} = 1.356\ \text{J}\ [\text{Note: } 1\ \text{lb}_f = 4.448\ \text{N}] \tag{4.6}$$

$$1\ \text{therm} = 100{,}000\ \text{BTU} = 1.055 \times 10^8\ \text{J}$$

$$1\ \text{atm·L} = 101.3\ \text{J}$$

Please note that the terms of "small calorie (calorie)" and "large calorie (Calorie)" are used in health science and food industry. One "Calorie" (with C in uppercase) = 1 kilocalories (kcal) = 1,000 calories.

Example 4.1: Conversion between KE and PE

A ball (1 kg, 2.2 lb_m) is falling from an elevation 20 m (66 ft) above the ground. Estimate its velocity before hitting the ground.

Solution:

a. Using Eq. 4.2 to find PE of the ball before falling off,

$$\text{PE} = mg\Delta h = (1\ \text{kg})\left(9.81\frac{m}{s^2}\right)(20\ \text{m}) = 196.2\ \text{kg}\cdot\frac{m^2}{s^2} = 192.6\ N\cdot m = 192.6\ J$$

b. Assuming all the PE is converted to KE (with no loss in the form of heat) and using Eq. 4.1,

$$\text{PE} = 196.2\ \text{kg}\cdot\frac{m^2}{s^2} = \text{KE}\ = \frac{1}{2}(1\ \text{kg})(v)^2$$

v = 19.8 m/s.

Discussion:

1. The question can also be solved using the given values in the US customary units.
2. The estimate is based on two assumptions: (i) the energy is conserved, and (ii) the PE is totally converted to the KE. The first assumption is always valid. However, when the ball is falling, it is subjected to the gravitational force, the buoyancy force from air, and the friction force between the ball and air. Consequently, the actual velocity should be <19.8 m/s.

4.3 THE ZEROTH LAW OF THERMODYNAMICS

As we have learned from physics, there are three Newton's laws of motion that describe the relationship between the motion of an object and the forces acting on it. Similarly, there are four fundamental laws of thermodynamics. These laws serve as the foundation to define a group of thermodynamic properties and to characterize systems in thermodynamic equilibrium.

The *zeroth law of thermodynamics* establishes that temperature is a fundamental and measurable property of a system. Ways to describe the zeroth law of thermodynamics include:

- Systems, that are in thermo-equilibrium, are at the same temperature.
- If each of two systems is in thermo-equilibrium with a third system, then all three systems are in thermo-equilibrium with each other.
- If systems are in thermal equilibrium, no heat flow will take place.

4.3.1 Temperature

As mentioned earlier, temperature (T) is a thermodynamic state function. Temperature of a substance is related to the average KE of its particles. At a higher temperature, kinetic energies of molecules within a substance would be larger. Since the KE is equal to ½ mv^2 (Eq. 4.1), a molecule would move at a larger velocity at a higher temperature because its mass stays the same.

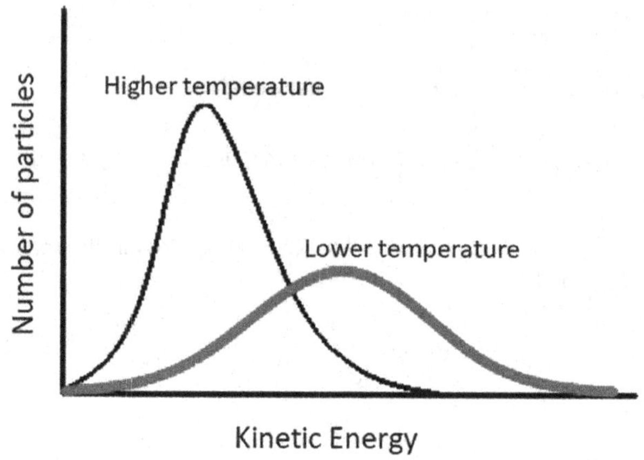

FIGURE 4.1 Distribution of kinetic energy as a function of temperature.

At a given temperature, the particles of any substance have the same average KE. However, at a specific temperature, not all the particles within the substance have the same KE (i.e., not uniform); they display a wide distribution instead. Figure 4.1 illustrates that most of the particles have their kinetic energies close to their average for both temperatures. However, at a higher temperature, the average KE becomes larger, and the distribution curve becomes narrower.

4.3.2 Units of Temperature

The temperature scales used in our daily uses are in Fahrenheit (°F) or Celsius (°C). The Fahrenheit scale sets the freezing and boiling points of water at the mean sea level at 32°F and 212°F; and those in the Celsius scale are at 0°C and 100°C, respectively. Conversions between these two temperature scales are:

$$T\ (\text{in }°F) = \frac{9}{5} \times T\ (\text{in}°C) + 32 \tag{4.7a}$$

$$T\ (\text{in }°C) = \frac{5}{9} \times \left[T\ (\text{in }\ °F) - 32\right] \tag{4.7b}$$

The absolute temperatures are more commonly used in science and engineering and their temperature scales are in Rankine (°R) and in Kelvin (K). Both absolute temperature scales start with zero (the "*absolute zero*"), and the conversions among all temperature scales are:

$$T\ (\text{in K}) = T\ (\text{in }°C) + 273.15 \approx T\ (\text{in }°C) + 273 \tag{4.8a}$$

$$T\ (\text{in }°R) = T\ (\text{in }°F) + 459.67 \approx T\ (\text{in }°F) + 460 \tag{4.8b}$$

$$T\ (\text{in K}) = \frac{5}{9} \times T\ \left(\text{in }°R\right) \tag{4.8c}$$

The physical meaning of the absolute zero will be described later in this chapter.

4.4 THE FIRST LAW OF THERMODYNAMICS

The *first law of thermodynamics* is essentially the law of energy conservation. Ways to describe the first law of thermodynamics include:

- The total energy of the universe remains constant.
- Energy can neither be created or destroyed; however, it may be exchanged between the system and its surroundings.
- A thermodynamic system possesses a state function, known as internal energy; and the total change of the internal energy (ΔU) of a closed system is equal to the total heat added into the system minus the total work done by the system.

Thus, the change of the total energy of a system (ΔU_{system}) should be equal to that of its surroundings ($\Delta U_{surroundings}$) as

$$\Delta U_{universe} = \Delta U_{system} + \Delta U_{surroundings} = 0 \tag{4.9a}$$

$$\Delta U_{system} = -\Delta U_{surroundings} \tag{4.9b}$$

Equation (4.9a) implies that the energy of the universe remains constant (i.e., $\Delta U_{universe} = 0$). The negative sign in Eq. (4.9b) means the energy gain of the system will be equal to the energy loss of its surroundings or vice versa.

4.4.1 Internal Energy

The *internal energy (U) of a molecule* is the sum of its KE and PE:

$$\text{Internal Energy} = (\text{KE})_{total} + (\text{PE})_{total} \tag{4.10}$$

The *internal energy of a system*, microscopically, refers to the energy of all the molecules making up the system. The internal energy of a system is the energy contained within it; however, it may not include the energy resulted from its movement (or its position) as a whole.

Determination of the total internal energy of a substance or a system is not possible; in other words, its absolute value is not available. Internal energy is also a thermodynamic state function. Only changes of the internal energy before and after a reaction/process is needed in thermodynamic analyses.

4.4.2 Heat and Work

Heat and work are two different ways of transferring energy between a system and its surroundings. In thermodynamics, *heat* is the transfer of thermal energy and *work* is the transfer of mechanical energy between the system and its surroundings.

For a closed system, the change in its internal energy (ΔU) is related to the transfers of heat (q) and work (w) between the system and its surroundings as:

$$\Delta U_{\text{system}} = q + w \quad (4.11)$$

If the system releases energy to its surrounding, the value of q is negative (its internal energy decreases). If a chemical reaction releases heat to its surroundings, the reaction is called *exothermic*. Conversely, an *endothermic* reaction receives heat from its surroundings to complete the reaction. The value of w is negative, if the system does work on its surroundings (its internal energy decreases). Conversely, w would be a positive value (and the internal energy increases) if the surroundings do work on the system.

A few concepts about internal energy are worthwhile to mention here:

- Although an absolute value of internal energy (U) is not available, change of internal energy (ΔU) can be determined from the values of q and w, which are usually measurable.
- Both internal energy and change of internal energy are state functions, while q and w are *path functions*, depending on the path(s) they went through. For example, a specific change in internal energy (ΔU) can come from many different combinations of q and w values (i.e., resulted from going through different paths), provided their sums are the same.
- Heat and work are both energy and they are transferrable. Work can be completely converted into heat; however, heat can only be partially converted into work. In other words, a complete conversion of heat into work is thermodynamically impossible.
- Different from internal energy which is contained/stored in the system, q and w are energy in transit (i.e., they move between the system and its surroundings and resulted in a change of U; and they are not stored by either the system or its surroundings).

Pressure-volume work (PV-work) is the work that is often associated with a system in which its volume changed due to formation or disappearance of gaseous substance [Note: From the Ideal Gas Law ($PV = nRT$), V is proportional to n (i.e., the number of moles present) at constant P & T]. It is often used in describing one of the most important thermodynamic state functions – that is enthalpy (H). Under a constant pressure (Note: Most chemical reactions occur under a constant pressure), the pressure-volume work is:

$$w = -P\Delta V \quad (4.12)$$

where ΔV = change of the system volume. The negative sign implies that the internal energy of the system decreases if its volume increases (i.e., the system did work on the environment through expansion). Combining Eqs. (4.11) and (4.12),

$$\Delta U_{\text{system}} = q + w = q - P\Delta V \quad (4.13)$$

4.4.3 Units of Pressure

Pressure is another important thermodynamic state function and an important parameter affecting the properties of gases. *Pressure/stress* is a force divided by the area where the force is applied:

$$\text{Pressure } (P) = \frac{\text{Force } (F)}{\text{Area } (A)} \tag{4.14}$$

Standard barometric/atmospheric pressure is the average atmospheric pressure at the sea level, and it is equal to 1 atmosphere (atm). The barometric pressure at a location would depend on its altitude, latitude, and weather conditions. The most commonly used SI and US customary units for pressure are Newton/m^2 (N/m^2 or Pascal (Pa)), bar, lb/in^2 (psi), mm-Hg, Torr, ft-H_2O, and m-H_2O, respectively. Common units of pressure and their unit conversions are given below.

$$\begin{aligned}
&1 \text{ atmosphere (atm)} \\
&= 1.01325 \times 10^5 \text{ N/m}^2 = 1.01325 \times 10^5 \text{ Pa} = 101.325 \text{ kN/m}^2 \approx 101.3 \text{ kN/m}^2 \\
&= 1.013 \text{ bar} = 1{,}013 \text{ milli-bar (mbar)} \\
&= 14.696 \text{ lb/in}^2 \text{ (psi)} \approx 14.7 \text{ psi} \qquad (4.15) \\
&= 760 \text{ millimeters of mercury (mm-Hg)} = 29.92 \text{ in-Hg} = 760 \text{ Torr} \\
&= 33.9 \text{ ft-H}_2\text{O} = 407 \text{ in-H}_2\text{O} = 10.33 \text{ m-H}_2\text{O}
\end{aligned}$$

Standard temperature and pressure (STP) are a set of conditions for experimental measurements, so that comparisons can be made between data taken under different conditions. In chemistry, the STP is defined by International Union of Pure and Applied Chemistry (IUPAC) as $T = 273.15$ K and $P = 100$ kPa. However, the standard conditions used in practice are often different. For example, the USEPA generally has 760 mm-Hg (or 1 atm) as the standard pressure, but the standard temperature for stack sampling is 20°C (68°F) and that for ambient air sampling and combustion analysis is 25°C (77°F). The South Coast Air Quality Management District (SCAQMD) in southern California has 60°F or 68°F as the standard temperatures in their regulations. It is always a good practice to state the conditions of the STP in data reporting.

Example 4.2: Relationship between Internal Energy, Heat, and Work

A system containing 30 L of air is compressed to 10 L under a constant pressure of 2 atm. During this process, the system releases 2,000 Joules of heat to its surroundings. What is the change in internal energy of this system?

Solution:

a. Using Eq. (4.12) to find the work of air compression,

$$w = -P\Delta V = -\ (2\ \text{atm})(10 - 30\ L) = 40\ \text{atm} \cdot L = 4{,}052J$$

b. Using Eq. (4.11) to find ΔU,

$$\Delta U = q + W = -2{,}000 + 4{,}052 = 2{,}052J$$

Discussion:

1. The conversion factor between atm·L and J is given in Eq. 4.6.
2. The air compression resulted in a negative ΔV, thus w is a positive value (work is done on the system).
3. Heat is released to the surroundings, so it is a negative value.

4.4.4 Enthalpy and Internal Energy

Enthalpy is a measurement of energy in a thermodynamic system, which represents the total heat content of a system. It is a state function. A *fluid* (e.g., liquid and gas) is a substance that has no fixed shape and yields readily to external pressure. For a fluid in a system, enthalpy is defined as the sum of its internal energy plus the product of the pressure (P) and the volume (V) of the system; that is:

$$H = U + PV \tag{4.16}$$

Then the change in the enthalpy of the system during a chemical reaction is then equal to the change in its internal energy plus the change in the product of P and V of the system as:

$$\Delta H = \Delta U + \Delta(PV) = \Delta U + P\Delta V + V\Delta P \tag{4.17}$$

Below are changes in enthalpy and in internal energy under three specific conditions:

1. <u>No change in the product of *P* and *V*</u>: From Eq. (4.17), the change in enthalpy is essentially the change in internal energy:

$$\Delta H = \Delta U \tag{4.18}$$

2. <u>Process at a constant pressure (i.e., an *isobaric* process)</u>: Since ΔP = 0 and assuming the only work done is compression/expansion (i.e., $P\Delta V$), then from Eqs. (4.13) and (4.17):

$$\Delta H = \Delta U + P\Delta V = (q + w) + P\Delta V = (q - P\Delta V) + P\Delta V = q \tag{4.19}$$

It means that the heat given off or absorbed during a chemical reaction at a constant pressure is equal to the change in the enthalpy of the system.

3. Process at a constant volume (i.e., *isochoric* process): Since there is no work done ($P\Delta V = 0$), then

$$\Delta U = q + w = q \tag{4.20}$$

It means that the heat given off or absorbed during a chemical reaction at a constant volume is equal to the change in the internal energy of the system.

4.4.5 Extensive Properties and Intensive Properties

Different substances can be characterized by their chemical and physical properties. A *physical property* is a characteristic of a substance that is not associated with a change in its chemical composition. Common examples of physical properties include mass, volume, density (the ratio of mass and volume), color, boiling point, and melting point. A *chemical property* describes the ability of a substance to go through a chemical change/reaction, and it can be observed during or after a reaction since its composition would be changed. Common examples of chemical properties include ignitability, corrosivity, reactivity, toxicity, radioactivity, enthalpy of formation, and heat of combustion. Ignitibility, corrosivity, reactivity, and toxicity are the four chemical characteristics that the USEPA uses to categorize hazardous wastes.

Properties of substance also fall into two categories: extensive and intensive. If a property depends on the amount of the substance present, it is an *extensive* property (e.g., mass and volume). On the other hand, if a property does not depend on the amount of the substance present, it is an *intensive* property (e.g., temperature, boiling point, ignitibility, and corrosivity). Physical properties can be extensive or intensive. For example, density (= mass/volume) is an intensive property, while both mass and volume are extensive properties. Similarly, chemical properties can be extensive or intensive. For example, heat of reaction is an extensive property because it is often expressed as the amount of heat released or absorbed by a reaction for a specific quantity of reactant(s) participated in the chemical reaction.

4.4.6 Heat Capacity and Specific Heat Capacity

Heat capacity. *Heat capacity* is the amount of heat required to raise the temperature of an object (or a system) by one degree unit of temperature (°C, K, °F, or °R). For example, the SI unit for the heat capacity is J/K. Since a temperature increase by 1°C is the same as an increase of 1 K, 1 J/K = 1 J/°C. The heat capacity in the US customary unit can be expressed in BTU/°R (which is the same as BTU/°F).

Specific heat capacity. *Specific heat capacity* is the amount of heat required to raise a specific quantity of substance by one degree of temperature. For example, the SI unit for the specific heat capacity is J/K/kg; and the common US customary unit for specific heat capacity is BTU/°F/lb.

One calorie is the energy needed to raise the temperature of 1 g of water by 1°C; while one BTU is the energy needed to raise the temperature of one pound of water by 1°F. The specific heat capacity of water is 1 cal/°C/g (or 1 BTU/°F/lb). Conversions for common heat capacity and specific heat capacity units are:

$$1 \text{ kJ/kg} = 0.43 \text{ BTU/lb} \tag{4.21a}$$

$$1 \text{ cal/°C/g} = 1 \text{ BTU/°F/lb} = 4.184 \text{ kJ/K/kg} \tag{4.21b}$$

Different substances have different specific heat capacities. Water has a larger specific heat capacity than most other substances (e.g., 4.184 vs. 0.45 kJ/K/kg for iron). The specific heat capacity is an extensive property; thus, 1 kg of water can absorb a specific amount of heat with a smaller temperature rise, when compared to 1 kg of iron. That is the reason why water is commonly used as a coolant for machinery. It should be noted that the specific heat capacity for a given substance is a function of temperature as well as of its phase (i.e., solid, liquid, or gas). The specific heat capacity values of water, ice, and water vapor are approximately 4.18, 2.06, and 1.87 kJ/K/kg, respectively; and the specific heat capacity values of water increase with temperature.

For a system having only physical processes occurring (i.e., no chemical reactions), the changes of internal energy and enthalpy depend only on the temperature. The heat transfer (q) of a system can be estimated as:

$$q = \Delta U = C_v \Delta T \text{ (for an isochoric process)} \tag{4.22}$$

$$q = \Delta H = C_p \Delta T \text{ (for an isobaric process)} \tag{4.23}$$

Where C_v is the *specific heat of a substance at a constant volume*, while C_p is the *specific heat at a constant pressure*. In other words,

$$C_v = \Delta U / \Delta T \text{ (for an isochoric process)} \tag{4.24}$$

$$C_p = \Delta H / \Delta T \text{ (for an isobaric process)} \tag{4.25}$$

For processes involving liquids and solids only, the difference between ΔU and ΔH is often small because ΔPV is small. However, the difference can be relatively larger for processes/reactions that involve gases, because the ΔPV value can be significant. For gases, its value of C_p is larger than that of C_v. The *specific heat ratio* (γ) is the ratio of C_p and C_v. The values of γ are 5/3, 7/5 for mono-atomic ideal gases and diatomic ideal gases (and air), respectively.

Under a constant temperature, the energy change (q) for a specific object with a specific heat capacity of C_p and a mass of m can be estimated by:

$$q = \Delta H = m \times C_p \times \Delta T \tag{4.26}$$

Example 4.3: Specific Heat Capacity (SI Units)

A 1-kg iron pipe was heated from 20°C to 90°C. It was then put into a 20-L water bath ($T = 20$°C) to cool down. Estimate:

a. the energy needed to heat up the iron pipe.
b. the final temperature of the water bath.

Solution:

a. Assuming the iron pipe has a specific heat capacity of 0.45 kJ/kg/K, use Eq. (4.26):

$$q = m \times C_p \times \Delta T = 1 \times 0.45 \times (90 - 20) = 31.5 \text{ kJ}$$

b. Assuming the system reaches an equilibrium temperature of T and water has a constant specific heat capacity of 4.18 kJ/kg/K, then

$$q = m \times C_p \times \Delta T = 1 \times 0.45 \times (90 - T) = 20 \times 4.18 \times (T - 20)$$

$$\text{T} = 20.38°\text{C}$$

Discussion:

The water temperature increased slightly, ~0.38°C. A few things should be noted:

1. The density of water was assumed to be 1 kg/L, so that 20 L of water has a mass of 20 kg.
2. Although the specific heat capacity values are in kJ/kg/K, the temperature changes (ΔT) were calculated using temperatures in °C because a change in K is the same as in °C.
3. The process was assumed to be *adiabatic* (no heat change between the system and its surroundings). The system of concern is just the water and the iron pipe, and the container was considered as part of the surrounding. There is no heat loss to (or heat gain by) the surroundings. All the heat released by the iron pipe was picked up by the water.

Example 4.4: Specific Heat Capacity (US Customary Units)

A 1-lb aluminum pipe was heated to 200°F. It was then put into a 5-gallon water bath ($T = 68$°F) to cool down. After reaching thermal equilibrium, the temperature of water bath went up 0.92°F, estimating the specific heat capacity of the aluminum pipe.

Solution:

Assuming the water has a constant heat capacity value of 1 BTU/lb/°R:

$$q = m \times C_p \times \Delta T = 1 \times C_p \times (200 - 68.92) = (5 \times 8.34) \times 1 \times (0.92)$$

$$C_p = 0.293 \text{ BTU/lb/°R } (= 1.224 \text{ kJ/kg/K})$$

Discussion:

A few things should be noted in the solution:

1. The final temperature of the system = 68 + 0.92 = 68.92 °F.
2. It was assumed the density of water is equal to 8.34 lb/gal (= 62.4 lb/ft^3) [Note: 1 gallon = 3.785 L; 1 m^3 = 35.3 ft^3], so that 5 gallons of water has a mass of 41.7 lb.
3. Although the specific heat capacity values are in BTU/lb/°R, the temperature changes (ΔT) were calculated using temperatures in °F because a change in °R is the same as in °F.
4. The process was assumed to be adiabatic (i.e., no heat change between the system and its surroundings).
5. 1 BTU/lb/°R = 4.184 kJ/kg/K.

4.5 THE SECOND LAW OF THERMODYNAMICS

4.5.1 Spontaneous and Non-Spontaneous Processes

Spontaneous vs. non-spontaneous. A *spontaneous process* is a process that occurs on its own, specifically without the need for a continuous input of energy from its surroundings. For example, ice will melt in a glass under ambient conditions, thus it is a spontaneous process. On the other hand, the water cannot go back to ice without putting the glass into a freezer. A *non-spontaneous* process will not take place by itself, and it needs a continuous input of external energy. A process can be spontaneous in one direction under a particular set of condition; however, it would be non-spontaneous in the opposite direction under the same set of condition (as shown in the case of ice melting).

Reversible vs. Irreversible. A process is *reversible* if the system and its surroundings can return to the exact states, that they were in, by following the reverse path. Conversely, an *irreversible* process is one in which the system and its surroundings cannot return to their exact states, that they were in, before the process started. An example of a *reversible* process is expansion and compression of air in a container against a piston. The air can expand when the applied pressure of the piston decreases; however, it can be compressed back to the original state, if the pressure of the piston returns to the original value. Combustion of methane gas (CH_4) with oxygen into water and CO_2 is an example of an *irreversible* process because it is impossible to convert the produced water and CO_2 back to original methane and oxygen. Most of the processes that we encounter in nature are irreversible so that they are called *natural processes*. As we know, heat can transfer naturally from a hotter object to a colder one (i.e., the process is *spontaneous*), but the reverse is not true. A spontaneous process is an irreversible process.

4.5.2 Entropy

The second law of thermodynamics is related to the "quality" of energy. Ways to describe the *second law of thermodynamics* include the following:

- Heat will not flow spontaneously from a colder object to a hotter object.
- When energy is transferred or transformed, some of it will be "wasted".
- If a system is isolated, any natural process that occurs in the system will be in the direction of increasing disorder (or in the direction of increasing entropy) of the system.
- Entropy is a measure of the unavailability of a system's energy to do work; it can be considered as a measure of disorder; the larger the entropy the greater the disorder.
- The entropy of the entire universe, as an isolated system, always increases over time.

Entropy is a thermodynamic quantity that represents the randomness or disorder of a system (or the extent how the energy is distributed). For example, the entropy of ice (in which the water molecules are not free to move) is less than that of water (in which the water molecules can move around within a container). Similarly, the entropy of ambient water vapor will be even larger. Thus, for a substance

$$S_{\text{solid}} < S_{\text{liquid}} < S_{\text{gas}} \tag{4.27}$$

For a reversible process, the combined entropy of the system and its surroundings remains a constant, which is:

$$S_{\text{final}} = S_{\text{initial}};\ \Delta S = 0\ (\text{for a reversible process}) \tag{4.28}$$

On the other hand, for an irreversible process, the combined entropy of the system and its surroundings will increase, which is:

$$S_{\text{final}} > S_{\text{initial}};\ \Delta S > 0\ (\text{for an irreversible process}) \tag{4.29}$$

Thus, when we consider the universe (i.e., a combination of a system and its surroundings), its entropy always increases over time,

$$\Delta S_{\text{universe}} = \Delta S_{\text{system}} + \Delta S_{\text{surroundings}} \geq 0 \tag{4.30}$$

$$\Delta S_{\text{universe}} = \Delta S_{\text{system}} + \Delta S_{\text{surroundings}} = 0\ (\text{only for a reversible process}) \tag{4.31}$$

A few concepts about entropy are worthwhile to mention here:

- Although $\Delta S_{\text{universe}}$ always ≥ 0, ΔS_{system} can be <0 (i.e., in this case, $\Delta S_{\text{surroundings}}$ should have a positive value to make $\Delta S_{\text{universe}}$ always ≥ 0).

- For a given substance, phase changes from solid to liquid (i.e., melting), liquid to gas (i.e., evaporation/vaporization), or solid to gas (i.e., sublimation), will incur an increase in entropy of the system.
- Dissolving a solute into a solution will increase the entropy of the solute (the solute becomes more random); while dissolution of gaseous compounds into a solution will decrease the entropy of the gas (the gas molecules become more in order).
- Increasing the volume that a gas occupies will increase the disorder/entropy of the gas.
- A chemical reaction among gas molecules, the entropy of the system will increase if the total number of gas molecules increases at the end of reaction.
- Entropy of a substance increases with temperature.

Similar to internal energy (U) and enthalpy (H), entropy (S) is a thermodynamic state function and ΔS depends only on the initial and final states of the system. For an *isothermal process* (i.e., the process occurs at a constant temperature), the change in entropy (ΔS) of a system from its initial state to the final state is equal to the change in heat (q) transferred divided by the absolute temperature (T) as:

$$\Delta S = \frac{q}{T} \tag{4.32}$$

Phase changes (e.g., ice melting and water vaporization) are reversible processes. The magnitude of ΔS for a reversible process such as a phase change is often reported as J/K/mole.

Ice melting and water evaporation are two good examples of reversible processes. *Latent heat of fusion* (or *enthalpy of fusion*) is the amount of energy that needs to be supplied to a solid substance (e.g., ice) to convert it to a liquid (e.g., water) when the pressure and temperature are kept constant. For example, when $P = 1$ atm and T is kept at the melting point of ice (0°C), the latent heat of fusion of ice is ~334 kJ/kg. *Heat of solidification* is the opposite of heat of fusion. Similarly, the *heat of vaporization* (or *enthalpy of vaporization*) is the amount of heat needed to convert a specific quantity of liquid to its vapor when pressure and temperature are kept constant. For example, the heat of vaporization of water at $P = 1$ atm and $T = 100$°C (the boiling point of water at $P = 1$ atm) is ~2,260 kJ/kg. It should be noted that the value of heat of vaporization of water is much larger than its heat of solidification. *Heat of condensation* is the opposite of heat of vaporization.

As mentioned in the first law of thermodynamics, the energy is conserved, and the change of the internal energy can be in the form of heat and work. However, not all the heat (thermal energy) can be converted into work. Entropy represents the portion of the thermal energy of a system that is not available for doing useful work. Taking groundwater pollution as an example, once pollutants enter a groundwater aquifer, some of them will get adsorbed onto the aquifer matrix while the rest will get dissolved and move downgradient with the groundwater flow. The randomness increases so that the entropy increases. In remediation, we need to spend energy to remove/recover/destroy the pollutants from the contaminated aquifer.

Example 4.5: Entropy Changes of Ice Melting and Water Evaporation

Estimate (a) the entropy change of 1 mole of ice to water; and (b) the entropy change of 1 mole of water to vapor. What are the entropy changes of the surroundings during these phase changes?

Solution:

a. With the heat of fusion of 334 kJ/kg, the molar heat of fusion of ice:

$$\Delta H_{\text{fusion}} = 334\frac{\text{kJ}}{\text{kg}} = 334\frac{J}{g} = \left(334\frac{J}{g}\right)\left(18\frac{g}{\text{mole}}\right) = 6{,}012\ \text{J/mole}$$

From Eq. (4.29), the entropy change from ice to water at 0°C:

$$\Delta S_{\text{ice to water}} = \frac{6{,}012}{(273+0)} = 22.0\left(\frac{\text{kJ}}{K}\right)/\text{mole}$$

b. With the heat of vaporization of 2,260 kJ/kg, the molar heat of vaporization of water:

$$\Delta H_{\text{vaporization}} = 2{,}260\frac{J}{g} = \left(2{,}260\frac{J}{g}\right)\left(18\frac{g}{\text{mole}}\right) = 40{,}680\ \text{J/mole}$$

From Eq. (4.29), the entropy change from water to water vapor at 100°C:

$$\Delta S_{\text{water to vapor}} = \frac{40{,}680}{(273+100)} = 109.1\left(\frac{J}{K}\right)/\text{mole}$$

c. Since both phase changes are reversible, the entropy changes of the surroundings are opposite to those of the systems (see Eq. 4.28).

Discussion:

1. Energy/heat needs to be added to ice (the system) from the surroundings for ice melting, so the value of q is positive (similarly for water evaporation).
2. For water, $S_{\text{vapor}} > S_{\text{water}} > S_{\text{ice}}$ (also see Eq. 4.27).
3. $\Delta S_{\text{water to vapor}} > \Delta S_{\text{ice to water}}$.

Example 4.6: Enthalpy Changes with Phase Changes

Estimate the energy required to turn 1 m^3 of ice (specific heat capacity = 2.093 kJ/K/kg; density = 0.91 g/cm^3 = 910 kg/m^3) from −10°C to 100°C water vapor at $P = 1$ atm.

Solution:

To find the energy requirement for converting the ice at -10°C to 100°C water vapor, the process can be viewed as four distinct steps: (a) raising the ice temperature from −10°C to the melting point (i.e., 0°C), (b) converting ice to water at 0°C, (c) raising the water temperature from 0°C to its boiling point (100°C), and (d) converting water to vapor at 100°C. The heat required for each step is calculated below:

a. $$\Delta H_{\text{temperature rise of ice}} = \left(1\ \text{m}^3 \times 910 \frac{\text{kg}}{m^3}\right)\left(\frac{2.093 \frac{\text{kJ}}{K}}{\text{kg}}\right)(10\ K) = 19{,}046\ \text{kJ}$$

b. $$\Delta H_{\text{fusion}} = \left(334 \frac{\text{kJ}}{\text{kg}}\right)(910\ \text{kg}) = 3.04 \times 10^5\ \text{kJ}$$

c. $$\Delta H_{\text{temperature rise ofwater}} = (910\ \text{kg})\left(\frac{4.184 \frac{\text{kJ}}{K}}{\text{kg}}\right)(100\ K) = 3.81 \times 10^5\ \text{kJ}$$

d. $$\Delta H_{\text{vaporization}} = \left(2{,}260 \frac{\text{kJ}}{\text{kg}}\right)(910\ \text{kg}) = 2.06 \times 10^6\ \text{kJ}$$

The total heat required is the sum of the above four values = 2.75 $\times 10^6$ kJ.

Discussion:

1. The amount of ice was given in volume, and it needs to be converted to mass first to match the units of specific heat capacity.
2. The specific heat of water was assumed to be constant, as 4.184 kJ/K/kg.
3. Heat of vaporization is the major portion of the total heat requirement.

4.6 THE THIRD LAW OF THERMODYNAMICS

The *third law of thermodynamics* is about the properties of closed systems in thermodynamic equilibrium. Ways to describe it include the following:

- The entropy of a system approaches a constant value when its temperature approaches the absolute zero.
- The entropy of a perfect crystal is zero when its temperature is equal to the absolute zero.

From the third law of thermodynamics, the entropy of any substance at T = the absolute zero is equal to zero. At any temperature above the absolute zero will have a positive value. While we cannot determine the absolute value of internal energy (only its changes between two states), the absolute values of entropy can be determined experimentally. Effects of temperature, entropy, enthalpy on the spontaneity of a chemical rection will be discussed in Chapter 5 – Chemical Reaction Equilibria.

REFERENCE

US EIA (2023). "What is Energy? Forms of Energy", United States Energy Information of Administration, www.eia.gov/energyexplained/what-is-energy/forms-of-energy.php, last updated August 16, 2023.

EXERCISE QUESTIONS

1. The Niagara Fall has a vertical drop of approximately 50 m (164 ft).
   a. Estimate the PE of 1 m^3 (35.3 ft^3) of water on top of the crest possesses, relative to the surface of the lake water, in J (or in ft-lb).
   b. Assuming all the PE is converted to KE, estimate the water velocity approaching the ground, in m/s (or in ft/s).
2. The temperature of a system is equal to 20°C. Express the system temperature in the following units: (a) Kelvin (K), (b) Fahrenheit (°F), and (c) Rankine (°R).
3. The temperature of a system is equal to 900°R. Express the system temperature in the following units: (a) Kelvin (K), (b) Fahrenheit (°F), and (c) Celsius (°C).
4. At what temperature, the numerical values of that temperature in °C and in °F are equal?
5. A system containing 2.0 ft^3 of air is compressed to 0.8 ft^3 under a constant pressure of 30 psi (lb_f/in^2). During this process, the system releases 3.0 BTU of heat to its surroundings.
   a. Convert the pressure of 30 psi to the unit of psf (lb_f/ft^2).
   b. What is the amount of the *PV*-work (in ft-lb_f, in J, and in BTU)?
   c. What is the change of internal energy of this system (in ft-lb_f, in J, and in BTU)?
6. Which of the following are extensive properties? Which of the following are state functions?
   (a) enthalpy, (b) entropy, (c) viscosity, (d) density, (e) heat of reaction, (f) heat capacity, and (g) specific heat capacity.
7. A 100-kg copper pipe (specific heat capacity = 0.385 kJ/K/kg) was heated from 20°C to 100°C. It was then put into a 500-L water bath (T = 20°C) to cool down. Estimate
   a. the energy needed to heat up the copper pipe, and
   b. the final temperature of the water bath.
8. A 200-lb lead pipe (specific heat capacity = 0.031 Btu/°R/lb) was heated to 300°F. It is then put into a 100-gallon water bath (T = 68°F) to cool down. Estimate the rise of the water bath temperature after the system reached a thermo-equilibrium.

5 Chemical Reaction Equilibria

5.1 INTRODUCTION

As mentioned in Chapter 4, a treatment process can be physical, chemical, biological, or thermal. A physical treatment process will just change the physical forms of substances, whereas a chemical, biological, or thermal treatment process involves the rearrangement of the molecular or ionic structure of a substance. A *chemical reaction* is a process in which one or more chemical compounds (i.e., the reactants) are converted to different compounds (i.e., the products). In that reaction, the original constituent atoms of the reactants are removed, rearranged, or combined with others to form the reaction products.

5.2 CONSERVATION OF MASS IN CHEMICAL REACTIONS

5.2.1 BALANCED REACTION EQUATION

A *chemical reaction equation* (e.g., Eq. 5.1 below) is often used to provide general information of a chemical reaction. The direction of the arrow shows the direction of the reaction, where A & B are the reactants and C & D are the products.

$$\text{Forward Reaction: } aA + bB \rightarrow cC + dD \tag{5.1}$$

Based on the "*Law of Mass Conservation*", the matter will not be created or destroyed in a chemical reaction. In other words, the total amount of each element in the reactants and that in the products will stay the same (but they may be in different combinations with other elements). A *stoichiometric coefficient* in Eq. (5.1) (i.e., *a*, *b*, *c*, and *d*) is the number written in front of a reactant or a product in that chemical reaction to balance the number of each element on both sides of the reaction equation. With proper stoichiometric coefficients, the reaction equation is balanced. For example, the balanced reaction equation for the combustion of heptane (C_7H_{16}) with oxygen in air is:

$$C_7H_{16(l)} + 11\ O_{2(g)} \rightarrow 7\ CO_{2(g)} + 8\ H_2O_{(g)} \tag{5.2}$$

For a complete combustion of hydrocarbons, the final products would be carbon dioxide (CO_2) and water (H_2O). The abbreviations in paratheses in a reaction equation indicate the physical states of the species (i.e., "*l*" for liquid, and "*g*" for gas). In this reaction, heptane is a liquid, while the other three species are in a gaseous form [Note: In a burner, a liquid fuel needs to be atomized into tiny droplets to facilitate better combustion].

DOI: 10.1201/9781003502661-5

In addition to mass balance (i.e., the conservation of matter), the reaction equation should also be charge-balanced. For example, the dissolution of calcium chloride ($CaCl_2$) in water into calcium and chloride ions:

$$CaCl_{2(s)} \overset{+H_2O}{\rightarrow} Ca^{+2}_{(aq)} + 2Cl^{-}_{(aq)} \tag{5.3}$$

The abbreviations, (*s*) and (*aq*), are for solid and for species in an aqueous solution, respectively. Equation (5.4) shows a half reaction for the oxidation of Zn (more in Chapter 9) and "e^-" is the symbol for an electron:

$$Zn_{(s)} \rightarrow Zn^{+2}_{(aq)} + 2e^{-} \tag{5.4}$$

Both equations (i.e., Eqs. 5.3 and 5.4) are balanced with regard to mass and charge.

The reaction equations above having an arrow of "→", which indicates that the directions of the reactions are all moving from left to right (i.e., forward reaction). However, many reactions are reversible (i.e., they can move in the opposite direction). Equation (5.5) shows the backward (the reverse) of the reaction shown in Eq. (5.1) [Note: The only difference is the direction of the arrow].

$$\text{Backward Reaction}: aA + bB \leftarrow cC + dD \tag{5.5}$$

In principle, all chemical reactions are reversible, meaning both forward and reverse processes are occurring at the same time. The rates of the forward and the backward reactions may not be the same at a specific time. A system would reach a state of *chemical equilibrium* in which the rate of the forward reaction is equal to the rate of the reverse reaction. At chemical equilibrium, no further changes in the concentrations of all the reactants and the products will occur. However, the reversibility of a chemical reaction may not be observable if the backward reaction is very slow or the fractions of the products in the equilibrium mixture are very small. Equation (5.6) shows an overall reaction by having a double-headed arrow (i.e., "⟷" or "⇆").

$$\text{Overall Reaction}: aA + bB \leftrightarrow cC + dD \tag{5.6}$$

A reaction is said to be *complete* when the equilibrium mixture contains insignificant amounts of certain reactants. For example, if all the heptane in the system is combusted (see Eq. 5.2), the reaction is a complete reaction.

5.2.2 Quantitative Information from Balanced Reaction Equations

$$C_7H_{16(l)} + 11\ O_{2(g)} \rightarrow 7\ CO_{2(g)} + 8\ H_2O_{(g)} \tag{5.7}$$

A balanced reaction equation (e.g., Eq. 5.2) shows qualitative information such as the molecular formula and the physical states of the reactants and the products. We can also obtain some quantitative information from a balanced reaction equation. The equation tells one unit of heptane reacts with 11 units of oxygen to form seven units of carbon dioxide and eight units of water. That unit is "mole" in chemistry.

In other words, in this reaction, 11 moles of O_2 will react with one mole of C_7H_{16} to produce seven moles of CO_2 and eight moles of H_2O. It should be noted that "mole" in chemistry is the gram-mole, or g-mole (i.e., molar mass of 1 g-mole of O_2 is 32 g). In engineering practices, "kg-mole" and "lb-mole" are also often used (i.e., the molar mass of 1 kg-mole of O_2 is 32 kg, and that of 1 lb-mole of O_2 is 32 pounds).

A *molar ratio* (*mole ratio*, or *mole-to-mole ratio*) is the ratio between the amounts (in moles) of any two compounds involved in a balanced chemical reaction, which can be found in the ratio of the corresponding stoichiometric coefficients.

Gaseous compounds participate in many chemical reactions as reactants and products; and their amounts are often expressed in volume or in pressure, instead of mass. An *ideal gas* is a gas in which there are no inter-molecular forces between gas molecules and these molecules occupy no volume. There is no such a thing as an ideal gas. However, behaviors of real gases deviate insignificantly from those of an ideal gas under ambient pressures and temperatures and in most environmental engineering applications. The ideal gas law is derived from four basic laws:

- *Boyle's Law* states that the volume (V) of a specific quantity of an ideal gas ("n" below represents the number of moles present) is inversely proportional to the system pressure (P) when the system absolute temperature (T) is held constant:

$$V \propto 1/P \text{ @ constant } n \text{ \& } T \tag{5.7}$$

- *Charles's Law* states that V is proportional to T when P is held constant:

$$V \propto T \text{ @ constant } n \text{ \& } P \tag{5.8}$$

- *Gay-Lussac's Law* states that P is proportional to T when V is held constant:

$$P \propto T \text{ @ constant } n \text{ \& } V \tag{5.9}$$

- *Avogadro's Law* states that V is proportional to the quantity of the gas (n) when T & P are held constant:

$$V \propto n \text{ @ constant } T \text{ \& } P \tag{5.10}$$

Combining these laws one can obtain:

$$PV = \left(\frac{m}{MW}\right)RT = nRT \tag{5.11}$$

The *ideal gas law* (or the *ideal gas equation of state*) relates P, V, T, and n of an ideal gas as

$$PV = \left(\frac{m}{MW}\right)RT = nRT \tag{5.12}$$

where m = mass, MW = molecular weight, and R = the universal gas constant. Below are the values of R in some commonly used units:

$$R = 0.082\ \frac{\text{L}\cdot\text{atm}}{\text{g-mole}\cdot\text{K}} = 8.314\frac{\text{m}^3\cdot\text{Pa}}{\text{g-mole}\cdot\text{K}} = 10.731\frac{\text{ft}^3\cdot\text{psi}}{\text{lb-mole}\cdot{}^\circ\text{R}} \tag{5.13}$$

Example 5.1: Ideal Gas Law

Find the volume of 1 g-mole and 1 lb-mole of an ideal gas at $T = 0°C$ and $P = 1$ atm.

Solution:

a. For 1 g-mole of an ideal gas,

$$V = \frac{nRT}{P} = \left[(1\ \text{g-mole})\left(0.08206\ \frac{\text{L}\cdot\text{atm}}{\text{g-mole}\cdot\text{K}}\right)(273\ \text{K})\right] \div 1\ \text{atm} = 22.4\ \text{L}$$

b. For 1-lb mole of an ideal gas,

$$V = \left[(1\ \text{lb-mole})\left(10.731\frac{\text{ft}^3\cdot\text{psi}}{\text{lb-mole}\cdot{}^\circ\text{R}}\right)(492\ {}^\circ\text{R})\right] \div 14.7\ \text{psi} = 359\ \text{ft}^3$$

Discussion:

1. Using a similar approach, one can readily find that volumes of 1 g-mole ideal gas @$T = 20°C$ and 25°C ($P = 1$ atm) are 24.05 and 24.46 L, respectively; while those of 1 lb-mole ideal gas would be 385 and 392 ft^3 @$T = 68$ and 77°F, respectively.
2. It may not be a bad idea to have these values readily available or memorized.
3. The molar volume of an ideal gas at STP ($T = 273.15$ K and $P = 100$ kPa) = 22.71 L, which is slightly larger than 22.4 L because of a smaller P (100 kPa versus 101.3 kPa, or 1 atm).

Example 5.2: Mass and Mole Relationships of Reactants and Products of a Chemical Reaction

A furnace is combusting heptane (C_7H_{16}) at a rate of 100 kg/hr.

a. What is the molar flow rate of heptane into the furnace?
b. What would be the stoichiometric amount of oxygen needed, in kg-mole/hr?
c. What would be the stoichiometric flow rate of oxygen in m^3/hr ($T = 20°C$ & $P = 1$ atm)?
d. What would be the rate of CO_2 produced from this furnace, in kg/hr?
e. What would be rate of CO_2 produced in m^3/hr (T = <u>25</u>°C & $P = 1$ atm), assuming the reaction is complete?

Solution:

a. *MW* of heptane (C_7H_{16}) = 12 × 7 + 1 × 16 = 100

Molar flow rate of heptane = (100 kg/hr) ÷ (100 kg/kg-mole)
= 1.0 kg-mole/hr (or 1,000 g-mole/hr)

b. Molar flow rate of oxygen = (1.0 kg-mole/hr) × (11/1) = 11.0 kg-mole/hr

c. Volumetric flow rate of oxygen @T = 20°C and P = 1 atm

= (11.0 kg-mole/hr) × 24.05 m³/kg-mole = 264.55 m³/hr

d. Molar flow rate of CO_2 = (1.0 kg-mole/hr) × (7/1) = 7.0 kg-mole/hr

Mass flow rate of CO_2 = (7.0 kg-mole/hr) × 44 = 308 kg/hr

e. Volumetric flow rate of CO_2 @T = 25°C and P = 1 atm

= (7.0 kg-mole/hr) × 24.46 m³/kg-mole = 168.35 m³/hr

Discussion:

1. "kg-mole' was used; the results would be the same, if "g-mole" is used.
2. "24.05 m³/kg-mole" in part (c) is the molar volume of an ideal gas @T = 20°C & P = 1 atm, which is the same as 24.05 L/g-mole.
3. "44" in part (d) is the MW of CO_2.
4. The molar flow rate and the mass flow rate are not affected by the system T & P; while the volumetric flow rates in parts (c) & (e) depend on T & P. The values of molar volume of an ideal gas used in parts (c) & (e) are different because the temperatures are different.

5.2.3 Limiting Reactant

In a balanced reaction equation, the reactants are shown in stoichiometric quantities. If the reaction is complete and the reverse/backward reaction is insignificant, none of the reactants would be left at the end of the reaction. However, for various reasons, the reactants in a reactor are often not present in stoichiometric quantities. Consequently, one or more of the reactants will not be used up completely at the end of the reaction. In this situation, the limiting reactant is the reactant that is completely used up and it will determine how much the product(s) would be produced.

Example 5.3: Limiting Reactant

One hundred pounds of calcium carbonate ($CaCO_3$) and 100 pounds of hydrochloric acid (HCl) were put into a reactor and the reaction below occurred. Determine (a) the mass of CO_2 produced, and (b) the amount of HCl remained at the end of a complete reaction.

$$CaCO_{3(s)} + 2\ HCl_{(aq)} \rightarrow CaCl_{2(aq)} + CO_{2(g)} + H_2O_{(aq)}$$

Solution:

a. MW of $CaCO_3$ = 100; MW of HCl = 36.5
Moles of $CaCO_3$ present = (100 lbs) ÷ (100 lbs/lb-mole) = 1.0 lb-mole
Moles of HCl present = (100 lbs) ÷ (36.5 lbs/lb-mole) = 2.74 lb-moles
Since one mole of $CaCO_3$ will react with only two moles of HCl, HCl is in excess (i.e., $CaCO_3$ is the limiting reactant).

Assuming the reaction is complete, 1 lb-mole of CO_2 would be produced for each lb-mole of $CaCO_3$ reacted (the stoichiometric ratio of $CaCO_3$:CO_2 = 1:1).
Mass of CO_2 produced = (1.0 lb-mole) × 44 = 44.0 lbs

b. The amount of HCl left = 2.74 − (1.0)(2/1) = 0.74 lb-mole = 26.27 lbs (= 0.74 × 35.5)

Discussion:

1. US customary unit for mass is "slug" and 1 slug = 14.59 kg. Weight of one slug mass at the sea level = (1 slug)(32.2 ft/s^2) = 32.2 lb_f (or = 32.2 lb). In engineering practices, lb_m (or lb) is more commonly used than slug as the unit of mass. One slug = 32.2 lb_m and 1 lb_m (or 1 lb) = 454 g = 0.454 kg (or 1 kg = 2.2 lb).
2. Volume of CO_2 produced can be found by using an approach similar to that used in Example 5.2 [Note: At P = 1 atm, the molar volumes of one lb-mole ideal gas are 385 and 392 ft^3 @ T = 68 and 77°F, respectively].

5.2.4 Reaction Quotient and Equilibrium Constant

Equation (5.6) is a general expression for a reversible chemical reaction, in which A and B are the reactants and C and D are the products, while a, b, c, and d are the stoichiometric coefficients. Once the reaction gets started, two reactions (i.e., the forward reaction and the reverse reaction) are going simultaneously, as shown in Eqs. (5.1) and (5.5), but at different rates until the system reaches an equilibrium in which the two rates become equal.

The "*Reaction Quotient*" (Q) is a measure of the relative molar concentrations of the reactants and the products present in a chemical reaction at a given time, by taking the stoichiometric constants into account. For the reaction shown in Eq. (5.6), Q can be expressed as

$$Q = \frac{[C]^c [D]^d}{[A]^a [B]^b} \tag{5.14}$$

where $[C]$ is the molar concentration of product C at that specific time (similarly $[A]$, $[B]$, and $[D]$ are concentrations of compounds A, B, and D, respectively); these concentrations are also often expressed as C_A, C_B, C_C, and C_D. As the reaction proceeds with time, the reaction quotient changes with the concentrations of the reactants and the products.

It should be noted that the effective concentrations (i.e., *activities*) should be used, instead of the molar concentrations, in Eq. (5.14). The *activity coefficient* (γ) of an ion or a molecule is defined as the ratio of *activity* (a) and the actual concentration (C) of that ion or molecule as:

$$\gamma = a/C; \text{ or } a = \gamma \times C \tag{5.15}$$

For an ideal solution, the activity is equal to the actual concentration (i.e., $\gamma = 1.0$). The value of an activity is difficult to be determined with accuracy because many factors could affect its value in a solution. Consequently, molar concentrations are used in most practices (also in this book).

When a system is in equilibrium, the molar concentrations of all reactants and products remain the same, even with both forward and backward reactions still ongoing. Consequently, the reaction quotient becomes a constant, and it is called the "*equilibrium constant*" (K_{eq}) of the reaction, as

$$K_{eq} = \frac{[C]^c [D]^d}{[A]^a [B]^b} \tag{5.16}$$

The equilibrium constant (K_{eq}) is the reaction quotient at a specific condition (i.e., the system is in equilibrium). In practice, the value of K_{eq} relates the molar concentrations (in mole/L or M) of all the reaction participants.

From Eqs. (5.14) and (5.16), a reaction quotient and an equilibrium constant should come with a unit of $M^{[(c+d)-(a+b)]}$. However, the reported values of equilibrium constants usually have no units. It is because activities should be used in these two equations, but molar concentrations are usually used instead.

Three cases should be noted here: (i) if water is involved in a chemical reaction, it will not show up in Eqs. (5.14) and (5.16) because the molar concentration of water, $[H_2O]$, is a constant; and it is included in the reported K_{eq} values; (2) the concentration of a pure solid is constant and set to be equal to unity; and (3) the concentrations of gaseous compounds are related to pressure, so partial pressures of gases are also often used in Eqs. (5.14) and (5.16).

5.2.5 Molar Concentration and Mass Concentration

Molar concentration (or *molarity*) is a measure of the concentration of a soluble substance (i.e., solute) in a solution. In chemistry, the most commonly used unit for molarity is the number of moles per liter of solution (i.e., mole/L, mol/L, mole/dm^3, or mol/dm^3). A solution with a concentration of 1 mole/L is said to be 1 *molar*, designated as 1 M. Thus,

$$\text{Molarity } (M) = \frac{\text{Number of moles of a solute}}{\text{Liters of the solution}} = \frac{(\text{mass of the solute})/(MW \text{ of the solute})}{\text{Liters of the solution}} \tag{5.17}$$

$$\text{Moles of the solute} = \text{Molarity } (M) \times \text{Liters of the solution} \tag{5.18}$$

Molar concentrations are used in chemistry formulas/equations (e.g., in equilibrium constant and in reaction quotient), while mass concentrations (i.e., (mass of solute) ÷ (volume of the solution)) are more commonly used in engineering practices, such as mg/L. The conversion from a molar concentration to its corresponding mass concentration, and vice versa, can be done as:

$$\text{Mass concentration}\left(\text{in}\frac{\text{g}}{\text{L}}\right) = \text{Molarity }(M)\times MW \tag{5.19}$$

Example 5.4: Molar Concentration and Mass Concentration

a. Estimate the molar concentration of water.
b. If $[Ca^{+2}] = 10^{-3}$ M, what is its mass concentration in mg/L?
c. The current primary Maximum Contaminant Level (MCL) of benzene in drinking water is 0.005 mg/L (or 5 µg/L), express this concentration in a molar concentration.

Solution:

a. *MW* of water (H_2O) = 18.
   Assuming the density of water = 1 kg/L = 1,000 g/L, and use Eq. (5.17) to find the molarity of water:

$$\text{Molarity} = \frac{(1{,}000\ \text{g})/(18\ \text{g/mole})}{1\ \text{L}} = 55.5\frac{\text{mole}}{\text{L}} = 55.5\ \text{M}$$

b. *MW* of Ca^{+2} = 40. Use Eq. (5.19) to find the mass concentration of Ca^{+2}:

$$\text{Mass concentration } = 10^{-3}\frac{\text{mole}}{\text{L}}\times 40\frac{\text{g}}{\text{mole}} = 0.04\frac{\text{g}}{\text{L}} = 40\ \text{mg/L}$$

c. *MW* of benzene (C_6H_6) = 78. Use Eq. (5.19) to find the molar concentration of benzene:

$$0.005\frac{\text{mg}}{\text{L}} = \frac{5\times 10^{-6}\text{g}}{\text{L}} = C_{\text{benzene}}\times 78\frac{\text{g}}{\text{mole}}$$

$$C_{\text{benzene}} = 6.41\times 10^{-8}\ \text{mole/L} = 6.41\times 10^{-8}\,\text{M}$$

Discussion:

1. The MCL of benzene = 0.005 mg/L = 6.41×10^{-8} M, which is very small (i.e., the regulatory limit is very stringent).
2. A unit of "parts per million (ppm)" is often used to represent the solute concentration unit of "mg/L". 1 ppm = 1 mg/L = 1,000 µg/L = 1,000 parts per billion (ppb). Thus, the current MCL of benzene = 0.005 ppm = 5 ppb.

5.2.6 Le Chatelier's Principle

Le Chatelier's Principle states that "when a system is in equilibrium, changes in the process variables (i.e., temperature, pressure, volume, or concentrations of the reactants and/or products) will result in a predictable shift in the position of the equilibrium; and the shift will counteract the effects of changes". Taking concentrations as an example, if we increase the concentration of one reaction product, the backward reaction rate will increase so that the concentrations of all the reactants and products will change until a new equilibrium is reached. For this case, the concentrations of all reactants will increase, but the equilibrium constant would stay the same.

The value of the reaction quotient (Q) can tell us the state of a system with respect to equilibrium. It can tell if the reaction will proceed; and, if so, the direction of the reaction. There are three potential scenarios when we compare the values of Q and K (i.e., the equilibrium constant):

i. If $Q = K$, the system is in equilibrium.
ii. If $Q > K$, the backward reaction is more favored, because the concentrations of the products at that stage are relatively larger than those at equilibrium – more reactants will be produced until a chemical equilibrium is reached.
iii. If $Q < K$, the forward reaction is more favored (more products will be produced until a chemical equilibrium is reached).

Example 5.5: Equilibrium Constant and Reaction Quotient

Some carbonic acid (H_2CO_3) molecules will dissociate into proton ion (H^+) and bicarbonate ion $\left(HCO_3^-\right)$ in water as:

$$H_2CO_{3(aq)} \rightleftharpoons HCO^-_{3(aq)} + H^+_{(aq)} \quad K = 4.47\times10^{-7}\ (T = 25\ °C)$$

a. At $T = 25°C$, a system was found to be in equilibrium and the concentrations of carbonic acid and bicarbonate were equal, what is the molar concentration of the proton ion?
b. A concentrated acid was added to raise the proton concentration to 4.47×10^{-5} M to break the equilibrium. What is the reaction quotient immediately after the acid addition?
c. Which way the reaction would move immediately after the acid addition?
d. If the proton concentration is kept at 4.47×10^{-5} M and the system is allowed to reach a new equilibrium, what would be the ratio of $\left[HCO_3^-\right]/[H_2CO_3]$?

Solution:

a. For the given reaction, the equilibrium constant can be written as

$$K = \frac{\left[HCO_3^-\right]\left[H^+\right]}{[H_2CO_3]} = 4.47 \times10^{-7}$$

Since $\left[HCO_3^-\right] = [H_2CO_3]$, $[H^+]$ would be equal to 4.47×10^{-7} M.

b. Upon the acid addition ($[HCO_3^-]/[H_2CO_3] = 1.0$ and $[H^+] = 4.47 \times 10^{-5}$ M), the reaction quotient can be found by using Eq. (5.9):

$$Q = \frac{[HCO_3^-][H^+]}{[H_2CO_3]} = (1.0)(4.47 \times 10^{-5}) = (4.47 \times 10^{-5})$$

c. Since Q is 100 times of K (i.e., $Q > K$), the backward reaction is favored; more reactant (i.e., H_2CO_3) will be formed.

d. With $[H^+] = 4.47 \times 10^{-5}$ M and the same equilibrium constant value:

$$K = \frac{[HCO_3^-][H^+]}{[H_2CO_3]} = 4.47 \times 10^{-7} = \frac{[HCO_3^-](4.47 \times 10^{-5})}{[H_2CO_3]}$$

$$[HCO_3^-]/[H_2CO_3] = 0.01 = 1/100$$

Discussion:

1. Values of equilibrium constants usually depend on the system temperatures. $T = 25°C$ for this example (as specified in the problem statement).
2. The definition of pH is "pH = $-\log[H^+]$". Thus, the initial pH of water = $-\log(4.47 \times 10^{-7}) = 6.35$. After acid addition, pH decreased to 4.35 ($= -\log(4.47 \times 10^{-5})$).
3. At a lower pH, H_2CO_3 becomes the most dominant carbonate species [Note: Carbonate species are H_2CO_3, HCO_3^- and carbonate ($CO_3^{=}$) in water].

5.3 CONSERVATIONS OF ENERGY IN CHEMICAL REACTIONS

Energy transfer is involved in all chemical reactions. During a chemical reaction, energy is used to break the bonds of the reactants; and, at the same time, energy is released when new bonds of the products are formed. Some chemical reactions require less energy to break the bonds in the reactants, and more energy is released when new bonds are formed in the products. These reactions are called *exothermic reactions* that release energy to the surroundings (most often in the form of heat). In other reactions, the system needs to take in more energy to break the bonds than the energy released when new bonds are formed. These reactions are *endothermic reactions*, which receive/absorb energy from the surroundings.

Internal energy is the energy content of a system, while enthalpy can be considered as heat content of the system. The heat, released from the system to its surroundings or absorbed/taken from the surroundings into the system during a chemical reaction at a constant pressure, is the change of enthalpy (see Eq. 4.19). Since the energy is conserved, the energy loss of the system should be the gain of its surroundings; and vice versa.

5.3.1 Standard Enthalpy of Formation

Standard heat of formation (also called "*standard enthalpy of formation*") of a compound $\left(\Delta H_f^\circ\right)$ is the amount of heat absorbed or emitted when one mole of this compound is formed from its constituent elements under standard conditions (typically at $T = 25°C$ and $P = 1$ atm; or $T = 25°C$ and $P = 1$ bar). The standard enthalpy of formation of a pure element in its most stable form under standard conditions is arbitrarily assigned a value of zero.

Table 5.1 tabulates standard enthalpies of formation of some substances ($T = 25°C$ and $P = 1$ atm), along with values of some other thermodynamic functions (i.e., values of molar entropy and Gibbs free energy). The values of common thermodynamic functions can be found in chemistry and chemical engineering handbooks, for example:

- Perry's Chemical Engineers' Handbook (9th ed.) (2019) by Don W. Green and Marylee Z. Southland, McGraw-Hill.
- Lange's Handbook of Chemistry (17th ed.) (2017) by James G. Speight, McGraw-Hill.
- CRC Handbook of Chemistry and Physics (104th ed.) (2023) by John R. Rumble (editor-in-chief), CRC Publisher.

TABLE 5.1
Standard Molar Thermodynamic Values ($P = 1$ atm, $T = 25°C$)

Compound	State	Enthalpy (kJ)	Entropy (J)	Gibbs Free Energy (kJ)
Ag	(*s*)	0.0	42.6	0.0
Ag^+	(*aq*)	105.6	72.7	77.1
AgBr	(*s*)	−100.4	107.1	−96.9
AgCl	(*s*)	−127.1	96.2	−109.8
AgI	(*s*)	−61.8	115.5	−66.2
$AgNO_3$	(*s*)	−124.4	140.9	−33.4
Al	(*s*)	0.0	28.3	0.0
Al^{+3}	(*aq*)	−531.4	−321.7	−485.3
Al_2O_3	(*s*)	−1,675.7	50.9	−1,582.3
$AlCl_3$	(*s*)	−704.2	110.7	−628.8
Ba	(*s*)	0.0	67.0	0.0
Ba^{+2}	(*aq*)	−537.6	9.6	−560.7
$BaCl_2$	(*s*)	−858.6	123.7	−810.4
$BaCO_3$	(*s*)	−1,219.0	112.0	−1,139.0
$BaSO_4$	(*s*)	−1,473.2	132.2	−1,362.2
C(graphite)	(*s*)	0.0	5.7	0.0
C_2H_2	(*g*)	226.7	200.9	209.2
C_2H_4	(*g*)	52.3	219.6	68.2
C_2H_5OH	(*g*)	−235.1	282.7	−168.5

(*Continued*)

TABLE 5.1 (*Continued*)
Standard Molar Thermodynamic Values ($P = 1$ atm, $T = 25°C$)

Compound	State	Enthalpy (kJ)	Entropy (J)	Gibbs Free Energy (kJ)
C_2H_5OH	(*l*)	−277.7	160.7	−174.8
C_2H_6	(*g*)	−84.7	229.6	−32.8
C_3H_8	(*g*)	−103.8	269.9	−23.5
C_4H_{10}	(*g*)	−147.6	231.0	−15.1
C_5H_{12}	(*g*)	−146.4	349.0	−8.4
C_6H_{14}	(*g*)	−167.2	388.4	−0.3
C_6H_{14}	(*l*)	−198.8	296.0	−4.0
C_6H_6	(*l*)	49.0	172.8	124.5
C_7H_{16}	(*g*)	−187.8	427.9	8.0
C_7H_{16}	(*l*)	−224.4	326.0	1.8
C_8H_{18}	(*l*)	−250.0	361.2	6.7
Ca	(*s*)	0.0	41.4	0.0
Ca^{2+}	(*aq*)	−543.0	−55.2	−553.0
$Ca(OH)_2$	(*aq*)	−1,002.8	−74.5	−868.1
$Ca(OH)_2$	(*s*)	−986.1	83.4	−898.5
$CaCl_2$	(*s*)	−795.8	104.6	−748.1
$CaCO_3$	(*s*)	−1,206.9	92.9	−1,128.8
CaF_2	(*s*)	−1,219.6	68.9	−1,167.3
CaO	(*s*)	−635.1	39.8	−604.0
$CaSO_4$	(*s*)	−1,434.1	106.7	−1,321.8
CCl_4 (*g*)	(*g*)	−102.9	309.9	−60.6
CCl_4 (*l*)	(*l*)	−135.4	216.4	−65.2
CH_3COOH	(*g*)	−434.8	282.5	−376.7
CH^3COOH	(*l*)	−484.0	160.0	−389.0
CH_3OH	(*g*)	−200.7	239.8	−162.0
CH_3OH	(*l*)	−238.7	126.8	−166.3
CH_4	(*g*)	−74.8	186.3	−50.7
$CHCl_3$	(*g*)	−103.1	295.7	−70.3
$CHCl_3$	(*l*)	−134.5	201.7	−73.7
Cl^- (*aq*)	(*aq*)	−167.0	57.0	−131.0
Cl_2	(*aq*)	−23.0	121.0	7.0
Cl_2	(*g*)	0.0	223.1	0.0
ClO^-	(*aq*)	−101.7	41.8	−36.8
CN^-	(*aq*)	150.6	94.1	172.4
CO	(*g*)	−110.5	197.7	−137.2
CO_2	(*g*)	−393.5	213.7	−394.4
CO_2	(*aq*)	−413.8	117.6	−386.0
CO_3^{2-}	(*aq*)	−676.3	−53.1	−528.1
Cu	(*s*)	0.0	33.2	0.0
Cu^{+1}	(*aq*)	71.7	40.6	50.0

(*Continued*)

TABLE 5.1 (*Continued*)
Standard Molar Thermodynamic Values ($P = 1$ atm, $T = 25°C$)

Compound	State	Enthalpy (kJ)	Entropy (J)	Gibbs Free Energy (kJ)
Cu^{+2}	(*aq*)	64.8	−99.6	65.5
$Cu(OH)_2$	(*s*)	−450.0	108.0	−372.0
$CuCl_2$	(*s*)	−220.1	108.1	−175.7
F^-	(*aq*)	−332.6	−13.8	−278.8
F_2	(*g*)	0.0	202.8	0.0
Fe	(*s*)	0.0	27.8	0.0
Fe_2O_3	(*s*)	−824.2	87.4	−742.2
$FeCl_2$	(*s*)	−341.8	118.0	−302.3
$FeCl_3$	(*s*)	−399.5	142.3	−333.9
FeS	(*s*)	−95.0	67.0	−97.0
FeS_2	(*s*)	−178.2	52.9	−166.9
$FeSO_4$	(*s*)	−929.0	121.0	−825.0
H^+	(*aq*)	0.0	0.0	0.0
H_2	(*g*)	0.0	130.7	0.0
H_2CO_3	(*aq*)	−698.7	191.0	−623.4
H_2O	(*g*)	−241.8	188.8	−228.6
H_2O	(*l*)	−285.8	69.9	−237.1
H_2O_2	(*l*)	−187.8	109.6	−120.4
H_2S	(*g*)	−20.6	205.8	−33.6
H_2SO_4	(*aq*)	−909.3	20.1	−744.5
H_2SO_4	(*l*)	−814.0	156.9	−690.0
HBr	(*g*)	−36.4	198.7	−53.5
HCl	(*aq*)	−167.2	56.5	−131.2
HCl	(*g*)	−92.3	186.9	−95.3
HCO_3^-	(*aq*)	−691.1	95.0	−587.1
HF(*aq*)	(*aq*)	−332.6	−13.8	−278.8
HF(*g*)	(*g*)	−271.1	173.8	−273.2
HNO_3	(*aq*)	−207.4	146.4	−111.3
HNO_3	(*g*)	−135.1	266.4	−74.7
HNO_3	(*l*)	−174.1	155.6	−80.7
I^-	(*aq*)	−55.0	106.0	−52.0
I_2	(*aq*)	23.0	137.0	16.0
I_2	(*g*)	62.4	260.7	19.3
I_2	(*s*)	0.0	116.1	0.0
Mg^{+2}	(*aq*)	−466.8	−138.1	−454.8
Mg	(*s*)	0.0	32.7	0.0
$Mg(OH)_2$	(*s*)	−924.5	63.2	−833.5
$MgCl_2$	(*s*)	−641.3	89.6	−591.8
$MgCO_3$	(*s*)	−1,095.8	65.7	−1,012.1
MgS	(*s*)	−346.0	50.3	−341.8

(*Continued*)

TABLE 5.1 (*Continued*)
Standard Molar Thermodynamic Values ($P = 1$ atm, $T = 25°C$)

Compound	State	Enthalpy (kJ)	Entropy (J)	Gibbs Free Energy (kJ)
N_2	(*g*)	0.0	191.6	0.0
N_2O	(*g*)	82.1	219.9	104.2
N_2O_4	(*g*)	9.2	304.3	97.9
Na	(*s*)	0.0	51.2	0.0
Na^+	(*aq*)	−240.0	59.0	−262.0
Na_2CO_3	(*s*)	−1,130.7	135.0	−1,044.4
NaCl	(*aq*)	−407.3	115.5	−393.1
NaCl	(*s*)	−411.2	72.1	−384.1
NaOH	(*aq*)	−470.1	48.1	−419.2
NaOH	(*s*)	−425.6	64.5	−379.5
NH_3	(*aq*)	−80.0	111.0	−27.0
NH_3	(*g*)	−46.1	192.5	−16.5
$NH_4^+(aq)$	(*aq*)	−132.0	113.0	−79.0
NH_4Cl	(*aq*)	−299.7	169.9	−210.5
NH_4Cl	(*s*)	−314.4	94.6	−202.9
NH_4NO_3	(*aq*)	−339.9	259.8	−190.6
NH_4NO_3	(*s*)	−365.6	151.1	−183.9
NO	(*g*)	90.3	210.8	86.6
NO_2	(*g*)	33.2	240.1	51.3
O_2	(*g*)	0.0	205.1	0.0
O_3	(*g*)	142.7	238.9	163.2
OH^-	(*aq*)	−229.9	−10.5	−157.3
Pb	(*s*)	0.0	64.8	0.0
$PbCl_2$	(*s*)	−359.4	136.0	−314.1
SO_2	(*g*)	−296.8	248.2	−300.2
SO_3	(*g*)	−395.7	256.8	−371.1
SO_4^{2-}	(*aq*)	−909.0	20.0	−745.0

In addition to these handbooks, NIST Chemistry WebBook (https://webbook.nist.gov/), by National Institute of Standards and Technology, US Department of Commerce, provides access to data compiled and distributed by NIST under the Standard Reference Data Program. With regard to thermochemistry, the WebBook provides thermochemical data (including enthalpy of formation, entropy, enthalpy of combustion, heat capacity, phase transition enthalpies and temperatures, and vapor pressure) for over 7,000 organic and small inorganic compounds and reaction thermochemistry data (i.e., enthalpy of reaction and free energy of reaction) for over 8,000 reactions. In addition, the WebBook also provides thermophysical property (including density, heat capacity at constant pressure (C_p), heat capacity at constant volume (C_v), enthalpy, internal energy, entropy, viscosity, thermal conductivity, Joule-Thomson coefficient, and surface tension) data for 74 fluids. One can search for data on specific compounds in the WebBook based on name, chemical formula, CAS registry number, molecular weight, chemical structure, or selected ion energetics and spectral properties. (NIST, 2023).

5.3.2 Standard Enthalpy of Reaction

The standard heat/enthalpy of a reaction $\left(\Delta H^{\circ}_{\text{rxn}}\right)$ can be calculated from the standard heats/enthalpies of formation of the reactants and the products $\left(\Delta H^{\circ}_{\text{f}}\right)$:

$$\Delta H^{\circ}_{\text{rxn}} = \sum n\Delta H^{\circ}_{\text{f}}(\text{products}) - \sum m\Delta H^{\circ}_{\text{f}}(\text{reactants}) \tag{5.20}$$

where m and n are the stoichiometric coefficients in the reaction equation. The term "standard" means that reaction takes place under standard conditions (typically $T = 25°\text{C}$ & $P = 1$ atm).

Example 5.6: Heat of Reaction

Determine the heat of reaction of two reactions shown below:

a. $S_{(s)} + O_{2(g)} \rightarrow SO_{2(g)}$

b. $SO_{2(g)} + \frac{1}{2}O_{2(g)} \rightarrow SO_{3(g)}$

Solution:

a. From Table 5.1, $\Delta H^{\circ}_{\text{f}}$ of $SO_2 = -296.8$ kJ/mole and those of $S_{(s)}$ and $O_{2(g)} = 0$ (element in its most stable form), then

$$\Delta H^{\circ}_{\text{rxn}} = (1)(-296.8) - \left[(1)(0) + (1)(0)\right] = \underline{-296.8 \text{ kJ}}$$

b. From Table 5.1, $\Delta H^{\circ}_{\text{f}}$ of $SO_3 = -395.7$ kJ/mole, then

$$\Delta H^{\circ}_{\text{rxn}} = (1)(-395.7) - [(1)(-296.8) + (1/2)(0)] = \underline{-98.9 \text{ kJ}}$$

Example 5.7: Heat of Reaction

Methane (CH_4) and propane (C_3H_8) can be oxidized by oxygen as shown below:

$$CH_{4(g)} + 2\ O_{2(g)} \rightarrow CO_{2(g)} + 2\ H_2O_{(g)}$$

$$C_3H_{8(g)} + 5\ O_{2(g)} \rightarrow 3\ CO_{2(g)} + 4\ H_2O_{(g)}$$

a. Determine the heat of reaction of each reaction.
b. How much heat will be released if 1 g of methane is combusted?
c. How much heat will be released if 1 g of propane is combusted?
d. If water is the final product, instead of water vapor, what would be the heat of reaction for methane combustion?

Solution:

a. From Table 5.1, ΔH_f^o values of CH_4, CO_2, and $H_2O_{(g)}$ are −74.8, −393.5, and −241.8 kJ/mole, respectively and that of $O_2 = 0$ (element in its most stable form), then

$$\Delta H_{rxn}^o = [(1)(-393.5)+(2)(-241.8)]-[(1)(-74.8)+(2)(0)] = \underline{-951.9 \text{ kJ/mole } CH_4}$$

From Table 5.1, ΔH_f° of $C_3H_8 = -103.8$ kJ/mole, then

$$\Delta H_{rxn}^o = [(3)(-393.5)+(4)(-241.8)]-[(1)(-103.8)+(5)(0)] = \underline{-2{,}043.9 \text{ kJ/mole } C_3H_8}$$

b. The two heat reaction values in part (a) are for one mole of CH_4 and one mole of C_3H_8, respectively.

$$\Delta H_{rxn}^o \text{ for each gram of methane} = -952 \text{ kJ}/(18 \text{ g}) = \underline{52.9 \text{ kJ/g } CH_4}$$

c. ΔH_{rxn}^o for each gram of pentane = −2,044 kJ/(44 g) = $\underline{46.5 \text{ kJ/g } C_3H_8}$

d. If water, instead of vapor, is considered as the final product, then for methane combustion:

$$CH_{4(g)} + 2\ O_{2(g)} \rightarrow CO_{2(g)} + 2\ H_2O_{(l)}$$

From Table 5.1, ΔH_f^o values of $H_2O_{(l)}$ is −285.8 kJ/mole, then

$$\Delta H_{rxn}^o = [(1)(-393.5)+(2)(-285.8)]-[(1)(-74.8)+(2)(0)] = \underline{-1{,}039.9 \text{ kJ/mole } CH_4}$$

Discussion:

1. Both combustion processes are exothermic because the ΔH_{rxn}^o value for each process is negative.
2. Methane has a lower energy content than pentane on the molar basis, but larger on the mass basis.
3. The values of ΔH_{rxn}^o between parts (c) and (a) is −88 kJ/mole CH_4 (= −1,039.9 − (− 951.9)). It means that more energy will be released to the surroundings if water (instead of water vapor) is the final product. The heating value of a substance (usually a fuel) is the amount of heat released during the combustion of a specified amount of that substance. Heating values of fuels are often reported in *higher heating values* (HHVs) and/or in *lower heating values* (LHVs). The numerical difference between the LHV and HHV of a fuel is the amount of heat of vaporization of

water produced. Here, two moles of water are produced for each mole of methane combusted. The heat of water vaporization at 25°C is 44 kJ/mole, which is essentially half of the difference (i.e., 88 kJ/mole of methane combusted) in this example.

5.3.3 Hess's Law of Constant Heat Summation

Hess's law of constant heat summation (or *Hess's law*) states that "the heat of reaction of any chemical reaction is a fixed quantity; and it does not depend on the path of the reaction or the number of steps taken to come up with the reaction". In other words, if a reaction consists of *n* consecutive sub-steps, the overall heat of reaction (ΔH^o) is the sum of the heat of reaction of each step (ΔH_n):

$$\Delta H^o_{\text{overall reaction}} = \sum \Delta H_n \tag{5.21}$$

Example 5.8: Hess's Law of Constant Heat Summation

a. Determine the heat of reaction of the reaction shown below, by using standard enthalpies of formation:

$$S_{(s)} + \frac{3}{2}O_{2(g)} \rightarrow SO_{3(g)}$$

b. Use Hess's law and the values of heat of reaction for two reactions shown below to find the heat of reaction shown in part (a)

$$S_{(s)} + O_{2(g)} \rightarrow SO_{2(g)}$$

$$SO_{2(g)} + \frac{1}{2}O_{2(g)} \rightarrow SO_{3(g)}$$

Solution:

a. From Table 5.1, ΔH_f^o of $SO_{3(g)} = -395.7$ kJ/mole and those of S and $O_2 = 0$ (element in its most stable form), then
   ΔH^o_{rxn} for the formation of $SO_{3(g)}$ from $S_{(s)}$ and $O_{2(g)}$
   $= (1)(-395.7) - [(1)(0) + (3/2)(0)] = -395.7$ kJ

b. The summation of two reaction equations in part (b) is the same as the reaction equation in part (a). Thus, using the two heat of reaction values from Example 5.6 and the Hess's law of summation, then
   ΔH^o_{rxn} for the formation $SO_{3(g)}$ from $S_{(s)}$ and $O_{2(g)} = (-296.8)(1) + (-98.9)(1) = -395.7$ kJ

Discussion:

The results from using Eqs. (5.20) and (5.21) are the same (i.e., –395.7 kJ).

5.3.4 STANDARD MOLAR ENTROPY

As mentioned in Section 5.3.1, the standard enthalpy of formation of a pure element in its most stable form under standard conditions is arbitrarily assigned a value of zero; and the values are used to come up with the standard formation of a compound $\left(\Delta H_f^\circ\right)$. From the third law of thermodynamics, the entropy of any substance at $T =$ the absolute zero is equal to zero. At any temperature above the absolute zero, the entropy of a substance will have a positive value; and the entropy of the substance increases with temperature because of increases in molecular motions. While we cannot determine the absolute value of internal energy (only its changes between two states), the absolute values of entropy can be determined experimentally (see Section 4.7).

The absolute entropy of substances can be measured, and the *standard molar entropy* (S°) is the entropy of one mole of substance under standard conditions (typically $P = 1$ atm and $T = 25°C$) in the unit of J/K/mole. The standard molar entropies of some substances are given in Table 5.1.

Entropy change of a reaction. The entropy change of a reaction can be determined if the standard entropies of each participant (i.e., the reactants and the products) are known using the equation below. The equation is similar to that for enthalpies of reactions (i.e., Eq. 5.20):

$$\Delta S^\circ_{rxn} = \sum n\Delta S^\circ\left(\text{products}\right) - \sum m\Delta S^\circ\left(\text{reactants}\right) \tag{5.22}$$

Example 5.9: Entropy Change of a Process/Reaction

Use the standard molar entropy values to find the entropy change of the following process:

$$H_2O_{(l)} \rightarrow H_2O_{(g)}$$

Solution:

a. From Table 5.1, ΔS° of $H_2O_{(l)} = 69.9$ J/K/mole and that of $H_2O_{(g)} = 188.8$ J/K/mole.

b. Use Eq. (5.22) to find the entropy change of water evaporation (at $P = 1$ atm and $T = 25°C$):

$$\Delta S^\circ_{rxn} = (1)(188.8) - (1)(69.9) = \underline{118.9 \text{ J/K/mole}}$$

Discussion:

1. As expected, the standard molar entropy of water vapor is larger than that of water (188.8 vs. 69.9 J/K/mole). The entropy change of water vaporization is positive because of the larger entropy value of water vapor.

2. In Example 4.5, the entropy change for water evaporation at 100°C was estimated, using the heat of water vaporization (i.e., 40.68 kJ/mole), to be 109.1 J/K/mole, which is essentially the same as 118.9 J/K/mole ($T = 25°C$) found in this example.

Example 5.10: Entropy Change of a Reaction

Find the entropy change of propane (C_3H_8) combustion by using oxygen as shown below:

$$C_3H_{8(g)} + 5\ O_{2(g)} \rightarrow 3\ CO_{2(g)} + 4\ H_2O_{(g)}$$

Solution:

a. From Table 5.1, $\Delta S°$ values of $C_3H_{8(g)}$, $O_{2(g)}$, $CO_{2(g)}$, $H_2O_{(g)}$ are 269.9, 205.1, 117.6, and 188.8 J/K/mole, respectively.

b. Use Eq. (5.22) to find the entropy change of this reaction (at $P = 1$ atm and $T = 25°C$):

$$\Delta S^{o}_{rxn} = \left[(4)(188.8) + 3(117.6)\right] - \left[(1)(269.9) + (5)(205.1)\right] = \underline{-187.4\,\text{J/K/mole}}$$

5.4 THERMODYNAMICS OF CHEMICAL REACTIONS

5.4.1 Spontaneity of a Chemical Reaction

As mentioned in Chapter 4, chemical thermodynamics will tell if a process/reaction can occur and to what extent the reaction can reach. There are three potential scenarios:

- $\Delta S_{universe} > 0$: the process is spontaneous in the direction written, but non-spontaneous in the reverse direction.
- $\Delta S_{universe} = 0$: the reaction is in equilibrium.
- $\Delta S_{universe} < 0$: the reaction is non-spontaneous in the direction written.

However, $\Delta S_{universe}$ is the sum of ΔS_{system} and $\Delta S_{surroundings}$ (as shown in Eq. 4.27). Finding ΔS_{system} for a chemical reaction is doable; however, finding $\Delta S_{surroundings}$ requires some quantitative measurement of the surroundings, and it is often challenging and not practically feasible. A criterion of spontaneity of a chemical reaction, which is based only on the state functions of the system, is of necessity (see Gibbs free energy below).

5.4.2 Gibbs Free Energy

Gibbs free energy (G) is a thermodynamic function that combines enthalpy, entropy, and absolute temperature (T), as:

$$G = (U + PV) - TS = H - TS \tag{5.23}$$

The Gibbs free energy has a unit of energy, but it is not really an energy. It is a thermodynamic function; one of its uses is to calculate the maximum amount of work that can be produced by a closed system at a constant T and P. More importantly, it can tell if a chemical reaction will occur under specific conditions.

At a constant T and P, the above equation can be written for a chemical reaction/process as:

$$\Delta G = \Delta H - T\Delta S \tag{5.24}$$

Under standard conditions (typically $T = 25°C$ and $P = 1$ atm), the above equation would become

$$\Delta G^\circ = \Delta H^\circ - T\Delta S^\circ \tag{5.25}$$

The sign of ΔG can tell if a reaction is spontaneous (and so is the direction of the reaction). At a given temperature, there are three potential scenarios:

1. $\Delta G < 0$: the reaction is spontaneous in the direction written. The reaction is often termed "*exergonic*".
2. $\Delta G = 0$: the reaction is at equilibrium.
3. $\Delta G > 0$: the reaction is non-spontaneous in the direction written, but spontaneous in the reverse direction. The reaction is often termed "*endergonic*".

The effect of temperature on the spontaneity of a reaction can be categorized into the following four groups if ΔH and ΔS are independent of temperature:

1. $\Delta H < 0$ and $\Delta S > 0$: $\Delta G < 0$ at all temperatures (the reaction is spontaneous).
2. $\Delta H > 0$ and $\Delta S < 0$: $\Delta G > 0$ at all temperatures (the reaction is non-spontaneous).
3. $\Delta H < 0$ and $\Delta S < 0$: $\Delta G < 0$ at lower temperatures (the reaction is spontaneous); but $\Delta G > 0$ at higher temperatures (the reaction is non-spontaneous)
4. $\Delta H > 0$ and $\Delta S > 0$: the effects of temperature are opposite of the group #3 above (i.e., $\Delta G > 0$ at lower temperatures, but $\Delta G < 0$ at higher temperatures).

It should be noted that in the absence of a phase change, both ΔH and ΔS vary insignificantly with temperature.

5.4.3 Change of Standard Gibbs Free Energy of a Chemical Reaction

Similar to standard enthalpies of formation $\left(\Delta H_f^\circ\right)$, standard free energy of formation $\left(\Delta G_f^\circ\right)$ of a substance is the change of free energy for its formation from its elements under standard conditions. Consequently, they can be used to calculate ΔG° of a reaction, analog to the calculation of ΔH, as

$$\Delta G^\circ = \sum n\Delta G_f^\circ(\text{products}) - \sum m\Delta G_f^\circ(\text{reactants}) \tag{5.26}$$

Example 5.11: Gibbs Free Energy if a Chemical Reaction

Find the change of Gibbs free energy of propane (C_3H_8) combustion by using oxygen as shown below:

$$C_3H_{8(g)} + 5\ O_{2(g)} \rightarrow 3\ CO_{2(g)} + 4\ H_2O_{(g)}$$

Solution:

a. The ΔG_f^o values of $C_3H_{8(g)}$, $O_{2(g)}$, $CO_{2(g)}$, $H_2O_{(g)}$ are –23.5, 0, –394.4, and –228.6 kJ/mole, respectively.

Use Eq. (5.26) to find the change of Gibbs free energy of this reaction:

$$\Delta G_{rxn}^{o} = [(3)(-384.4) + 4(-228.6)] - [(1)(-23.5) + (5)(0.0)] = -2{,}074 \text{ kJ/mole}$$

b. Use the ΔH^o and ΔS^o values from Examples 5.9 and 5.12 and Eq. (5.25) to find ΔG_{rxn}^{o} as:

$$\Delta G^o = \Delta H^o - T\Delta S^o = (-2043.9) - (298)(-0.1874) = -1{,}988 \text{ kJ/mole}$$

Discussion:

1. The ΔS value is in J/K/mole, while the ΔH value is in kJ/mole.
2. The ΔG values from two approaches are slightly different (–2,074 vs. 1,988), but comparable. The difference plausibly comes from the differences in the reported thermodynamic values.
3. With the negative ΔG value, the reaction should be spontaneous.

5.5 FREE ENERGY AND THE EQUILIBRIUM CONSTANT

Chemical reactions do not always occur under standard conditions. The relationship between ΔG^o (the change of free-energy under standard conditions) and ΔG (the change of free-energy under any other conditions) is:

$$\Delta G = \Delta G^o + RT(lnQ) \tag{5.27}$$

where R = ideal gas constant and Q = reaction quotient [Note: $\Delta G = \Delta G^o$, when $Q = 1$].

When a system is in equilibrium, ΔG is equal to zero and the values of the reaction quotient (Q) and the equilibrium constant (K) are equal. Consequently, Eq. (5.27) would become:

$$\Delta G = 0 = \Delta G^o + RT(lnK) \tag{5.28}$$

Then

$$\Delta G^{\circ} = -RT(lnK) \tag{5.29}$$

Or

$$K = Exp\left[-\frac{\Delta G^{o}}{RT}\right] \tag{5.30}$$

Equations (5.29) or (5.30) relates the equilibrium constant of a chemical reaction to its change of Gibbs free energy.

Example 5.12: Equilibrium Constant and Reaction Quotient

Carbonic acid (H_2CO_3) molecules will dissociate into proton ion (H^+) and bicarbonate ion $\left(HCO_3^-\right)$ in water as:

$$H_2CO_{3(aq)} \rightleftharpoons HCO^-_{3(aq)} + H^+_{(aq)} \quad K = 4.47\times10^{-7}\ (T = 25\ °C)$$

a. Find the change of Gibbs free energy of the forward reaction.
b. Use the value found in part (a) to determine the equilibrium constant of the reaction at 25°C.

Solution:

a. The ΔG_f^o values of $H_2CO_{3(aq)}$, $HCO^-_{3(aq)}$, and $H^+_{(aq)}$ are −623.4, −587.1, and 0 kJ/K/mole, respectively
   Use Eq. (5.22) to find the change of Gibbs free energy of this reaction (at $P = 1$ atm and $T = 25$°C):

$$\Delta G^{\circ}_{rxn} = \left[(1)(-587.1)+1(0)\right]-\left[(1)(-623.4)\right] = \underline{36.3\ \text{kJ/mole}}$$

b. Use Eq. (5.22) to find K:

$$K = Exp\left[-\frac{\Delta G^{o}}{RT}\right] = Exp\left[-\frac{\left(36{,}300\frac{\text{J}}{\text{mole}}\right)}{\left(8.3145\frac{\text{J}}{(\text{mole}\cdot\text{K})}(298\ \text{K})\right)}\right] = 4.34\times10^{-7}$$

Discussion:

The calculated value of the equilibrium constant (4.34×10^{-7}) is very close to the reported value of 4.47×10^{-7}.

5.6 TYPES OF CHEMICAL REACTIONS

There are many ways to classify types of chemical reactions. Most of the classifications have chemical reactions into four types: (1) synthesis, (ii) decomposition, (iii) single-replacement, and (iv) double-replacement reactions.

A *synthesis reaction* occurs when two reactants interact to form one product. A typical example is the formation of water vapor from hydrogen and oxygen as:

$$2\ H_{2(g)} + O_{2(g)} \rightarrow 2\ H_2O_{(g)} \tag{5.31}$$

A *decomposition reaction* occurs when a reactant breaks down into simpler products. An example is the decomposition of hydrogen peroxide into oxygen and water as:

$$2\ H_2O_{2(l)} \rightarrow 2\ H_2O_{(l)} + O_{2(g)} \quad (5.32)$$

A *simple-replacement* (also called *simple-displacement* or *substitution*) *reaction* occurs when a single element replace/displace an element in a compound to produce a new compound. An example is that copper replaces silver in silver nitrate as:

$$2\ Ag(NO_3) + Cu \rightarrow Cu(NO_3)_2 + 2\ Ag \quad (5.33)$$

A *double-replacement* (also called *double-displacement*) *reaction* occurs when two ionic species in two compounds (i.e., the reactants) exchange ions between them to form two completely different compounds. An example is the reaction between silver nitrate and potassium chloride to form silver chloride and potassium nitrate as:

$$Ag(NO_3) + KCl \rightarrow AgCl + K(NO_3) \quad (5.34)$$

In addition to four fundamental types of chemical reactions, many consider acid-base reaction, precipitation, oxidation-reduction, combustion, and hydrolysis as types of chemical reactions.

An *acid-base* (or *neutralization*) *reaction* is the reaction between acids and bases. An example is the neutralization between hydrochloric acid and sodium hydroxide to form neutral sodium chloride and water as:

$$HCl + NaOH \rightarrow NaCl + H_2O \quad (5.35)$$

A *precipitation reaction* occurs when two soluble compounds (or two ionic species) react to form an insoluble solid (i.e., precipitate). The reaction equation between silver nitrate and potassium chloride shown in Eq. (5.34) can be rewritten as:

$$Ag(NO_3)_{(aq)} + KCl_{(aq)} \rightarrow AgCl_{(s)} + K(NO_3)_{(aq)} \quad (5.36)$$

$$Ag^+_{(aq)} + Cl^-_{(aq)} \rightarrow AgCl_{(s)} \quad (5.37)$$

The nitrate ion and potassium ion do not show up in Eq. (5.37) because they do not actually participate in this participation reaction. It should also be noted that the neutralization reaction shown in Eq. (5.35) as well as the precipitation reaction shown in Eq. (5.36) are also double-displacement reactions.

An *oxidation-reduction* (or *redox*) *reaction* is a reaction that involves the transfer of electrons between the reactants. Oxidation of methane into carbon dioxide and water is an example of a redox reaction. Through this reaction, the oxidation state of carbon in methane gas and that of oxygen in oxygen gas are changed because of the transfer of electrons from carbon to oxygen. Oxygen in oxygen gas is the electron-acceptor in this reaction (more later).

$$CH_{4(g)} + 2\ O_{2(g)} \rightarrow CO_{2(g)} + 2\ H_2O_{(g)} \quad (5.38)$$

A *combustion reaction* is a reaction in which a substance reacts with oxygen gas (also a reactant) at elevated temperatures such as the reaction equation shown in Eq. (5.38). It is also a redox reaction. The final products of complete hydrocarbon combustion are typically carbon dioxide and water, and this type of reaction are usually exothermic and releases heat to its surroundings.

A *hydrolysis* reaction is the breaking of a chemical bond through a reaction with water. For example, ethyl acetate ($CH_3C(=O)OC_2H_5$) reacts with water to form acetic acid (CH_3COOH) and ethanol (C_2H_5OH) as:

$$CH_3COOC_2H_5 + H_2O \rightarrow CH_3COOH + C_2H_5OH \tag{5.39}$$

Instead of discussing the chemical reactions type-by-type, this book takes a different approach. In the coming chapters, solutes, solvents, and solutions will be discussed first (Chapter 6), followed by ions and ionic equations (Chapter 7), acid-base equilibria (Chapter 8), and then oxidation-reduction reactions (Chapter 9).

REFERENCE

NIST (2023). "Welcome to the NIST Chemistry WebBook", National Institute of Standards and Technology, US Department of Commerce, https://webbook.nist.gov/.

EXERCISE QUESTIONS

1. Find the volume of 1 g-mole and 1 lb-mole of an ideal gas at $T = 30°C$ and $P = 0.95$ atm.
2. A furnace is combusting butane (C_3H_8) at a rate of 88 lb/hr.
   a. What is the molar flow rate of butane into the furnace?
   b. What would be the stoichiometric amount of oxygen needed, in lb/hr?
   c. What would be the stoichiometric flow rate of oxygen in ft^3/hr ($T = 68°F$ & $P = 1$ atm)?
   d. What would be the rate of CO_2 produced from this furnace, in lb/hr?
   e. What would be rate of CO_2 produced in ft^3/hr ($T = \underline{77}°F$ & $P = 1$ atm), assuming the reaction is complete?
3. Fe_2O_3 and CO are being fed to a blast furnace at a mass ratio of 2:1. The Fe_2O_3 feed rate is 5,000 kg/hr and assume the reaction below reaches completion.
   a. Which is the limiting reactant (i.e., Fe_2O_3 or CO)?
   b. What is the CO_2 emission rate (in kg/hr)?

$$Fe_2O_{3(s)} + 3\ CO_{(g)} \rightarrow 2\ Fe_{(s)} + 3\ CO_{2(g)}$$

4. a. A water sample contains 40 m/L of Ca^{+2} and 40 mg/L of Na^+, what are their molar concentrations?
   b. The current primary MCLs of carbon tetrachloride (CCl_4) and fluoride (F^-) in drinking water are 0.005 and 4.0 mg/L, respectively; express these concentrations in molar concentration.

5. Some carbonic acid (H_2CO_3) molecules will dissociate into proton ion (H^+) and bicarbonate ion $\left(HCO_3^-\right)$ in water as:

$$H_2CO_{3(aq)} \rightleftharpoons HCO^-_{3(aq)} + H^+_{(aq)} \quad K = 4.47 \times 10^{-7} \ (T = 25\ °C)$$

   a. At $T = 25°C$, a system was found to be in equilibrium and the concentrations of carbonic acid and bicarbonate were equal, what is the molar concentration of the proton ion?
   b. A concentrated acid was added to raise the proton concentration to 4.47×10^{-4} M to break the equilibrium. What is the reaction quotient immediately after the acid addition?
   c. Which way the reaction would move immediately after the acid addition?
   d. If the proton concentration is kept at 4.47×10^{-4} M and the system is allowed to reach a new equilibrium, what would be the ratio of $\left[HCO_3^-\right]/\left[H_2CO_3\right]$?

6. Use the standard enthalpies of formation data in Table 5.1 to determine the heat of reaction of the following reaction.

$$4\ FeS_{2(s)} + 11\ O_{2(g)} \rightarrow 2\ Fe_2O_{3(s)} + 8\ SO_{2(g)}$$

7. Ethane (C_2H_6) and ethylene/ethene (C_2H_4) can be oxidized by oxygen as shown below:

$$C_2H_{6(g)} + 3.5\ O_{2(g)} \rightarrow 2\ CO_{2(g)} + 3\ H_2O_{(g)}$$

$$C_2H_{4(g)} + 3\ O_{2(g)} \rightarrow 2\ CO_{2(g)} + 2\ H_2O_{(g)}$$

   a. Determine the heat of reaction of each reaction.
   b. How much heat will be released if 1 g of ethane is combusted?
   c. How much heat will be released if 1 g of ethylene/ethene is combusted?
   d. If water is the final product, instead of water vapor, what would be the heat of reaction for ethane combustion?

8. Given the heat of formation values (ΔH^o) for the reactions in the table below (note: at 298 K), Use Hess's Law to determine the heat of reaction at 298 K for the following reaction:

$$CS_{2(l)} + 3\ O_{2(g)} \rightarrow CO_{2(g)} + 2\ SO_{2(g)}$$

	ΔH^o (kJ/mole)
$C_{(s)} + O_{2(g)} \rightarrow CO_{2(g)}$	−395
$C_{(s)} + 2\ S_{(s)} \rightarrow CS_{2(l)}$	88
$S_{(s)} + O_{2(g)} \rightarrow SO_{2(g)}$	−298

9. Use the standard molar enthalpy and entropy values in Table 5.1 to find the entropy change and the heat of reaction of the following process:

$$HNO_{3(l)} \rightarrow HNO_{3(g)}$$

10. Use the standard molar entropy values in Table 5.1 to find the entropy change for the formation of water vapor from gaseous hydrogen and oxygen:

$$2\ H_{2(g)} + O_{2(g)} \rightarrow 2\ H_2O_{(g)}$$

11. Use the standard molar enthalpy values in Table 5.1 to find the heat of reaction for the formation of water vapor from gaseous hydrogen and oxygen:

$$2\ H_{2(g)} + O_{2(g)} \rightarrow 2\ H_2O_{(g)}$$

12. Use the standard Gibb's free energy values in Table 5.1 to find the change of Gibb's free energy for the formation of water vapor from gaseous hydrogen and oxygen:

$$2\ H_{2(g)} + O_{2(g)} \rightarrow 2\ H_2O_{(g)}$$

13. a. Use the values of change of entropy and heat of reaction from Question #10 and #11, respectively, and Eq. (5.25) to estimate the change of Gibb's free energy for the formation of water vapor from gaseous hydrogen and oxygen:

$$2\ H_{2(g)} + O_{2(g)} \rightarrow 2\ H_2O_{(g)}$$

b. Are the values from part (a) and from Question #12 comparable?

14. a. Use the value of the change of Gibb's free energy for the formation of water vapor from gaseous hydrogen and oxygen to estimate its equilibrium constant.

$$2\ H_{2(g)} + O_{2(g)} \rightarrow 2\ H_2O_{(g)}$$

b. Is the value of the equilibrium constant relatively large or small?

15. Use the values of entropy change and the heat of reaction found in Question #9 for the reaction shown below.
    a. Will the reaction be spontaneous in the forward direction?
    b. Estimate the equilibrium constant for this reaction @$T = 25°C$.

$$HNO_{3(l)} \rightarrow HNO_{3(g)}$$

6 Solutes, Solvents, and Solutions

6.1 INTRODUCTION

A mixture is *homogenous* if components of the mixture are uniformly distributed throughout the entire mixture. *Solutions* are homogeneous mixtures of two or more components, in which the *solvent* is the substance present in the largest amount, while the other components are the *solutes*. An *aqueous solution* is a solution in which water is the solvent.

6.1.1 Types of Solution

Solutions can be formed by many types/forms of solutes and solvents. Types of solutions with water or air as the solvent are of most environmental professionals' concern; they include:

1. Liquid-in-liquid solutions
2. Gas-in-gas solutions
3. Gases in aqueous solutions
4. Solids in aqueous solutions
5. Particulates in air
6. Solids with adsorbed substances in aqueous solutions
7. Partition of compounds among air, water, and soil grains in soil

6.1.2 Water as a Polar Solvent

Water is often referred as the universal solvent because it can dissolve many different substances due to the polar arrangement of oxygen and hydrogen atoms. A water molecule has a slightly negative end on the side of the oxygen atom and a slightly positive end on the side of the hydrogen atoms. The positive hydrogen end of one water molecule can attract the negative end of another water molecule (i.e., the hydrogen bonding). With the hydrogen bonding, an elaborate network of water molecules is formed (see Figure 1.7). In general, polar compounds are more soluble in polar solvents and less soluble in non-polar solvents.

6.1.3 The Solvation Process

Dissolution of solute(s) into a solvent to form a solution is a physical process (i.e., no changes in chemical forms of both solute(s) and the solvent). *Solvation* is the interaction between the solvent and the dissolved solute(s). In the case of an aqueous

DOI: 10.1201/9781003502661-6

solution, solvation is often referred to as *hydration*. Three steps are needed to make this solvation process happen. Firstly, energy is required to overcome the intermolecular attraction between the solutes ($\Delta H_{\text{solute-solute}}$). In the meantime, energy is also required to overcome the intermolecular attraction between the solvent molecules ($\Delta H_{\text{solvent-solvent}}$). These two steps are endothermic (i.e., $\Delta H > 0$). The third step is the solute–solvent interaction, and it is exothermic (i.e., $\Delta H_{\text{solute-solvent}} < 0$). The overall enthalpy for the formation of a solution would be:

$$\Delta H_{\text{solution}} = \Delta H_{\text{solute-solute}} + \Delta H_{\text{solvent-solvent}} + \Delta H_{\text{solute-solvent}} \tag{6.1}$$

The value of $\Delta H_{\text{solution}}$ is not always negative. Recall that the change of Gibb's free energy should be less than zero (i.e., $\Delta G < 0$) for a spontaneous process, and

$$\Delta G = \Delta H - T\Delta S \tag{5.24}$$

A large ΔS value can make ΔG turns negative, even with a positive ΔH value. For dissolving a solid substance into water, the degree of disorder of the solute should usually increase ($\Delta S > 0$). A larger temperature should also make ΔG less positive or more negative. Consequently, the solubility of a solid substance in water usually increases with temperature.

6.2 LIQUID-IN-LIQUID SOLUTIONS

The liquids are *miscible* if they can form a homogenous mixture in which the composition is uniform throughout the entire mixture. Otherwise, they are *immiscible*. Alcoholic beverages (ethanol in water) and gasoline are good examples of liquid-in-liquid solutions. Water is apparently the solvent in wines and beer because water is the largest component. Gasoline is often a mixture of more than 200 organic compounds, and it is difficult to name a single substance as the solvent. Virtually all common organic liquids are miscible with another. However, many organic solvents (e.g., trichloroethylene (TCE, C_2HCl_3) and tetrachloroethylene (PCE, C_2Cl_4)) and water are *immiscible*.

6.2.1 Miscibility and Solubility

Solubility is the maximum amount of a solute that can dissolve in a solvent at a specified temperature and pressure. At that point, the solution is termed "*saturated*". If a solution contains less than that maximum amount of solute, it is *unsaturated*. If a solution is *supersaturated* (i.e., containing a solute larger than the maximum amount that can be dissolved), the molecular solute will become immiscible with the solvent. In a supersaturated solution, if the solutes are in the ionic form, the dissolved species will tend to form a solid through precipitation or crystallization (more later).

Although TCE (and PCE) and water are immiscible, TCE and PCE are soluble in water to a limited extent (i.e., they are sparingly soluble in water). The solubilities of TCE and PCE in water are approximately 1,000 and 150 mg/L at 25°C, respectively. Chlorinated solvents such as TCE and PCE are common pollutants found in contaminated groundwater aquifers. The migration/spread of a dissolved pollutant plume is always of concern. Due to the limited solubility of TCE in groundwater, for example, the spread of dissolved TCE with groundwater flow should be limited. However, if alcohol is also present in groundwater, alcohol can work as a cosolvent and TCE will travel much faster and spread wider in the aquifer because of its greater solubility in alcohol (USEPA, 1989). A *cosolvent* is a substance added to a primary solvent in small amounts to increase the solubility of a poorly-soluble compound.

6.2.2 Octanol-Water Partition Coefficient

Water and n-octanol are immiscible. If a vessel contains n-octanol and water, the liquid content should be in two distinct layers with n-octanol on the top (the density of n-octanol is smaller than that of water). If a substance is added into the container, and the solution is then well mixed to allow the dissolution of the substance into water and n-octanol to reach an equilibrium. After that, the liquid content is allowed to settle back into two distinct layers again. The *octanol-water partition coefficient* (K_{ow}) is the partition coefficient of this substance in the water and n-octanol two-phase system as:

$$K_{ow} = \frac{C_{\text{octanol}}}{C_{\text{water}}} \tag{6.2}$$

where C_{octanol} = equilibrium concentration of an organic compound in octanol; C_{water} = equilibrium concentration of the organic compound in water.

Organic compounds that are very soluble in water are called *hydrophilic* and the opposite of hydrophilic is *hydrophobic*. The octanol-water partition coefficient (K_{ow}) serves as an indicator of how an organic compound will partition itself between an organic liquid and water. Values of K_{ow} range widely, from 10^{-3} to 10^{7}. Organic compounds having low K_{ow} values are hydrophilic (like to stay in water); while those having large K_{ow} values are more hydrophobic. A hydrophobic compound is generally more *lipophilic* (i.e., more soluble in fat); and it tends to accumulate in the fatty issues of organisms (*bioaccumulation*). Bioconcentration factor (BCF), bioaccumulation factor (BAF), and biomagnification factor (BMF) are terms often used to describe the bioaccumulation potential of a substance in an aqueous environment (EPA, 2023).

- *Bioconcentration factor (BCF)*, typically with a unit of L/kg, is the ratio of concentration of a substance in fish or other tissue (in mg/kg) and the concentration of the substance in water (in mg/L).
- *Bioaccumulation factor (BAF)*, typically dimensionless, is the ratio of the concentration in the test organism (e.g., fish) (in mg/kg) and the concentration of the substance in the surrounding medium (e.g., soil or sediment) (in mg/kg).
- *Biomagnification factor (BMF)*, typically dimensionless, is the ratio of concentration of a substance in a predator (in mg/kg) and the concentration of the substance in predator's prey (in mg/kg).

Values of BCFs have been found closely related to K_{ow} values. A substance with a large K_{ow} value would incur a larger BCF value. The USEPA used the K_{ow} as an approximation of the baseline BCF for development of national bioaccumulation factors (USEPA, 2016).

6.3 GAS-IN-GAS SOLUTIONS

Out of three phases of matter (i.e., solid, liquid, and gas), gas has the smallest density (i.e., the smallest amount of mass in a specific volume). Both water and gas/air are fluid because they can fill a confined space in which they are placed. However, under ambient conditions, gas/air is considered compressible (i.e., the volume decreases as the applied pressure increases), while water is incompressible. There are relatively large spaces among gas molecules, and the interactions among the gas molecules are relatively weak.

Air is a good example of gas-in-gas solution or a gas mixture, and it contains various gaseous components. Chemical makeup of our atmosphere, excluding water vapor, is: 78.084% N_2, 20.947% O_2, 0.934% argon (Ar), and 0.035% CO_2 by volume, and many other gases in trace amounts; and the top four components making up 99.998% of all gases (NOAA, 2023).

With regard to the thermodynamic nature of a gas mixture, the intermolecular interactions in gases are weak. In other words, all three ΔH values (i.e., $\Delta H_{\text{solute-solute}}$, $\Delta H_{\text{solvent-solvent}}$, and $\Delta H_{\text{solute-solvent}}$) are very small; consequently, Eq. (6.1) becomes:

$$\Delta H_{\text{solution}} \approx 0 \quad (\text{for a gas mixture}) \tag{6.3}$$

In addition, mixing of gases will increase the degree of disorder in the system. Consequently, all gases "dissolved" in each other to form solutions in all proportions readily.

6.3.1 Basic Properties of Air

This section discusses some basic air properties including molecular weight, density, viscosity, and specific heat capacity.

Molecular weight. Since air is a mixture of many gases, the molecular weight (MW) of air can be estimated from its composition. For most air pollution calculations, it is usually satisfactory to treat the composition of dry air as 21% O_2 and 79% N_2 by volume (also by mole). Thus,

- $MW_{\text{dry air}} \approx [(32)(21\%) + (28)(79\%)] \approx 29$ g/g-mole (lb/lb-mole)
- Molar ratio of N_2:O_2 in air $\approx (79\%) \div (21\%) \approx 3.76$
- Mass ratio of N_2:O_2 in air $\approx [(28)(79\%)] \div [(32)(21\%)] \approx 3.29$

For wet air, the apparent MW can be found by including the volume (or molar) fraction of water (f_{water}). Thus,

$$MW_{\text{wet air}} = (1 - f_{\text{water}})(MW_{\text{dry air}}) + (f_{\text{water}})(MW_{\text{water}}) \tag{6.4}$$

Density. *Density* (ρ) of a material is the ratio of its mass (m) and the volume (V) it occupies. For an ideal gas:

$$\rho = \left(\frac{m}{V}\right) = \frac{P \times MW}{R \times T} \tag{6.5}$$

Example 6.1: Molecular Weight/Mass of Air

Assuming that air is a binary mixture of nitrogen and oxygen (79% N_2 and 21% O_2 by volume), answer the questions below.

a. What is mole fraction of oxygen?
b. What is the molecular weight of air?
c. What is the density of air @ $P=1$ atm and $T=25°C$?
d. What is the partial pressure of oxygen @ $P=1$ atm?
e. What is the oxygen concentration in air (% by wt.)?
f. Will methane gas and chlorine gas be lighter or heavier than air?

Solution:

a. Assuming the air behaves as an ideal gas (i.e., the volume is proportional to its molar content), the mole fraction of oxygen is equal to 0.21.

b. *MW* of nitrogen (N_2)=28 and that of oxygen (O_2)=32, thus
MW of air=(28)(79%)+(32)(21%) ≈ 29 g/g-mole (or 29 lb/lb-mole)

c. The molar volume of an ideal gas=24.5 L @ $P=1$ atm and $T=25°C$ (from Example 5.1).
Density of air=29/24.5=1.21 g/L @ $P=1$ atm and $T=25°C$

d. Partial pressure of oxygen=0.21 atm @ $P=1$ atm.

e. %O_2 in air (by weight)=(32)(21%) ÷ 29=23.2%

f. *MW* of methane (CH_4)=16 and *MW* of chlorine gas (Cl_2)=71, so methane gas is lighter than air while chlorine gas is heavier than air.

Discussion:

1. 29 is a good estimate for *MW* of air.
2. The % by wt. is larger than % by volume for oxygen in the air because of its *MW*>*MW* of air.
3. Methane is generated in landfills under anaerobic conditions, collection of landfill gas is usually needed to prevent (i) methane emission to the atmosphere (because methane is one of the major greenhouse gases) and/or (ii) its accumulation within the landfill which may exert a potential of pushing the final cover upward.
4. The leakage of chlorine gas may form a toxic chlorine cloud on the ground because of its heavier density.

Example 6.2: Density of Air

Estimate the density of air for the following three conditions:

a. under the standard conditions ($T=20°C$ and $P=1$ atm)
b. $T=200°C$ and $P=1$ atm
c. $T=200°C$ and $P=2$ atm.

Solution:

a. Assuming *MW* of air = 29, then

$$\rho = \frac{(1\text{ atm})\left(29\dfrac{g}{\text{g-mole}}\right)}{\left(0.082\dfrac{L\cdot\text{atm}}{\text{g-mole}\cdot K}\right)(293\ K)} = 1.21\text{ g/L}$$

or

$$\rho = \frac{(14.7\text{ psi})\left(29\dfrac{\text{lb}}{\text{lb-mole}}\right)}{\left(10.731\dfrac{\text{ft}^3\cdot\text{psi}}{\text{lb-mole}\cdot{}^\circ R}\right)(528\ {}^\circ R)} = 0.075\text{ lb/ft}^3$$

b. At $T = 200°C$ and $P = 1$ atm:

$$\rho = \frac{(1)(29)}{(0.082)(473)} = 0.75\text{ g/L}$$

c. At $T = 200°C$ and $P = 2$ atm:

$$\rho = \frac{(2)(29)}{(0.082)(473)} = 1.50\text{ g/L}$$

Discussion:

1. The density of air @ $T = 20°C$ and $P = 1$ atm is around 1.21 g/L, which is ~800 times smaller than that of water (1 kg/L).
2. The volume of a gas increases with an increase in temperature, thus, its density will decrease with temperature. The ratio of densities at two temperatures is the inverse of the ratio of two absolute temperatures.
3. The density of an ideal gas is directly proportional to the system pressure.

Viscosity. *Viscosity* of a fluid (e.g., air or water) is a measure of its resistance to flow. This resistance is due to the intermolecular friction between layers of the fluid when they slide against each other. The *shearing stress* (τ_{shear}) between layers of a fluid is related to the velocity gradient *(du/dy)* between the liquid layers (see equation below). [Note: A *gradient* is the change in value of a quantity (e.g., elevation, velocity, concentration) to the change of a given variable (usually distance). It can be considered as a driving force to cause a movement from one location to another. In environmental engineering applications, temperature, pressure, concentration, and hydraulic gradients are commonly encountered.]

$$\tau_{shear} = \mu\left(\frac{du}{dy}\right) \tag{6.6}$$

where μ is the *dynamic/absolute viscosity* or just the *viscosity*. The common units of SI for viscosity are N-s/m^2 and Poise; while that of the US customary system is lb$_f$-s/ft^2. The conversions among commonly used viscosity units are [Note: 1 N = 1 kg·m/s^2]:

$$1\ \text{N·s/m}^2 = 1\ \text{kg/(m·s)} = 10\ \text{Poise} = 0.0209\ \text{lb}_f\text{-s/ft}^2 = 0.672\ \text{lb}_m\text{/ft-s} = 0.0209\ \text{slug/ft-s}$$

$$1\ \text{Poise} = 100\ \text{centi-Poise (cp)} = 0.1\ \text{N·s/m}^2 = 2.09\times10^{-3}\ \text{lb}_f\text{-s/ft}^2 = 0.0672\ \text{lb}_m\text{/ft-s} \quad (6.7)$$

In engineering applications, *kinematic viscosity* (ν) is also used, which is the ratio of the absolute viscosity (μ) and the mass density (ρ) of the fluid (i.e., $\nu = \mu/\rho$). The common units of SI for kinematic viscosity are m^2/s, Stoke (St), or centi-Stoke (cSt); while that of the US customary system is ft^2/s.

$$1\ \text{St} = 100\ \text{cSt} = 1\ \text{cm}^2\text{/s} = 10^{-4}\ \text{m}^2\text{/s} = 1.08\times10^{-3}\ \text{ft}^2\text{/s} \quad (6.8)$$

Because gas molecules are far apart for intermolecular cohesion to take place, gas is much less viscous than water. The (absolute/dynamic) viscosity values of water and air at 293 K are 1.01×10^{-3} and 1.81×10^{-5} N-s/m^2, respectively (Table 6.1). Thus, water is ~55 times more viscous than air under ambient conditions. However, it should be noted that the kinematic density of air at T = 20°C & P = 1 atm is larger than that of water (1.50×10^{-5} vs. 1.01×10^{-6} m^2/s), mainly because the density of air is ~800 times smaller.

Viscosity of a fluid is relatively independent of pressure but depends significantly on temperature. Liquid viscosity decreases when the temperature increases; but the gas has the opposite trend because the increasing temperature would cause more collisions among the gas molecules. The viscosity of air at any temperature can be estimated by using the viscosity value at a reference absolute temperature (μ_{ref}) and the equation below (USEPA, 2012). The viscosity value of 1.81×10^{-5} N·s/m^2 in Table 6.1 can be used as the reference value @ T = 293 K.

$$\frac{\mu}{\mu_{ref}} = \left(\frac{T}{T_{ref}}\right)^{0.768} \quad (6.9)$$

TABLE 6.1
Viscosities of Air and Water (T = 293 K & P = 1 atm)

	Viscosity (N·s/m^2)	Kinematic Viscosity (m^2/s)
Air	1.81×10^{-5}	1.50×10^{-5}
Water	1.01×10^{-3}	1.01×10^{-6}

Example 6.3: Viscosity of Air

Estimate the (dynamic/absolute) viscosity and the kinematic viscosity of air @ $T = 200°C$ and $P = 1$ atm.

Solution:

a. Use Eq. (6.6), the (dynamic/absolute) viscosity of air @ $P = 1$ atm and $T = 200°C$:

$$\frac{\mu}{\mu_{\text{ref}}} = \left(\frac{T}{T_{\text{ref}}}\right)^{0.768} = \frac{\mu}{1.81 \times 10^{-5}} = \left(\frac{473}{293}\right)^{0.768}$$

$$\mu_{473K} = 4.73 \times 10^{-5}\ \text{kg/ (m·s)} = 9.89 \times 10^{-7}\ \text{lb}_f\text{-s/ft}^2$$

b. The mass density of air ($P = 1$ atm and $T = 200°C$) = 0.75 g/L = 0.75 kg/m³ (Example 6.2),

$$\text{Kinematic viscosity} = \text{viscosity} \div \text{density}$$

$$= 4.73 \times 10^{-5}\ \text{kg/ (m·s)} \div 0.75\ \text{kg/m}^3 = 4.73 \times 10^{-5}\ \text{m}^2\text{/s}$$

Discussion:

As expected, viscosity of air increases with temperature.

Specific heat capacity. We disucssed energy, heat capacity, and specific heat capacity in Chapter 4. Since heat can be added to a gas mass while the volume or pressure of the system remains constant, there are specific heat at constant volume (C_v) and specific heat at constant pressure (C_p). For an ideal gas,

$$C_p = C_v + R \tag{6.10}$$

where R is the universal gas constant.

From the definitions of calorie and Btu (see Chapter 4), the specific heat capacity of water under room temperatures should be around 1 Btu/lb-°F, 1 cal/g·°C, 1 kcal/kg·°C, or 4.184 kJ/kg·K. The specific heat capacity of a gas stream depends on its composition and temperature. Table 6.2 tabulates the values of specific heat capacity of air, CO_2, and water vapor at different temperatures (68°F to 3,800°F; 20°C to 2,093°C). For a temperature not listed in this table, one can interpolate between two known values. As shown, the specific heat capacity increases with temperature. The specific heat capacity of water vapor is approximately twice as those of air and CO_2; thus, that of a gas stream containing a large amount of water vapor would have a larger value than that of the dry gas. Consequently, it would take more heat to raise the temperature of a wet gas stream than when it is drier.

Most of the energy-related calculations for gas combustion dealing with flue gases at or below 1,500°F (815°C). For these cases, the heat capacity of dry gas is about 0.25 Btu/lb-°F (1.05 kJ/kg·K) and that of water vapor is about 0.50 Btu/lb-°F (2.1 kJ/kg·K). These values can be used for most calculations without introducing significant errors (USEPA, 2012).

TABLE 6.2
Specific Heat Capacity (in Btu/lb-°F or kcal/kg·°C) of Air, CO_2, and Water Vapor (USEPA, 2012)

Temperature		Specific Heat Capacity		
(°F)	(°C)	Air	CO_2	Water Vapor
68	20	0.242	0.200	0.445
212	100	0.244	0.218	0.452
500	260	0.249	0.245	0.470
1,100	593	0.260	0.285	0.526
2,200	1,204	0.278	0.315	0.622
3,000	1,649	0.297	0.325	0.673
3,800	2,093	0.303	0.330	0.709

Example 6.4: Specific Heat Capacity of a Gas Stream

Estimate the specific heat capacity of an exhaust gas stream with 15% water vapor and the stack temperature is 700°F (371°C).

Solution:

a. Interpolating the values at 500 and 1,100°F, the specific heat capacities of dry air and water vapor at 700°F can be estimated as 0.253 and 0.489 Btu/lb-°F, respectively.

b. Heat capacity of the gas stream = (0.253)(1 – 15%) + (0.489)(15%)

= 0.289 Btu/lb-°F (= 1.21 kJ/kg·K)

6.3.2 Gas Mixtures

The atmosphere of the Earth is a mixture of many different gases held by the gravitational force. The ideal gas law is applicable to gas mixtures just as to pure gases. *Dalton's law of partial pressure* states that "the total pressure of a gas mixture (P_{system}, or P_{total}) is the sum of the partial pressure (P_i) of all its components", or "in a gas mixture the pressure exerted by each component gas is the same as which it would exert if it were alone":

$$P_{system} = P_{total} = \sum P_i \tag{6.11}$$

6.3.3 Units of Gaseous Concentrations

Concentration is to express the quantity of a solute contained in a specific amount of solution. Common units of gaseous concentration include pressure, mole fraction, % by volume, % by weight, mass concentration (mass/volume), and parts per million by volume ("ppmV", or just "ppm").

Conversions between the mass and volume concentrations are often needed. It can be easily done by using one of the two equations below. MW is the molecular weight of the compound and MV is the molar volume of the air (or the gas mixture) at that temperature and pressure:

$$1 \text{ ppm} = \left(\frac{MW}{MV\ (\text{in L})}\right)\frac{\text{mg}}{m^3} \tag{6.12}$$

$$1 \text{ ppm} = \left(\frac{MW}{MV\left(\text{in ft}^3\right)}\right)\times 10^{-6}\,\frac{\text{lb}}{\text{ft}^3} \tag{6.13}$$

The conversions between concentrations in lb/ft^3 and mg/m^3 are:

$$1\ \text{lb/ft}^3 = 1.603 \times 10^7\ \text{mg/m}^3 \tag{6.14a}$$

$$1\ \text{mg/m}^3 = 6.24 \times 10^{-8}\ \text{lb/ft}^3 \tag{6.14b}$$

Example 6.5: Concentrations of Hazardous Air Pollutants

Mercury vapor is toxic. The vapor pressure of mercury is 2.63×10^{-6} atm at $T=25°\text{C}$.

a. Convert the vapor pressure from atm to Pa (N/m^2).
b. Convert the vapor pressure value to mm-Hg and to Torr.
c. If mercury is allowed to evaporate to reach an equilibrium in an enclosed space, estimate the mercury concentration in the void space, in ppm and in % by volume ($P=1$ ppm & $T=25°\text{C}$).

Solution:

a. Using the conversion factors in Eq. (4.15),

$$P = \left(2.63\times10^{-6}\ \text{atm}\right)\left(\frac{1.013\times10^5\,\text{Pa}}{1\ \text{atm}}\right) = 0.27\ \text{Pa}$$

b. Similarly,

$$P = \left(2.63\times10^{-6}\ \text{atm}\right)\left(\frac{760\ \text{mmHg}}{1\ \text{atm}}\right) = 0.002\ \text{mmHg} = 0.002\ \text{Torr}$$

c. Since $P=2.430\times10^{-6}$ atm, the mercury concentration could be as high as 2.43 ppmV (ppm) under this condition, that is, 0.000243% by volume.

Discussion:

1. 1 mm-Hg = 1 Torr
2. 1 % by volume (or 1% by mole) = 10,000 ppmV.

Example 6.6: Molecular Weight/Mass of Gases

Carbon dioxide (CO_2) concentration in ambient air exceeds 400 ppmV in many locations.

a. Convert this volume concentration to mass concentration @ $T=20°C$ & $P=1$ atm.
b. What is the CO_2 concentration in % by volume?
c. What is the partial pressure of CO_2 in the air (in mm-Hg)?
d. What would be the equivalent mass concentration at a location ($P=23.4$ in-Hg and $T=55°F$)?

Solution:

a. Using Eq. (6.12) with $MV=24.05$ (or 385 ft³) and MW of $CO_2=44$,

$$1 \text{ ppm} = \left(\frac{44}{24.05}\right)\frac{\text{mg}}{m^3} = 1.83\ \frac{\text{mg}}{m^3}$$

$$1 \text{ ppm} = \left(\frac{44}{385}\right)\times 10^{-6}\,\frac{\text{lb}}{\text{ft}^3} = 1.14\times 10^{-7}\,\frac{\text{lb}}{\text{ft}^3}$$

Therefore, 400 ppm = (400)(1.83) = 732 mg/m³ = 4.56×10^{-5} lb/ft³.

b. 400 ppm = (400) ÷ (1,000,000) = 0.04%

c. 400 ppm = (400×10^{-6}) × (760 mm-Hg) = 0.304 mm-Hg.

d. The molar volume of CO_2 at this location can be determined by using the ideal gas law or

$$\left(\frac{V_2}{V_1}\right) = \left(\frac{T_2}{T_1}\right)\left(\frac{P_1}{P_2}\right) \tag{6.15}$$

$$MV = (24.05)\left(\frac{460+55}{460+68}\right)\left(\frac{29.92}{23.4}\right) = 30.0L$$

Then, $1 \text{ ppm} = \left(\frac{44}{30.0}\right)\frac{\text{mg}}{m^3} = 1.47\ \frac{\text{mg}}{m^3}$

400 ppm = (400)(1.47) mg/m³ = 588 mg/m³.

Discussion:

1. Mixed units were used in this example.
2. Like what was done in part (c), if the volume at one set of T & P is known, the volume at another set of T & P can be calculated by using Eq. (6.15) (shown in part (c) of this solution).
3. CO_2 concentration in our ambient air is ~ 0.04% (400 ppmV).
4. With the same air concentration, in ppmV, its mass concentration could be different at locations having different pressures and/or temperatures.

6.3.4 Humidity of Air

Humidity is the amount of water vapor present in the air. *Absolute humidity* is the total mass of water vapor in a specific volume of air, often in g/m^3, and it is a function of T and P. The amount of water vapor needed to achieve saturation increases with temperature. *Relative humidity* (*RH*) measures the existing humidity relative to the saturation humidity at that temperature. It can be defined as the ratio of the partial pressure of water vapor (P_{water}) to the vapor pressure in equilibrium with pure water under the same conditions:

$$RH = \left(\frac{P_{\text{water}}}{P_{\text{water}}^{\text{vap}}} \right) \times 100\% \qquad (6.16)$$

Dew point is the temperature when the water vapor starts to condense out of the air. Dry-bulb, wet-bulb, and dew-point temperatures are important parameters that describe the status of air with regard to humidity. The *dry-bulb temperature* is what we measure with a normal thermometer and is basically the temperature of the ambient air. The *wet-bulb temperature* can be measured by wrapping the bulb of the thermometer in a wet fabric. With the cooling of the bulb from adiabatic evaporation of water, the wet-bulb temperature is lower than the corresponding dry-bulb temperature and higher than the dew-point temperature [Note: "*Adiabatic*" denotes a process or condition in which heat does not leave or enter the system. In other words, no heat exchange between the system and its surroundings.]. When the humidity of air is high, the difference between the wet-bulb and the dry-bulb temperatures would be smaller because less tendency for water to evaporate from the wet fabric. In addition, a smaller difference between the dry-bulb and the dew-point temperatures indicates that the RH of the air is high. The relationships among them under standard atmospheric conditions can be retrieved from a *psychrometric chart*. Figure 6.1 shows an example psychrometric chart. Psychrometric charts and tutorials on how to use them are readily available on the Internet.

Example 6.7: Calculation of Dew-point Temperature

The flue gas from a combustion chamber contains 10% of water by volume. Estimate its dew-point temperature ($P = 1$ atm), using the Antoine equation below (for water vapor between 1°C and 100°C):

$$Log_{10}P\ (\text{in mm-Hg}) = 8.07131 - \frac{1{,}730.63}{T\ (\text{in °C}) + 233.426}$$

Solution:

a. Use Eq. (6.11) and the mole fraction of water vapor to find its vapor pressure:

$$P_{\text{water vapor}} = y_{\text{water vapor}} \times P = (10\%)(760) = 76.0 \text{ mm-Hg}$$

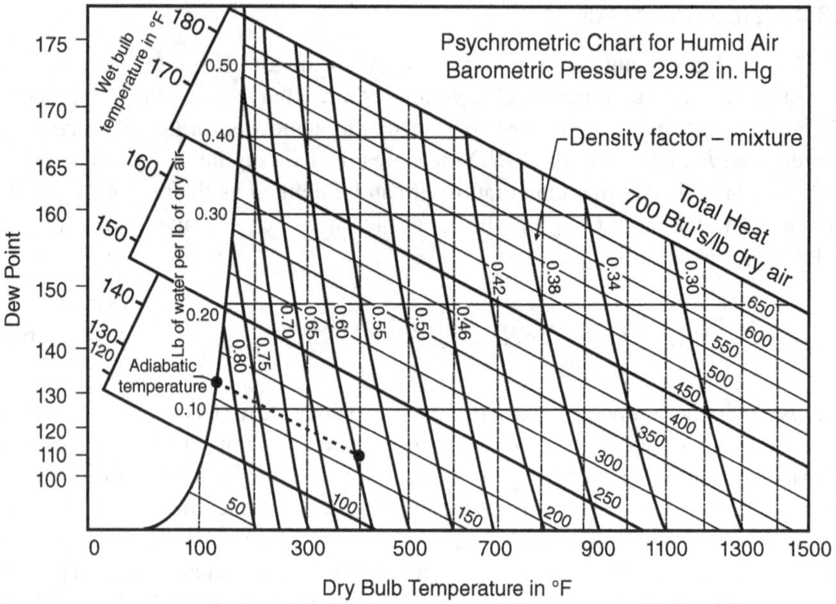

FIGURE 6.1 A psychrometric chart. (USEPA, 2012.)

b. $$Log_{10}(76.0) = 8.07131 - \frac{1{,}730.63}{T\ (\text{in }°C) + 233.426}$$

$$T = 46.1°C$$

Discussion:

1. Antonine constants are readily available on the Internet. However, we need to pay attention to the units associated with the parameters since the equation is empirical. For example, "Log_{10}" is in the equation above, not the "natural log"; and T is in °C, not in Kelvin.
2. The vapor pressure of water @ 46°C is 75.7 mm-Hg (from the Internet).

6.3.5 Heating Value of a Gas Mixture

In combustion, the combined influent to the combustion chamber should possess sufficient energy to sustain burning. Below are some of the terms that are commonly used with regard to energy/heat.

Sensible heat (H_s) is the amount of heat needed to cause a change in temperature of a unit mass of a substance without any phase change. It is a multiplication product of specific heat capacity (C_p) and the temperature change (ΔT) as

$$H_s = C_p \times \Delta T \tag{6.17}$$

Latent heat, or *heat of vaporization*, (H_v) is the heat/energy required for a unit mass of a substance to transform itself from a liquid state to a vapor state at its boiling point. It involves a change of phases, but the temperature stays the same. For example, water needs to absorb heat to evaporate; and the latent heat of water vaporization at its boiling point (i.e., 100°C @ $P=1$ atm) is 970.3 Btu/lb, or 2,255 kJ/kg.

Enthalpy (or *heat content*) is the sum of sensible and latent heat present in a substance minus that at the reference conditions. The common reference conditions are $T=20$°C and $P=1$ atm.

Higher heating value (HHV), also called the *gross heating value*, is the total amount of heat released from complete combustion of a fuel at 15.56°C (60°F) and its combustion products cooled down to 15.56°C (60°F). The *lower heating value* (LHV), or the *net heating value* (NHV), is the HHV minus the latent heat of water formed from combustion. Heating values of gases are often expressed in energy per unit mass or energy per unit volume. Common units for heating value per unit mass are shown in Eq. (4.21a):

$$1 \text{ kJ/kg} = 0.43 \text{ Btu/lb (or 1 Btu/lb} = 2.324 \text{ kJ/kg)} \tag{4.21a])}$$

Common units for heating value per unit volume are:

$$1 \text{ kJ/m}^3 = 0.02685 \text{ Btu/ft}^3 \text{ (or 1 Btu/ft}^3 = 37.24 \text{ kJ/m}^3\text{)} \tag{6.18}$$

If the heating value per unit mass of a gas stream is known, its corresponding heating value per unit volume can be readily calculated by multiplying the known value with the density of the gas stream.

Organic compounds generally contain higher heating values, and they can serve as energy sources for combustion. The higher the organic concentration in a gas stream, the higher the heat content; and, consequently, the less the requirement of the auxiliary fuel. Dulong's formula has been commonly used to estimate heat values of fuels or organic substances. If the heating value of a compound of concern (COC) is not available, the Dulong's formula can be used:

$$\text{Heating Value}\left(\text{in } \frac{\text{Btu}}{\text{lb}}\right) = 145.4\ C + 620\left(H - \frac{O}{8}\right) + 41S \tag{6.19}$$

$$\text{Heating Value}\left(\text{in } \frac{\text{kJ}}{\text{kg}}\right) = 337.9\ C + 1{,}441\left(H - \frac{O}{8}\right) + 95S \tag{6.20}$$

where C, H, O, and S are the percentages (by weight) of these elements in the compound.

According to U.S. Energy Information Administration (EIA), the average heat content of natural gas in 2015 is 1,037 Btu/ft^3 (38,600 kJ/m^3) and that of the motor gasoline in retail is 120,476 Btu/gallon (33,580 kJ/L). It should be noted that "*therm*" is a common energy unit for natural gas; and 1 therm = 100,000 Btu.

Example 6.8: Estimate the Heating Value of Methane Gas

Using Dulong's formula to estimate the heating value of methane (CH_4) gas.

Solution:

a. *MW* of $CH_4 = 12 \times 1 + 1 \times 4 = 16$

$$\text{Weight percentage of C in methane} = 12 \div 16 = 75\%$$

$$\text{Weight percentage of H in methane} = 4 \div 16 = 25\%$$

$$\text{Heating Value } = 145.4\ (75) + 620(25) = 26{,}405 \frac{\text{Btu}}{\text{lb}}$$

$$\text{Heating Value}\left(\text{in } \frac{\text{kJ}}{\text{kg}}\right) = 337.9(75) + 1{,}441(25) = 61{,}325 \frac{\text{kJ}}{\text{kg}}$$

b. Use Eq. (6.5) to find the density of methane at $T = 20°C$ (68°F):

$$\rho = \left(\frac{m}{V}\right) = \frac{P \times MW}{R \times T} = \left(\frac{16}{24.05}\right) = 0.665 \frac{\text{kg}}{m^3}$$

$$\rho = \left(\frac{m}{V}\right) = \left(\frac{16}{383}\right) = 0.0418 \frac{\text{lb}}{\text{ft}^3}$$

c. Heating value = (61,325 kJ/kg)(0.665 kg/m^3) = 41,080 kJ/m^3

= (26,405 Btu/lb) × (0.0418 lb/ft^3) = 1,103 Btu/ft^3

Discussion:

1. Methane is the principal constituent of natural gas, and it represents about 95% of the mixture (Kuo et al., 2015). The heating value of methane calculated from Dulong's formula, 1,103 Btu/ft^3 or 41,080 kJ/m^3, is very close to the reported average heating value of the natural gas, 1,037 Btu/ft^3 (38,600 kJ/m^3).
2. The weight percentage of C in methane is 75%. A value of 75, not 75%, should be used in the Dulong's formula.
3. Methane gas is lighter than air. Its density can also be found by multiplying the air density with the molecular weight ratio of methane and air (i.e., 16/29).

6.4 GASES IN AQUEOUS SOLUTIONS

In an environmental system, water, air, and soil are often present and in contact. In this section, we will work on gases in aqueous solutions.

6.4.1 Basic Properties of Water

Water is the major component in an aqueous solution, and it serves as the solvent of the solution. Some basic physicochemical properties of water include:

1. Water is an essential ingredient for the existence of life; all known biochemical processes occur in aqueous environments. The composition of most living things is overwhelmingly water, by weight.
2. Three states of water (i.e., snow/ice, water, and water vapor) are commonly co-present in our living environment.
3. Water is a polar compound and can form hydrogen bonding among water molecules. It can dissolve many polar molecular compounds and ionic compounds.
4. The specific heat capacity of water is relatively high, due to the hydrogen bonding among water molecules.
5. Water has an amphoteric nature (i.e., water can act as an acid and as a base).

Density of water. As defined in Eq. (6.5), density of a substance is its mass per unit volume. The SI unit for density is kg/m^3; however, g/cm^3 is also often used. The US customary unit for density is slug/ft^3, but lb$_m$/ft^3 is more often used [Note: 1 slug = 32.174 lb$_m$ ≅ 32.2 lb$_m$]. The conversions among them are:

$$1 \text{ g/cm}^3 = 1{,}000 \text{ kg/m}^3 = 62.4 \text{ lb}_m/\text{ft}^3 = 1.938 \text{ slug/ft}^3 = 8.34 \text{ lb}_m/\text{gallon} \quad (6.21)$$

The melting point and the boiling point of pure water at $P = 1$ atm are 0°C and 100°C (or 32°F to 212°F), respectively. The density values of water from 0°C to 100°C are tabulated in Table 6.3. As shown, the density of water is the largest around 4°C with a value of 999.97 kg/m^3, and it decreases as temperature increases (the density of water at 100°C = 958.4 kg/m^3). Under ambient conditions, the density of water is very close 1,000 kg/m^3 (1 g/cm^3, or 62.4 lb/ft^3), this value is often used, instead of more exact values (as those shown in Table 6.3) in typical engineering calculations.

The ice is lighter than water, and its density is about 917 kg/m^3 (or ~0.92 g/cm^3).

Specific gravity. Specific gravity (SG) is a dimensionless unit, defined as the ratio of the density of a substance to the density of a reference substance. For liquid and solid, the reference is usually water

$$\text{Specific Gravity } (SG) = \frac{\rho}{\rho_{\text{water}}} \quad (6.22)$$

where ρ_{water} = density of water at 4°C [Note: The SG of water = 1.0]. For example, the SG of sand is equal to 2.65; thus, its density is equal to 2,650 kg/m^3 (i.e., 2.65 g/cm^3, or 165.4 lb/ft^3). For SG of gases, the reference is usually the air and

$$\text{Specific Gravity } (SG) = \frac{\rho}{\rho_{\text{air}}} \quad (6.23)$$

TABLE 6.3
Density of Pure Water as a Function of Temperature

Temperature (°C)	Density (kg/m³)
0	999.8
4	999.97
10	999.7
15	999.1
20	998.2
25	997.0
30	999.8
40	992.2
50	988.0
60	983.2
80	971.8
100	958.4

where ρ_{air} = density of air, normally at T = 20°C and P = 1 atm, which is 1.21 g/L (see Example 6.2). The SG of chlorine gas is larger than one, while that of methane gas is less than unity under ambient conditions.

Viscosity of water. As mentioned in Section 6.3, viscosity of a fluid is a measure of its resistance to flow. The viscosity of water decreases as the temperature increases (Table 6.4).

Surface tension of water. *Surface tension* can be defined as the property of the liquid surface that, due to the cohesive nature of the liquid, allows it to resist an external force. Because of its hydrogen bonding, water is more cohesive

TABLE 6.4
Viscosity of Water as a Function of Temperature

Temperature	Dynamic Viscosity		Kinematic Viscosity
(°C)	(N·s/m²)	(lb_f-s/ft²)	(m²/s)
10	1.31×10^{-3}	2.73×10^{-5}	1.31×10^{-6}
20	1.00×10^{-3}	2.09×10^{-5}	1.00×10^{-6}
25	8.90×10^{-4}	1.86×10^{-5}	8.93×10^{-7}
30	7.97×10^{-4}	1.67×10^{-5}	8.01×10^{-7}
40	6.53×10^{-4}	1.36×10^{-5}	6.58×10^{-7}
50	5.46×10^{-4}	1.14×10^{-5}	5.53×10^{-7}
60	4.66×10^{-4}	9.73×10^{-6}	4.74×10^{-7}
70	4.04×10^{-4}	8.43×10^{-6}	4.13×10^{-7}
80	3.54×10^{-4}	7.39×10^{-6}	3.64×10^{-7}
90	3.14×10^{-4}	6.56×10^{-6}	3.26×10^{-7}
100	2.82×10^{-4}	5.88×10^{-6}	2.94×10^{-7}

TABLE 6.5
Surface Tension of Water as a Function of Temperature

Temperature [°C]	Surface Tension (N/m)
10	7.40×10^{-2}
15	7.33×10^{-2}
20	7.25×10^{-2}
25	7.18×10^{-2}
30	7.10×10^{-2}

> than many other liquids. Consequently, the surface tension of water is among the highest. The droplets of water tend to be pulled into a spherical shape due to the cohesive force of the surface layer. Surface tension is often expressed as an amount of force exerted on the surface perpendicular per unit length, in N/m.

The surface tension of water decreases as temperature increases, and its values at 20°C is equal to 7.25×10^{-2}N/m (Table 6.5).

6.4.2 Vapor Pressure

Vapor pressure (P^{vap}) of a compound is the equilibrium pressure of a vapor with its liquid/solid in a closed container. As the temperature increases, the vapor pressure increases. Intermolecular forces among molecules in a substance determine its vapor pressure. Therefore, some compounds are more volatile than the others. When $P=1$ atm, the P^{vap} of water is equal to 1 atm @$T=373$ K (i.e., 100°C, the boiling point of water), while it is only 18 mm-Hg @$T=273$ K (i.e., 0°C). The Antoine equation is commonly used to estimate P^{vap} at a specific temperature as:

$$ln\left(P^{vap}\right) = A - \frac{B}{T+C} \tag{6.24}$$

where A, B, and C are the Antoine constants; and their values can be readily found on the Internet or in chemistry handbooks. It should be noted these Antonie constants are empirical constants; and, for a specific compound, their values may be different for different temperature ranges. In addition, the reported values of these constants can be different from different studies. One should also pay attention to the units of T and P^{vap} that go with these constants (This suggestion applies to all empirical equations).

The Clausius-Clapeyron equation correlates the P^{vap} and the absolute temperature. It assumes that enthalpy of vaporization (ΔH^{vap}) is independent of temperature:

$$ln\frac{P_1^{vap}}{P_2^{vap}} = -\frac{\Delta H^{vap}}{R}\left[\frac{1}{T_1} - \frac{1}{T_2}\right] \tag{6.25}$$

where R = the universal gas constant and T = absolute temperature. Values of ΔH^{vap} can be readily found on the Internet or in chemistry handbooks. The unit of R should match those of T and ΔH^{vap} used in the equation.

For an ideal solution, the vapor-liquid equilibrium follows the *Raoult's law* as:

$$P_A = P^{vap} \times x_A \tag{6.26}$$

where P_A = partial pressure of compound A in the vapor phase and x_A = mole fraction of compound A in the solution. Raoult's law holds well only for ideal solutions. In dilute aqueous solutions that are commonly encountered in environmental engineering applications, Henry's law is often more applicable (detail later).

The *partial pressure* is the pressure that a compound would exert if all the other gases were not present. This is equivalent to the mole fraction of the compound in the gas phase (y) multiplied by the total pressure (P_{total}):

$$P_A = P_{total} \times y_A \tag{6.27}$$

The total pressure is the sum of all the partial pressures (see Eq. 6.11). For example, the ambient air contains 21% by mole (or by volume) of oxygen. Thus, the mole fraction of oxygen in the air is 0.21; and its partial pressure is 0.21 atm (= 0.21 × 1 atm). It should be noted that 1% is $10{,}000 \times 10^{-6}$; that is 10,000 parts per million by volume (ppmV). [Note: The ppmV, or ppm (v/v), is determined by comparing the volume of one constituent with the total volume of the substance. Gas concentrations are always expressed in ppm (v/v) as opposed to the ppm (w/w) format that is often used for liquid or solid. In this book, ppm will be used as an abbreviation of ppmV, or ppm (v/v)]. Thus, the oxygen concentration in our ambient air is about 210,000 ppm. If CO_2 concentration at a location is 400 ppm, then its mole fraction is 0.0004. The corresponding CO_2 concentration in percentage is 0.04% by mole (or by volume). Its partial pressure would be 4×10^{-4} atm or 0.304 mm-Hg (= 760 mm-Hg × (4×10^{-4})) – see Example 6.6.

Example 6.9: Estimate Partial Pressures of a Mixed Solvent

An above-ground storage tank contains 50% (by wt.) of toluene (C_7H_8, MW = 92) and 50% of ethyl benzene (C_8H_{10}; MW = 102). Estimate the maximum toluene and ethyl benzene concentrations (in ppm) in the void space of the tank. The ambient temperature is 20°C.

Solution:

a. At 20°C, the vapor pressure of toluene = 22 mm-Hg and that of ethyl benzene = 7 mm-Hg.

b. Using a total mass of 100 g as the basis for calculations; for 50% by weight of toluene, its mole fraction = (moles of toluene) ÷ [(moles of toluene) + (moles of ethyl benzene)]

$$= (50/92) \div [(50/92) + (50/106)] = 0.535 = 53.5\%$$

c. Using Eq. (6.31), the partial pressure of toluene in the void

$$= (22)(0.535) = 11.78 \text{ mm-Hg} = 0.0155 \text{ atm} = 15{,}500 \text{ ppm}$$

d. The partial pressure of ethyl benzene in the void

$$= (7)(1 - 0.535) = 3.25 \text{ mm-Hg} = 4{,}300 \text{ ppm}$$

Discussion:

The vapor concentrations are those in equilibrium with the solvent. An equilibrium between the liquid and vapor phases can occur in a closed container or in a stagnant space.

Example 6.10: Using the Antoine Equation to Estimate the Vapor Pressure

The empirical constants of the Antoine equation for benzene are: $A = 15.9008$, $B = 2{,}788.51$; $C = -52.36$ for P^{vap} in mm-Hg and T in Kelvin (Reid et al., 1987). Estimate the vapor pressures of benzene at $T = 20$ and 25°C and $P = 1$ atm using the Antoine equation.

Solution:

a. Using Eq. (6.24), at 20°C (293 K):

$$ln\left(P^{vap}\right) = A - \frac{B}{T + C} = 15.9008 - \frac{2{,}788.51}{(293 - 52.36)} = 4.322$$

$$P^{vap} = 75.3 \text{ mm-Hg}$$

b. Use Eq. (6.24), at 25°C (298 K):

$$ln\left(P^{vap}\right) = 15.9008 - \frac{2{,}788.51}{(298 - 52.36)} = 4.557$$

$$P^{vap} = 95.3 \text{ mm-Hg}$$

Discussion:

As expected, the vapor pressure of benzene at 25°C (95.3 mm-Hg) is larger than that at 20°C (75.3 mm-Hg).

Example 6.11: Using the Clausius-Clapeyron Equation to Estimate the Vapor Pressure

Using the vapor pressure value of benzene at 20°C (obtained from Example 6.10) and the Clausius-Clapeyron Equation to estimate its vapor pressure at 25°C. Note: Enthalpy of vaporization of benzene = 33,830 J/g-mole (Lide, 1992).

Solution:

Using Eq. (6.25),

$$\ln\frac{75.3}{P_2^{vap}} = -\frac{33{,}840}{8.314}\left[\frac{1}{293}-\frac{1}{298}\right].$$

P^{sat} of benzene at 25°C = <u>95.1 mm-Hg</u>

Discussion:

The value (95.1 mm-Hg), derived in this example, by using the Clausius-Clapeyron Equation is essentially the same as the value (95.3 mm-Hg) in Example 6.10 by using the Antoine equation.

6.4.3 Henry's Law

In addition to gaseous components, air also contains moisture and some liquid droplets. Gaseous components have tendencies to move across the interfacial boundaries between the air and the liquid droplets to get absorbed/dissolved. On the other hand, dissolved species would also have tendencies to leave the liquid droplets and move into the air. *Absorption* is a common air pollution control process that uses liquid to remove gaseous COCs. Consequently, it is important to understand the interactions between vapor and liquid.

As mentioned earlier, Raoult's law describes the vapor-liquid equilibrium (see Eq. 6.26). In aqueous solutions of low COC concentrations, which are commonly encountered in environmental engineering applications, Henry's law has been found more applicable. *Henry's law* states that, at equilibrium, the partial pressure of a compound A (P_A) in the gas phase above a liquid is proportional to the concentration of that compound in the liquid (C_A) as

$$P_A = H_A \times C_A \tag{6.28}$$

where H_A = *Henry's constant* (or the *Henry's law constant*) of compound A. This equation shows a linear relationship between the liquid and gas concentrations. The larger the liquid concentration, the larger the gas concentration will be. A compound with a large Henry's constant means that it prefers staying in the air. It should be noted that in some air pollution books or literature, Henry's law is written in an opposite way as $C_A = H_A P_A$; Henry's constant in this format is the inverse of the one shown in Eq. (6.28).

Henry's law can also be expressed in the following form:

$$G = H \times C \tag{6.29}$$

where C is the COC concentration in the liquid phase and G is the corresponding COC concentration in the gas phase. Henry's law has been widely used in various disciplines to describe the distribution of a compound between the gas and the liquid

TABLE 6.6
Unit Conversions for Henry's Constant (Kuo & Cordery, 1988)

Desired Unit for Henry's Constant (H)	Conversion Equation
atm/M, or atm·L/mole	$H = H^{*}RT$
atm·m³/mole	$H = H^{*}RT/1{,}000$
M/atm	$H = 1/(H^{*}RT)$
atm/(mole fraction in liquid), or atm	$H = (H^{*}RT)[1{,}000\ \gamma/W]$
(mole frac. in vapor)/(mole frac. in liquid)	$H = (H^{*}RT)[1{,}000\ \gamma/W]/P$

γ = specific gravity of the solution (1 for dilute solution); H^{*} = Henry's constant in the dimensionless form; M = solution molarity in (g mol/L); P = system pressure in atm (usually =1 atm); R = 0.082 atm/(K)(M); T = system temperature in Kelvin; W = equivalent MW of solution (18 for dilute aqueous solution)

phases. The units of Henry's constant reported in the literature vary considerably. The commonly encountered units for Henry's constant include atm/mole fraction, atm/M, M/atm, atm/(mg/L), and dimensionless. When inserting a value of Henry's constant into one of the two equations above, it is important to check if its units match with the other two variables. Engineers often use the units they are familiar with and have difficulties in performing the necessary unit conversions. Uses of Henry's constant in dimensionless form have been increasing because it is more straightforward and easier. Please note that it is not a "(mole fraction)/(mole fraction)" dimensionless unit. The actual meaning of Henry's constant in the dimensionless format is "(concentration in the gas phase)/(concentration in the liquid phase)", which can be [(mg/L)/(mg/L)]. To be more precise, it has a unit of (unit volume of liquid)/(unit volume of air). Table 6.6 tabulates conversions among commonly-use units for Henry's constant.

The Henry's constant of a compound is practically the ratio of its vapor pressure and solubility, provided that both are measured at the same temperature; that is,

$$H = (\text{Vapor Pressure}) \div (\text{Solubility}) \tag{6.30}$$

This equation implies that the higher the vapor pressure, the larger the Henry's constant is. It also indicates that the lower the solubility, the larger the Henry's constant. For most organic compounds, the vapor pressure increases with the increase in temperature and the solubility decreases with the increase in temperature. Consequently, Henry's constant, as defined in Eq. (6.30), should increase as the system temperature increases.

Example 6.12: Unit Conversions for Henry's Constant

Henry's constant of benzene (C_6H_6) in water at 25°C is 5.55 atm/M. Convert this value to a value with (a) the dimensionless units and (b) a unit of atm.

Solution:

a. From Table 6.6,

$$H = H^{*}RT = 5.55 = H^{*}(0.082)(273 + 25)$$

$$H^{*} = \underline{0.227} \text{ (dimensionless)}$$

b. Also from Table 6.6,

$$H = (H^{*}RT)[1{,}000\ \gamma/W]$$

$$= [(0.227)(0.082)(273+25)][(1{,}000)(1)/(18)] = \underline{308.3 \text{ atm}}$$

Discussion:

1. Benzene is a VOC of concern and appears in most, if not all, databases containing Henry's constant. It may not be a bad idea to memorize Henry's constant value of benzene, which is equal to approximately 0.23 under ambient conditions.
2. To convert Henry's constant of another COC in the database, just multiply the ratio of the Henry's constants (in any units) of COC and benzene by a multiple of 0.23. For example, to convert the Henry's constant of methylene chloride (CH_2Cl_2), 2.03 atm/M to dimensionless, just multiply the ratio by 0.23 as $[(2.03)/(5.55)] \times (0.23) = 0.084$ (dimensionless).

Example 6.13: Estimate Henry's Constant from Solubility and Vapor Pressure

Vapor pressure of benzene (C_6H_6) and its solubility in water at 298 K are 95.2 mm-Hg and 1,780 mg/L, respectively. Estimate its Henry's constant (in atm/M) from the given data.

Solution:

a. $P^{vap} = 95.2/760 = 0.125$ atm

b. C = 1,780 mg/L = 1.78 g/L

$$= (1.78\ \text{g/L}) \div (78.1\ \text{g/g-mole}) = 0.0228\ \text{mole/L} = 0.0228\ \text{M}$$

c. $H = (0.125 \text{ atm}) \div (0.0228 \text{ M}) = \underline{5.48 \text{ atm/M}}$

Discussion:

The calculated value, 5.48 atm/M, is essentially the same as 5.55 atm/M, given in Example 6.12.

Example 6.14: Use Henry's Law to Estimate Equilibrium Concentrations

An air sample was taken from the head space of a landfill leachate collection tank. The tetrachloroethylene (PCE, CCl_4) concentration in the air sample was found to be 1,250 ppm. Estimate the PCE concentration in the leachate. Note: $T = 298$ K, Henry's constant of PCE = 25.9 atm/M, and MW of PCE = 165.8).

Solution:

a. 1,250 ppm $V = 1{,}250 \times 10^{-6}$ atm $= 1.25 \times 10^{-3}$ atm $= P_A$

b. From Eq. (6.28),

$$P_A = 1.25 \times 10^{-3} \text{ atm} = H_A \times C_A = (25.9 \text{ atm/M}) \times (C_A)$$

c. $C_A = 4.82 \times 10^{-5}$ M $= (4.82 \times 10^{-5}$ mole/L)(165.8 g/mole)

$$= 8 \times 10^{-3} \text{ g/L} = \underline{8 \text{ mg/L}} = \underline{8 \text{ ppm}}$$

We can also use the dimensionless Henry's constant to solve this problem.

a. $H = H^* RT = 25.9 = H^*(0.082)(273 + 20)$

$$H^* = 1.08 \text{ (dimensionless)}$$

b. Use Eq. (6.12) to convert ppm to mg/m^3:

$$1{,}250 \text{ ppm} = (1{,}250) \times [(165.8/24.05)] \text{ mg/m}^3$$

$$= 8{,}620 \text{ mg/m}^3 = 8.62 \text{ mg/L}$$

c. Use Eq. (6.29):

$$G = 8.62 \text{ mg/L} = H \times C = 1.08 \times C$$

$$\text{So, } C = \underline{8 \text{ mg/L}} = \underline{8 \text{ ppm}}$$

Discussion:

1. These two approaches yield identical results.
2. Henry's constant of PCE is five times larger than that of benzene (1.08 vs. 0.227).
3. A concentration of 8 mg/L of PCE in the leachate is in equilibrium with a gas concentration of 1,250 ppmV.
4. The numeric value of the gas concentration (1,250 ppm) is much higher than that of the corresponding liquid concentration (8 ppm).

When a gas dissolved in water dissociates (i.e., does not stay in its molecular form), then Henry's law does not apply. They will be more soluble in water than predicted by Henry's law. For example, $NH_{3(g)}$ will form ammonium ion and $H_2S_{(g)}$ will dissociate into ions when dissolved into water as:

$$NH_{3(g)} \overset{+H_2O}{\rightarrow} NH_{3(aq)} + H_2O_{(l)} \rightarrow NH^+_{4(aq)} + OH^-_{(aq)} \quad (6.31)$$

$$H_2S_{(g)} \overset{+H_2O}{\rightarrow} H_2S_{(aq)} \rightarrow HS^-_{(aq)} + H^+_{(aq)} \quad (6.32)$$

6.4.4 Dissolved Oxygen

The dissolved oxygen (*DO*) in water is a very important water quality parameter. Many aquatic species live on the dissolved oxygen present in water. Although the partial pressure of oxygen in our ambient air = 0.21 atm, the *DO* concentrations in ambient water are only about 10 mg/L or less, depending on the temperature. The *saturated DO concentration* (DO_{sat}) is the maximum amount of oxygen that can be dissolved in water at that temperature. The *DO deficit* is defined as the difference between DO_{sat} and the measured *DO* of a water sample as:

$$DO\ Deficit = DO_{\text{deficit}} - DO \tag{6.33}$$

Example 6.15: Dissolved Oxygen Concentration

Henry's constant of O_2 in water at 20°C is 1.35 × 10^{-3} M/atm, estimate the DO_{sat} value @T = 20°C.

Solution:

Partial pressure of oxygen in air = 0.21 atm,

$$0.21 \text{ atm} \times \left(1.35 \times 10^{-3} \frac{M}{\text{atm}}\right) = C_A$$

$C_A = 2.835 \times 10^{-4}$M

Discussion:

1. This is a good example for properly using the reported Henry's constant values. Henry's law is in the form of "$P_A \times H_A = C_A$"; Henry's constant is on the opposite side of Eq. (6.28).
2. The DO_{sat} value found here is for pure water; the actual DO_{sat} values for ambient water would be slightly different.
3. The *DO* concentration is more commonly expressed in mg/L. For this example,

$DO_{sat} = 2.835 \times 10^{-4}$ M = 9.07 mg/L [= (2.835 ×10^{-4} M) × 32,000 mg/mole].

4. Since the solubility of oxygen in water decreases with temperature, the DO_{sat} values will be smaller at higher temperatures.

6.5 SOLIDS IN AQUEOUS SOLUTIONS

Up to this point, a solution is considered as a homogeneous mixture; and the solutes are considered "dissolved" in the solvent. However, not all the constituents are dissolved in aqueous solutions that we have encountered or may encounter. There are three types of solids present in aqueous solutions.

6.5.1 Types of Solids in Aqueous Solutions

Based on the approximate size of particles, solids present in aqueous solutions are grouped into:

- *Dissolved solids.* dissolved in water in an ionic form or in a molecular form.
- *Suspended solids.* suspended in water that can be retained by a filter paper. They typically having a size larger than approximately one micrometer (1 micron, 1 μm, or 10^{-6}m). Suspended solids may settle to the bottom of a container due to gravitational force if the water is allowed to stand still for a sufficient period of time.
- *Colloidal particles.* not dissolved, but they are able to remain evenly distributed throughout the solution. Colloidal particles are commonly considered having diameters between 10^{-3} to 1 micron.

6.5.2 Solubility

Solvation is the interaction between the dissolved species and the solvent. If water is the solvent, the term *hydration* is often used for solvation. As mentioned, the dissolved species in water can be in an ionic form or in a molecular form. *Dissociation* is a chemical reaction in which a compound breaks into two or more ionic species in water, carrying charges; however, the solution remains neutral in charge (see Eq. 6.34). Molecular compounds are dissolved in water due to dipole-dipole and hydrogen bond attractions, and no ionic species will be formed (see Eq. 6.35).

$$NaCl_{(s)} \overset{H_2O}{\rightarrow} Na^+_{(aq)} + Cl^-_{(aq)} \tag{6.34}$$

$$CH_3OH_{(l)} \overset{H_2O}{\rightarrow} CH_3OH_{(aq)} \tag{6.35}$$

Terms "soluble" and "insoluble" in water are qualitative. *Solubility* is the maximum quantity of a substance that can be dissolved in water (i.e., the solvent). An aqueous solution is *saturated* when it contains the maximum concentration of a solute dissolved in water. Before reaching the status of saturated, the solution is called *under-saturated.* Beyond saturation (i.e., the maximum amount of solute can be dissolved at a specific condition), the solution is *supersaturated*; the tendency of precipitation of ionic species increases when the solution is supersaturated.

Factors affecting solubility in water include:

- <u>Types of chemical compounds</u>. Salts are generally more soluble than molecular compounds.
- <u>Polarity of compounds</u>. Polar compounds are more soluble in water because water is polar.
- <u>Temperature</u>. Solubility generally increases with temperature. Calcium carbonate ($CaCO_3$) is an exception; its solubility decreases as the temperature of water increases. That explains the calcium carbonate deposits on the surface of the heat exchanger inside a water heater.

- Pressure. Increasing pressure would force more solutes dissolved into a solution; this is especially true for dissolving gaseous compounds in water.

More discussion on solubility in water and the effects of solids in aqueous solutions on water quality will be given in subsequent chapters.

6.5.3 Units of Solid Concentrations in Water

Commonly used units for dissolved solid concentrations include molarity, normality, mole fraction, and mass concentration (mass/volume); those for suspended solids are often in mass concentration (mass/volume) and % by weight.

Molarity. Molarity (M) is used by chemists to express the solid concentration in aqueous solution. *Molarity* (i.e., the molar concentration) is the number of moles of a solute per liter of the solution [Note: not per liter of water]. Thus,

$$\text{Molar Concentration } (M) = \frac{\text{Moles of solute}}{\text{Liter of solution}} = \frac{\left(\dfrac{\text{Mass of solute}}{MW \text{ of the solute}}\right)}{\text{Liter of solution}} \tag{6.36}$$

$$1\ M = 1 \text{ mole/L} \tag{6.37}$$

Normality. *Equivalent weight* is commonly used in chemistry for acid-base reactions as well as for alkalinity and hardness concentrations (more in the chapter for water quality later). *Equivalent weight* (EW) is the molecular weight of a substance divided by its equivalency as:

$$\text{Equivalent Weight} = \frac{\text{Molecular weight}}{\text{Equivalency}} \tag{6.38}$$

The *equivalency* is (i) the number of proton ions that a compound can give or receive, (ii) the number of electrons transferred in an oxidation-reduction reaction, or (iii) the valence state of ions in precipitation reactions. For example, the equivalencies of carbonic acid (H_2CO_3) and bicarbonate (HCO_3^-) are 2 and 1, respectively; and those of Ca^{+2} and PO_4^{-3} are 2 and 3, respectively. The EW of H_2CO_3 is 31 (= 62/2) and that of HCO_3^- is 61 (= 61/1). *Normality* (N) is the number of equivalents per liter of solution as:

$$\text{Normality } (N) = \frac{\text{\# of Equivalents}}{\text{Liter of solution}} = \frac{\left(\dfrac{\text{Mass of solute}}{\text{Equivalent Weight}}\right)}{\text{Liter of solution}} \tag{6.39}$$

$$\text{Normality } (N) = \text{Molar Concentration } (M) \times \text{Equivalency } (n) \tag{6.40}$$

It should be noted that the numeric values of molar concentration and normality are temperature-dependent because the volume of the solution is a function of temperature.

Mole fraction. *Mole fraction* of a substance i (X_i) in a solution is defined as:

$$X_i = n_i \Big/ \sum n_j \tag{6.41}$$

where n_i = the number of moles of solute i, and n_j = the number of moles of substance j in the solution including all the solutes and water in this case. The sum of all mole fractions in a solution should be equal to unity.

Molality. *Molality* (m) is the number of moles of a solute per kg of the solvent [Note: It is "per kg of water", not "per kg of the solution"]. Thus,

$$\text{Molality} = \frac{\text{Moles of solute}}{\text{kg of water}} = \frac{\left(\dfrac{\text{Mass of solute}}{MW \text{ of the solute}}\right)}{\text{kg of water}} \tag{6.42}$$

ppm, ppb, ppt. In environmental engineering applications, the COC concentrations in aqueous solutions are relatively low; for example, in the mg/L range. Since the mass of 1-L water is essentially 1 kg under ambient concentrations, 1 mg/L concentration in water is approximately 1 mg/kg. With 1 kg = 1,000,000 mg, then 1 mg/kg = 1 mg/(1,000,000 mg) = 1/1,000,000 = 1 part per million (ppm). Thus, a reported concentration of 1 mg/L is often shown as having a concentration of 1 ppm; 1 µg/L = 1 part per billion (ppb); and 1 nanogram/L = 1 part per trillion (ppt). Thus,

$$1 \text{ ppm} = 1\frac{\text{mg}}{L}; 1 \text{ ppb} = 1\frac{\mu\text{g}}{L}; 1 \text{ ppm} = 1{,}000 \text{ ppb} = 1{,}000{,}000 \text{ ppt} \tag{6.43}$$

% by wt. When the solid concentrations in aqueous solutions are relatively high (e.g., water/wastewater sludge), they can be expressed in "% by wt." as:

$$\% \text{ by wt.} = \left(\frac{\text{Weight of the solid}}{\text{Weight of the solution}}\right) \times 100\% \tag{6.44}$$

Under ambient conditions, if the density of the solution is close to that of water (i.e., 1 kg/L), then the following relationship is valid:

$$1\% \text{ by wt. solids} = 10{,}000\frac{\text{mg}}{\text{kg}} \cong 10{,}000\frac{\text{mg}}{L} = 10{,}000 \text{ ppm} \tag{6.45}$$

Example 6.16: Solute Concentrations in Aqueous Solutions

Express the concentration of 3% by weight $CaSO_4$ solution in water in terms of: (a) mg/L; (b) ppm; (c) molarity; (d) normality, (e) molality, and (f) mole fraction.

Solution:

a. Assume the density of the solution is the same as water (i.e., 1 kg/L), and use 1 L of solution as the basis for the calculations in this example.

$$MW \text{ of } CaSO_4 = 136$$

$$\text{Mass of } CaSO_4 \text{ in the solution} = (1\text{ kg})(3\%) = 0.03\text{ kg} = 30\text{ g} = 30{,}000\text{ mg}$$

$$\text{Concentration} = 30{,}000\text{ mg} \div 1\text{ L} = \underline{30{,}000\text{ mg/L}}$$

b. 30,000 mg/L = <u>30,000 ppm</u>

c. $\text{Molarity }(M) = \dfrac{\left(\dfrac{30}{136}\right)\text{moles}}{1\text{ L}} = 0.22\ M$

d. The equivalency of $CaSO_4 = 2$ (because calcium and sulfate ions are divalent)

Normality $(N) = 0.22 \times 2 = \underline{0.44\text{ N}}$

e. $\text{Molality} = \dfrac{\left(\dfrac{30}{136}\right)\text{moles}}{0.97\text{ kg of water}} = 0.23\text{ mole/kg}$

f. $\text{Mole fraction of } CaSO_4 = \dfrac{\text{mole of } CaSO_4}{\text{mole of } CaSO_4 + \text{mole of } H_2O}$

$$= \frac{\left(\frac{30}{136}\right)}{\left(\frac{30}{136}\right) + \left(\frac{970}{18}\right)} = 4.08 \times 10^{-3}$$

Discussion:

1. The molar concentration of pure water =1,000 g ÷ 18 g/mole = 55.5 mole/L = 55.5 M
2. The mole fraction of water in this example $= 1 - 4.08 \times 10^{-3} = 0.996$.

Example 6.17: Solute Concentrations in Aqueous Solutions

A batch reactor contains 200 ft^3 of water. The initial chloride concentration in the reactor is 0 mg/L as chloride. A slug dose of sodium chloride (NaCl) solution (V = 5 gallons and C = 30 g/L as chloride) is added to the reactor. Assuming the mixing is complete, estimate the chloride concentration after the dose, in mg/L.

Solution:

a. Total mass of chloride in the reactor = (30 g/L)[(5 gal)(3.785 L/gal)] = 567.75 g
b. Total volume of the solution in the reactor = [(200 ft^3)(7.48 gal/ft^3) + 5 gal] = (1,501 gal)(3.785 L/gal) = 5,681 L
c. Concentration = mass/volume = (567.75 g)/(5,681 L) = 0.10 g/L = <u>100 mg/L</u>

Discussion:

1. The procedure used to solve this question is straightforward, but it requires some unit conversions. With multiple-choice questions as the norm for the FE and PE exams, unit conversions are often part of the tasks needed to solve many questions.
2. 1 gallon = 3.785 L; and 1 ft^3 = 7.48 gallons.

6.5.4 Boiling Point Elevation and Freezing Point Depression

The *boiling point* of a solution is the temperature at which its vapor pressure is equal to the pressure of its surroundings. When a non-volatile solute is added to a solution, the vapor pressure of the new solution is lower than that of the pure solvent. Consequently, more heat needs to be supplied to the solution for it to boil. The increase in the boiling point of the solution is the *molal boiling point elevation* (ΔT_b), which is proportional to the molality of the solute in the solution (m) as:

$$\Delta T_b = k_b \times m \tag{6.46}$$

where k_b = *molal boiling point elevation constant* (in K/m typically).

Similarly, the freezing point of a solvent will decrease when an additional amount of solute is added. The *molal freezing point depression* (ΔT_f) is also proportional to the molality of the solute in the solution as:

$$\Delta T_f = k_f \times m \tag{6.47}$$

where k_f = *molal freezing point depression constant* (in K/m typically).

Here are a couple of real-life examples related to the boiling point elevation and the freezing point depression. Salts are commonly used for deicing roads, sidewalks, parking lots and driveways because the deicing salts reduce the melting point and prevent ice from forming. If salts are added to a vessel containing boiling water, the water will cease boiling because the boiling point will go higher. The author found his rice cooker could not cook the rice well in Laramie, Wyoming, because the boiling point is lower due to the higher elevation (elevation ≅ 7,220 ft). A viable solution is to add salts into the rice cooker to raise the boiling point of water to get the rice well cooked.

Example 6.18: Boiling Point Elevation and Freezing Point Depression

For water, the molal boing point elevation constant is 0.512 K/m and the molal freezing point depression constant is −1.86°C/m. A seawater sample contains 3.5% by wt. of dissolved solids, estimate the boiling point and the freezing point of this seawater sample.

Solution:

a. Assume all the dissolved solids are in the form of sodium chloride (MW = 58.5), then:

Moles of NaCl in 1 kg of seawater = (1,000 g)(3.5%) ÷ (58.5 g/mole) = 0.599 moles

Molality of NaCl in this seawater sample = 0.599 moles ÷ 0.965 kg water = 0.62 mole/kg

b. $\Delta T_b = k_b \times m = (0.512 \times 0.62) = 0.32$ K

Thus, the boiling point of this seawater sample = 100 + 0.32 = 100.32°C.

c. $\Delta T_f = k_f \times m = (-1.86 \times 0.62) = -1.15\ K$

Thus, the freezing point of this seawater sample = 0 – 1.15 = –1.15°C.

Discussion:

1. The total dissolved solids (TDS) concentration of typical seawater is around 35,000 mg/L; and the major ion concentration are 18,980 mg/L (Na^+), 10,560 mg/L (Cl^-), 2,650 mg/L $\left(SO_4^=\right)$, 1,260 mg/L (Mg^{+2}), 400 mg/L (Ca^{+2}), 380 mg/L (K^+), and 140 mg/L $\left(HCO_3^-\right)$.
2. Using NaCl as the dissolved species in this example appears to be acceptable, based on its dominance as the dissolved species in typical seawater.

6.6 ADSORPTION OF SOLUTES ONTO SOLIDS IN AQUEOUS SOLUTIONS

Many toxic organic substances and heavy metals are hydrophobic, and they tend to get adsorbed onto solids. Consequently, they are often present in high concentrations on the suspended solids as well as on the sediments of waterbodies. This section discusses the adsorption of solutes onto solids in aqueous solutions.

Adsorption. *Adsorption* is the process in which a compound moves from the liquid phase (or from the air – more in the next section) to the surface of the solids across the interfacial boundary. Adsorption is caused by interactions among three distinct components:

- adsorbent (e.g., vadose zone soil, aquifer matrix, and activated carbon)
- adsorbates (e.g., the COCs)
- solvent (e.g., soil moisture and groundwater)

In adsorption, the adsorbate is removed from the solvent and taken by the adsorbent. Adsorption is an important mechanism governing the COC's fate and transport in the environment.

Adsorption Isotherms. For a system where a solid phase and a liquid phase coexist, an adsorption isotherm describes the equilibrium relationship between the liquid and the solid phases. The "isotherm" indicates that the relationship is for a constant temperature.

The most popular isotherms are the Langmuir isotherm and the Freundlich isotherm. Both were derived in the early 1900s. The Langmuir isotherm (see Eq. 6.48) has a theoretical basis that assumes a mono-layer coverage of the adsorbent surface by the adsorbates; while the Freundlich isotherm is a semi-empirical relationship. For a Langmuir isotherm (see Eq. 6.49), the concentration on the solid increases with increasing concentration in liquid until the maximum concentration on the solid is reached. The Langmuir isotherm can be expressed as follows:

$$S = S_{max}\frac{KC}{1+KC} \tag{6.48}$$

where S = the adsorbed concentration on the solid surface, C = the dissolved concentration in liquid, K = the equilibrium constant, and S_{max} = the maximum adsorbed concentration. On the other hand, the Freundlich isotherm can be expressed in the following form:

$$S = K_F C^{1/n} \tag{6.49}$$

Both K_F and $1/n$ are the empirical constants.

The constants in the above two isotherms are different for different combinations of adsorbates, adsorbents, and solvents. For a given compound, the values will also be different for different temperatures. When using the isotherms, we should ensure that the units among the parameters and the constants are in consistency.

Both isotherms are nonlinear. Incorporating the nonlinear Langmuir or Freundlich isotherm into a mass balance model to evaluate the COC's fate and transport in environmental media will make the computer simulation more difficult or more time-consuming. However, it was found that, in many environmental engineering applications, the linear form of the Freundlich isotherm applies. It is called the *linear adsorption isotherm* with $1/n = 1$, thus

$$S = K_F C \tag{6.50}$$

which simplifies the mass balance equation in a fate and transport model.

Partition Coefficient. For soil-water systems, the linear adsorption isotherm is often written in the following form:

$$S = K_p C; \qquad \text{or } K_p = \frac{S}{C} \tag{6.51}$$

where K_p is called the *partition coefficient* that measures the tendency of a compound to be adsorbed onto the surface of soil or sediment from a liquid phase. It describes how a COC distributes (partitions) itself between the two media (i.e., solid and liquid). Henry's constant, which was discussed earlier, can be viewed as the vapor-liquid partition coefficient.

For a given organic chemical compound, the partition coefficient is not the same for every soil. The dominant mechanism of organic adsorption is the hydrophobic bonding between the compound and the natural organics associated with the soil. It was found that K_p increases linearly with the fraction of organic carbon (f_{oc}) in soil, thus

$$K_p = f_{oc} K_{oc} \tag{6.52}$$

The *organic carbon partition coefficient* (K_{oc}) can be considered as the partition coefficient for the organic compound into a hypothetical pure organic carbon phase. Since a soil surface is not completely covered by organics, the partition coefficient is discounted by the factor, f_{oc}. Clayey soil is often associated with more natural organic matters and, thus, has a stronger adsorption potential for organic COCs.

The octanol-water partition coefficient (K_{ow}) serves as an indicator of how an organic compound will partition between an organic phase and water (see Section 6.2). Values of K_{ow} range widely, from 10^{-3} to 10^7. Organic compounds with low K_{ow} values are hydrophilic (like to stay in water) and have low soil adsorption. There are many correlation equations between K_{oc} and K_{ow} (or solubility in water, S_w) reported in the literature. Table 6.7 tabulates the ones mentioned in an EPA Handbook (USEPA, 1991). As shown, K_{oc} increases linearly with increasing K_{ow} or with decreasing S_w on a log-log plot. The following simple correlation is also commonly used (LaGrega et al., 1994):

$$K_{oc} = 0.63K_{ow} \tag{6.53}$$

Example 6.19: Solid-Liquid Equilibrium Concentrations

The aquifer underneath a site is impacted by tetrachloroethylene (PCE). A groundwater sample contains 200 ppb of PCE (K_{ow}=398). Estimate the PCE concentration adsorbed onto the aquifer material, which contains 1% of organic carbon. Assume the adsorption isotherm follows a linear model.

Solution:

a. From Table 6.7, for PCE (a chlorinated hydrocarbon)

$$\text{Log } K_{oc} = 1.00(\text{Log } K_{ow}) - 0.21$$

$$= 2.6 - 0.21 = 2.39$$

$$K_{oc} = 245 \text{ mL/g} = 245 \text{ L/kg}$$

b. From Eq. (6.53),

$$K_{oc} = 0.63K_{ow} = 0.63(398) = 251 \text{ mL/g} = 251 \text{ L/kg}$$

c. Use Eq. (6.52) to find K_p:

$$K_p = f_{oc}\, K_{oc} = (1\%)(251) = 2.51 \text{ mL/g} = 2.51 \text{ L/kg}$$

d. Use Eq. (6.51) to find S:

$$S = K_pC = (2.51 \text{ L/kg})(0.2 \text{ mg/L}) = 0.50 \text{ mg/kg}$$

Discussion:

1. Eq. (6.53) (K_{oc}=0.63 K_{ow}) looks very simple, but it yields an estimate of K_{oc} (251 kg/L), that is very comparable to the value (245 L/kg) from using the correlation equation in Table 6.7.
2. Most technical articles do not give the units of K_p. K_p has a unit of "(volume of solvent)/(mass of adsorbent)"; and it is equal to mL/g or L/kg in most, if not all, of the correlation equations.

6.7 LIQUIDS AND SOLIDS IN AIR

In addition to gaseous constituents, our ambient air also contains particulate matter (*PM*), which includes suspended solid and liquid, such as dust, soot, smoke, and liquid droplets as well as living matters such as pollen, bacteria, spores, mold, and fungus. *Dust* is a loose term applied to solid particles, predominantly larger than colloids and capable of temporary suspension in air or other gases. *Soot* is a term for particles of amorphous carbon and tars generated from incomplete combustion of hydrocarbons. An *aerosol* is a mixture of fine particles, which can be either liquid or solid, and air. Aerosols can be natural (e.g., fog and geyser steam) or artificial (e.g., haze). *Fog* is a loose term applied to visible aerosols in which the dispersed phase is liquid, mainly formed by condensation. *Smoke* is an aerosol which is a visible mixture of gases and fine particles generated from combustion, and the particles are present in a sufficient quantity to be observable independently of the presence of other solids. *Smog* is a term derived from smoke and fog, and it was first used around 1950 to describe the combination of smoke and fog in London. *Photochemical smog* is an air pollution caused by photochemical reactions among nitrogen oxides and hydrocarbons under the action of sunlight; and ozone is its main component.

6.8 PARTITION OF COMPOUNDS AMONG AIR, WATER, AND SOIL GRAINS IN A SOIL FORMATION

Once a non-aqueous phase liquid (e.g., organic solvents) enters subsurface such as soil, it may end up in four different phases: (1) dissolved in soil moisture, (2) adsorbed onto soil grains, (3) volatilized into the pore space of the soil matrix, and (iv) stayed in its pure liquid form (i.e., "*free product*"). We have discussed the equilibrium systems of liquid and its vapor (i.e., Raoult's Law), liquid and gas (i.e., Henry's Law) and liquid and soil (i.e., adsorption onto solid). Now we move one step further to discuss the system including liquid, vapor, and solid (and free product in some of the applications).

The soil moisture in the vadose zone is in contact with both the soil grains and the air in the void, and the COCs in each of these phases can travel to the other phases. The dissolved concentration in the soil moisture, for example, is affected by the concentrations in the other phases (i.e., soil, vapor, and free product). If the entire system is in equilibrium, these concentrations are related by the equilibrium equations mentioned earlier. In other words, if the entire system is in equilibrium and the COC concentration in one phase is known, the concentrations at other phases can be estimated by using the equilibrium relationships. Although in real applications, the equilibrium condition does not always exist; the estimate from such a condition serves as a good starting point or as the upper or the lower limit of the actual values.

Example 6.20: Solid-Liquid-Vapor-Free Product Equilibrium Concentrations

Free-product phase of 1,1,1-trichloroethane (1,1,1-TCA, $C_2H_3Cl_3$) was found in the subsurface at a site. The soil is silty with an organic content of 2%. The subsurface temperature is 20°C. Estimate the maximum concentrations of TCA in the void, in the soil moisture, and on the soil grains.

Solution:

a. Since free product is present, the maximum vapor concentration will be the vapor pressure of 1,1,1-TCA at that temperature (= 100 mm-Hg at 20°C):

$$100 \text{ mmHg} = (100 \text{ mm-Hg}) \div (760 \text{ mm-Hg/atm}) = 0.132 \text{ atm.}$$

$$G = 0.132 \text{ atm} = 132{,}000 \text{ ppmV}$$

Use Eq. (6.9) to convert ppmV to mg/m³, (*MW* of = 133.5)

$$132{,}000 \text{ ppmV} = (132{,}000) \times [(133.5/24.05)] \text{ mg/m}^3$$

$$G = 732{,}200 \text{ mg/m}^3 = \underline{732.2 \text{ mg/L}}$$

b. Henry's constant of 1,1,1-TCA at 20°C = 14.4 atm/M
Convert it to dimensionless, using Table 6.6:

$$H = H^*RT = 14.4 = H^* \times (0.082) \times (273 + 20)$$

$$H^* = 0.60 \text{ (dimensionless)}$$

Use Eq. (6.29) to find the concentration in soil moisture:

$$G = H \times C = 732.2 \text{ mg/L} = (0.60) \times C$$

$$\text{So, } C = 1{,}220 \text{ mg/L} = \underline{1{,}220 \text{ ppm}}$$

c. K_{ow} of 1,1,1-TCA at 20°C = 309 L/kg (Kuo, 2014)

From Table 6.7, for TCA (a chlorinated hydrocarbon)

$$\text{Log } K_{oc} = 1.00(\text{Log } K_{ow}) - 0.21 = 2.49 - 0.21 = 2.28$$

$$K_{oc} = 191 \text{ mL/g} = 191 \text{ L/kg}$$

TABLE 6.7
Some Correlation Equations between K_{oc} and K_{ow} (or S_w) (USEPA, 1991)

Equation	Data Base
$\log K_{oc} = 0.544 (\log K_{ow}) + 1.377$; or $\log K_{oc} = -0.55 (\log S_w) + 3.64$	Aromatics, carboxylic acids and esters, insecticides, ureas and uracils, triazines, miscellaneous
$\log K_{oc} = 1.00 (\log K_{ow}) - 0.21$	Polycyclic aromatics, chlorinated hydrocarbons
$\log K_{oc} = -0.56 (\log S_w) + 0.93$	PCBs, pesticides, halogenated ethanes & propanes, PCE, 1,2-dichlorobenzene

Note: S_w is the solubility in water, in mg/L.

Or, from Eq. (6.53),

$$K_{oc} = 0.63 K_{ow} = 0.63(309) = 195 \text{ mL/g} = \underline{195 \text{ L/kg}}$$

Use Eq. (6.52) to find K_p:

$$K_p = f_{oc} K_{oc} = (2\%)(191) = 3.82 \text{ mL/g} = 3.82 \text{ L/kg}$$

Use Eq. (6.50) to find the concentration on the soil grain, S:

$$S = K_p C = (3.82 \text{ L/kg})(1{,}220 \text{ mg/L}) = \underline{4{,}660 \text{ mg/kg}}$$

Discussion:

1. The calculated liquid concentration, 1,220 mg/L, is smaller than its solubility, 4,400 mg/L.
2. The simple equation, Eq. (6.53), again yields an estimate value of K_{oc} (195 L/kg) that is very comparable to the value (191 L/kg) from using the correlation equation in Table 6.7.
3. The calculated concentrations are the maximum possible values; the actual values would be lower if the system is not in equilibrium and is not a confined system.

Example 6.21: Solid-Liquid-Vapor Equilibrium Concentrations (Absence of Free Product)

For a subsurface impacted by 1,1,1-trichloroethane (1,1,1-TCA), the soil vapor concentration at a location was found to be 1,320 ppmV. The soil is silty with an organic content of 2%. The subsurface temperature is 20°C. Estimate the maximum concentrations of TCA in the soil moisture and on the soil grains.

Solution:

a. With a concentration of 1,320 ppmV, this is 100 times smaller than that in Example 6.20:

$$G = 1{,}320 \text{ ppmV} = 7{,}320 \text{ mg/m}^3 = \underline{7.32 \text{ mg/L}}$$

b. With the dimensionless Henry's constant of 0.60 (from Example 6.20):

$$G = HC = 7.32 \text{ mg/L} = (0.60) \times C$$

$$\text{So, } C = 12.2 \text{ mg/L} = \underline{12.2 \text{ ppm}}$$

c. With $K_p = 3.82$ L/kg (from Example 6.20),

$$S = K_p C = (3.82 \text{ L/kg})(12.2 \text{ mg/L}) = \underline{46.6 \text{ mg/kg}}$$

Discussion:

1. The equilibrium relationships ($G = HC$ and $S = K_p C$) are linear. With a vapor concentration of 1,320 ppmV, which is 100 times smaller than 132,000 ppmV in Example 6.20, the corresponding liquid and solid concentrations are correspondingly smaller by 100 times. It should be noted that this is only valid when two systems have the same characteristics (i.e., the same H and K_p values).
2. The concentrations are based on an assumption that the system is in equilibrium; the actual values would be different if the system is not in equilibrium.

REFERENCES

Kuo, J.F. and Cordery, S.A. (1988). "Discussion of Monograph for Air Stripping of VOC from water, *J. Environ.* Eng., V. 114, No. 5, p. 1248–50.

Kuo, J.F. (2014). *Practical Design Calculations for Groundwater and Soil Remediation*, 2nd edition, CRC Press, Boca Raton, Florida.

Kuo, J., Hicks, T.C., Drake, B., Chan, T.F. (2015) "Estimation of Methane Emission for California Natural Gas Industry", *J. Air & Waste Management Association* 65(7), 844–55.

LaGrega, M.D., Buckingham, P.L. and Evans, J.C. (1994). *Hazardous Waste Management*, McGraw-Hill, New York.

Lide, D.R. (1992). *Handbook of Chemistry and Physics*, 73rd edition, CRC Press, Boca Raton, Florida.

NOAA (2023). "The Atmosphere", National Oceanic and Atmospheric Administration, https://www.noaa.gov/jetstream/atmosphere, last updated July 28, 2023.

Reid, R.C., Prausnitz, J.M. and Poling, B.F (1987). *The Properties of Liquids and Gases*, 4th edition, McGraw-Hill, Inc., New York, NY.

USEPA (1989). "Transport and Fate of Contaminants in the Subsurface", EPA 625/4-89/019, US EPA, Washington, DC.

USEPA (1991). "Site Characterization for Subsurface Remediation", EPA 625/R-91/026, US EPA.

USEPA (2012). *APTI 427: Combustion Source Evaluation - Student Manual*, 3rd edition, prepared by ICES Ltd. For Air Pollution Training Institute, USEPA.
USEPA (2016). "Development of National Bioaccumulation Factors: Supplemental Information for EPA's 2015 Human Health Criteria Update", EPA 822-R-16-001, United States Environmental Protection Agency, https://www.epa.gov/sites/default/files/2016-01/documents/national-bioaccumulation-factors-supplemental-information.pdf, last updated January 2016.
USEPA (2023). "KABAM Version 1.0 User's Guide and Technical Documentation", https://www.epa.gov/pesticide-science-and-assessing-pesticide-risks/kabam-version-10-users-guide-and-technical, last updated September 5, 2023.

EXERCISE QUESTIONS

1. If air is a binary mixture of nitrogen and oxygen (80% N_2 and 20% O_2 by volume), answer the questions below.
   a. What is the molecular weight of air?
   b. What is the density of air @ $P=1$ atm and $T=25°C$?
   c. What is the density of air @ $P=1$ atm and $T=0°C$?
   d. What is the partial pressure of oxygen @ $P=1$ atm?
   e. What is the oxygen concentration in air (% by wt.)?
2. Estimate the density of air (in g/L and in lb/ft^3):
   a. $T=300°F$ and $P=14.0$ psi
   b. $T=300°F$ and $P=28.0$ psi
   c. $T=500°F$ and $P=14.0$ psi.
3. Estimate the viscosity and the kinematic viscosity of air @ $T=500°F$ and $P=1$ atm.
4. Estimate the specific heat capacity of an exhaust gas stream with 10% water vapor and the stack temperature = 250°C.
5. Benzene vapor is toxic. The vapor pressure of benzene at $T=25°C$ is 95.1 mmHg ($P=1$ atm).
   a. Convert the vapor pressure from mm-Hg into atm.
   b. Convert the vapor pressure from mm-Hg to Pa and in Torr.
   c. If benzene is allowed to evaporate to reach an equilibrium in an enclosed space, estimate the benzene concentration in the air space, in ppm and in % by volume ($P=1$ ppm & $T=25°C$).
6. One liter of gasoline ($MW=100$; density = 0.85 kg/L) is spilled in a room (10 m × 20 m × 5 m).
   a. Determine the maximum gasoline concentration in the room air in mg/m^3.
   b. Express the maximum concentration in ppmV ($T=15°C$ and $P=0.95$ atm)
7. The flue gas from a combustion chamber contains 15% of water by volume. Estimate its dew-point temperature ($P=1$ atm), using the Antoine equation below (for water vapor between 1°C to 100°C):

$$Log_{10}P\ (\text{in mm-Hg}) = 8.07131 - \frac{1,730.63}{T\ (\text{in °C}) + 233.426}$$

8. a. Using Dulong's formula to estimate the heating value of propane (C_3H_8) gas.
   b. Compare your result with the heating value of methane found in Example 6.8.
9. An above-ground storage tank contains 60% (by wt.) of xylenes and 40% of toluene. Estimate the maximum xylenes and toluene concentrations (in ppm) in the void space of the tank. The ambient temperature is 20°C [Note: The vapor pressure of xylenes = 10 mm-Hg and that of toluene = 22 mm-Hg at T = 20°C].
10. a. The empirical constants of the Antoine equation for benzene are: A = 6.90565, B = 1211.033; C = 220.79 for P^{vap} in mm-Hg and T in °C. Estimate the vapor pressures of benzene at T = 20 and 25°C and P = 1 atm using the Antoine equation below [Note: The operator in the equation below is "Log_{10}", while that in Eq. (6.24) or in Example 6.10 is "natural log". In addition, the T in the equation below is in °C].

$$Log_{10}\left(P^{vap}\right) = A - \frac{B}{T + C}$$

    b. Compare your results with those of Example 6.10.
11. a. Using the vapor pressure of benzene at 20°C (obtained from Question #10) and the Clausius-Clapeyron Equation to estimate its vapor pressure at 15°C and 25°C. Note: Enthalpy of vaporization of benzene = 33,830 J/g-mole (Lide, 1992).
    b. Compare the vapor pressures at these temperatures (i.e., 15°C, 20°C, and 25°C).
12. Henry's constant of toluene (C_7H_8) in water at 20°C is 6.7 atm/M. Convert this value to (a) dimensionless and (b) with a unit of atm.
13. a. Vapor pressure of toluene (C_7H_8) and its solubility in water at 293 K are 22 mm-Hg and 515 mg/L, respectively. Estimate its Henry's constant (in atm/M) from the given information.
    b. Compare your result with the value given in Question 12.
14. An air sample was taken from the head space of a landfill leachate collection tank. The toluene concentration in the air sample was found to be 400 ppm. Estimate the toluene concentration in the leachate. Note: T = 293 K and Henry's constant of toluene = 6.7 atm/M.
15. The dissolved oxygen (DO) concentration at a location of a river immediately after a wastewater discharge is 5.0 mg/L (T = 25°C). The Henry's constant for O_2 in water at 25°C is 1.50×10^{-3} M/atm.
    a. Estimate the DO_{sat} value of this river water when T = 25°C.
    b. Compare your result with the value found in Example 6.15.
    c. What is the DO deficit at this location?
16. Express the concentration of 5% by weight $MgCO_3$ solution in water in terms of: (a) mg/L; (b) ppm; (c) molarity; (d) normality, (e) molality; and (f) mole fraction.

17. A batch reactor contains 100 ft^3 of water. The initial chloride concentration in the reactor is 0 mg/L as chloride. A slug dose of sodium chloride (NaCl) solution (V = 10 gallons and C = 20 g/L as chloride) is added to the reactor. Assuming the mixing is complete, estimate the chloride concentration after the dose, in mg/L.
18. For water, the value of molal boing point elevation constant is 0.512 K/m and the molal freezing point depression constant is –1.86°C/m.
    a. A seawater sample contains 3.0% by wt. of dissolved solids. Estimate the boiling point and the freezing point of this seawater.
    b. Compare your results with those of Example 6.18.
19. The aquifer underneath a site (T = 20°C) is impacted by 1,2-dichloroethane (1,2-DCA, $C_2H_4Cl_2$). A groundwater sample contains 300 ppb of 1,2-DCA (K_{ow} = 63). Estimate the 1,2-DCA concentration adsorbed onto the aquifer material, which contains 1.5% of organic carbon. Assume the adsorption isotherm follows a linear model.
20. Free-product phase of 1,2-dichloroethane (1,2-DCA, $C_2H_4Cl_2$) was found in the subsurface at a site. The soil is silty with an organic content of 1.2%. The subsurface temperature is 20°C. Estimate the maximum concentrations of 1,2-DCA in the air void, in the soil moisture, and on the soil grains [Note: The vapor pressure of 1,2-DCA = 180 mm-Hg, K_{ow} = 63, and its Henry's constant = 4.26 atm/M @T = 20°C].
21. A subsurface was impacted by 1,2-dichloroethane ($C_2H_4Cl_2$), the soil vapor concentration at a location was found to be 2,000 ppmV (no presence of free product). The soil is silty with an organic content of 1.2%. The subsurface temperature is 20°C. Estimate the maximum concentrations of 1,2-DCA in the soil moisture and on the soil grains [Note: the vapor pressure of 1,2-DCA = 180 mm-Hg, K_{ow} = 63, and its Henry's constant = 4.26 atm/M @ T = 20°C].

7 Ions in Aquatic Solutions

7.1 INTRODUCTION

As mentioned in Section 1.3, when atoms are connected by covalent bonds, a molecule is formed. On the other hand, if the atoms are connected with ionic bonds, they are ionic compounds. When an ionic compound is added to water, some or all of the added ionic compounds will be dissolved and dissociated into ions to form a solution. The ions present in the solution will affect the properties of the solution.

Alcohols are molecular compounds, and they are miscible with water; while some molecular compounds (e.g., organic solvents) are not miscible with water, but are soluble in water to an extent. Similarly, some ionic compounds are more soluble in water than the others.

7.2 IONIC EQUATIONS

Electrolytes are substances that, when dissolved in water, break into positive-charge ions (cations) and negative-charge ions (anions). Table salt (NaCl) is an electrolyte; when added into water, it will dissolve and dissociate into positive sodium ion and negative chloride ion as:

$$NaCl_{(s)} \xrightarrow{+H_2O} Na^+_{(aq)} + Cl^-_{(aq)} \tag{7.1}$$

Equation (7.1) is an ionic equation in which the electrolytes in aqueous solution are expressed as dissociated ions. Similar to a balanced molecular equation (in which compounds are expressed in molecules), the number and type of atoms on both sides of the reaction are the same. In addition, the net charge should be the same on both sides of the balanced equation.

The dissociated ions present in aqueous solutions are stabilized by ion–dipole interaction with water molecules. The charged ions allow water to conduct electricity. In general, ionic compounds are usually electrolytes, whereas molecular substances are not, with the exceptions of acids and bases.

7.3 SOLUBILITY AND STRENGTH OF ELECTROLYTES

7.3.1 Strength of Electrolytes

An electrolyte (e.g., NaCl), when dissolved in water will dissociate into ions, can conduct an electric current. While many molecular compounds (e.g., sugar and alcohol) will not dissociate when dissolved in water; they cannot conduct an electric current in the solution, and they are *non-electrolyte*. Electrolytes are grouped into two types: strong and weak.

DOI: 10.1201/9781003502661-7

Strong Electrolytes. *Strong electrolytes* ionize almost completely in water. For example, sulfuric acid (H_2SO_4) is a strong acid, and it will almost completely dissociate (except under very low pHs) into protons and sulfate ions as:

$$H_2SO_{4(aq)} \rightarrow 2H^+_{(aq)} + SO^=_{4(aq)} \quad (7.2)$$

Below are examples of strong electrolytes:

- Strong acids: HCl (hydrochloric acid), HBr (hydrobromic acid), HI (hydroiodic acid), nitric acid (HNO_3), and sulfuric acid (H_2SO_4)
- Strong bases: NaOH (sodium hydroxide), KOH (potassium hydroxide), $Ca(OH)_2$ (calcium hydroxide), and $Mg(OH)_2$ (magnesium hydroxide)
- Salts: NaCl (sodium chloride), KBr (potassium bromide), $MgSO_4$ (magnesium sulfate), $CaCl_2$ (calcium chloride), Na_2CO_3 (sodium carbonate), and NH_4NO_3 (ammonium nitrate) [Note: Most salts are made of metal cations and non-metal anions. This statement applies to all the compounds in this paragraph, except NH_4NO_3].

Weak Electrolytes. *Weak electrolytes* only have a small amount of dissociation in solution. The molecules of a weak electrolyte are in equilibrium with its ions in a solution. Therefore, the ionization equation of a weak electrolyte is presented with a double-head arrow (i.e., the reaction is reversible); for example,

$$NH_{3(aq)} + H_2O \leftrightarrow NH^+_{4(aq)} + OH^-_{(aq)} \quad (7.3)$$

Ammonia (NH_3) is a weak electrolyte; it means that only a small percentage of ammonia molecules will become ionized when dissolved in water. Below are examples of weak electrolytes:

- Weak acids: HF (hydrofluoric acid), HCN (hydrogen cyanide), CH_3COOH (acetic acid)
- Weak bases: NH_3 (ammonia)

Example 7.1: Electrolytes and Nonelectrolytes

Classify the following compounds as (i) molecular or ionic compound and (ii) strong, weak, or non-electrolyte.

a. Benzene (C_6H_6) and 1,2-DCA ($C_2H_4Cl_2$)
b. Methanol (CH_3OH) and phenol (C_6H_5OH)
c. Acetic acid (CH_3COOH) and carbonic acid (H_2CO_3)
d. Potassium chloride (KCl) and sodium flouride (NaF)

Solution:

a. Both benzene (C_6H_6) and 1,2-DCA ($C_2H_4Cl_2$) are molecular compounds, and they are non-electrolyte.

b. Both methanol (CH_3OH) and phenol (C_6H_5OH) are molecular compounds, but methanol is a non-electrolyte and phenol is a weak electrolyte. Phenol is a weak acid and it can dissociate into a phenolate ion ($C_6H_5O^-$) and a proton.
c. Both acetic acid (CH_3COOH) and carbonic acid (H_2CO_3) are molecular compounds. They are weak acids and weak electrolyte.
d. Potassium chloride (KCl) and sodium fluoride (NaF) are ionic compounds, and they are strong electrolytes.

7.3.2 Solubility of Ionic Compounds

Solubility of ionic compounds in water is the measure of the maximum amount of solute that can be dissolved in a given quantity of water at a given temperature and pressure. When an ionic compound dissolved in water, it dissociates into ions. If the energy released from the interaction of the ions with water molecules is sufficient to compensate the energy required to break the ionic bonds in the solid and to separate the water molecules, the ions can then be inserted among water molecules. The solubility is usually expressed in molarity (M) in chemistry, but more commonly in mg/L in environmental engineering applications.

The solubility of ionic compounds in water depends on the types of ions (i.e., cations and anions) that form the compounds. Below is the general classification to tell if an ionic compound is soluble, insoluble, or slightly soluble:

- A salt is considered "*soluble in water*" if it yields a solution with a concentration of ≥0.1 M with water under ambient conditions.
- A salt is considered "*insoluble in water*" if it yields a solution with a concentration of <0.001 M with water under ambient conditions.
- A salt is considered "*slightly (sparingly) soluble in water*" if it yields a solution with a concentration of between 0.001 and 0.1 M with water under ambient conditions.

The solubility (or insolubility) of an ionic compound in water can be predicted by a set of rules as described below.

The solubility rules for "soluble" ionic substances which contain:

1. All alkali metals (i.e., Li^+, Na^+, K^+, Rb^+, Cs^+)
2. All ammonium $\left(NH_4^+\right)$
3. All nitrate $\left(NO_3^-\right)$, perchlorate $\left(ClO_4^-\right)$, chlorate $\left(ClO_3^-\right)$, and acetate (CH_3COO^-)
4. Most chlorides (Cl^-), bromides (Br^-) and iodides (I^-)
    - Exceptions: when with Ag^+, Cu^+, Tl^+, Pb^{+2}, and Hg_2^{+2}.
5. Most sulfates $\left(SO_4^=\right)$
    - Exceptions: when with Ca^{+2}, Sr^{+2}, Ba^{+2}, Ag^+, and Pb^{+2}.

The solubility rules for "insoluble" ionic substances which contain:

1. Most hydroxides (OH^-)
   - Exceptions: when with (i) alkali metals (i.e., Li^+, Na^+, K^+, Rb^+, and Cs^+), (ii) ammonium (NH_4^+), and (iii) Ca^{+2}, Sr^{+2}, Ba^{+2}.
2. Most carbonates $(CO_3^=)$, oxalates $(C_2O_4^=)$, and phosphate (PO_4^{-3})
   - Exceptions: when with (i) alkali metals (i.e., Li^+, Na^+, K^+, Rb^+, and Cs^+) and (ii) ammonium (NH_4^+).
3. Most sulfides ($S^=$)
   - Exceptions: when with (i) alkali metals (i.e., Li^+, Na^+, K^+, Rb^+, Cs^+), (ii) ammonium (NH_4^+), and (iii) Ca^{+2}, Sr^{+2}, Ba^{+2}.
4. All oxides ($O^=$)
   - Exceptions: when with (i) alkali metals (i.e., Li^+, Na^+, K^+, Rb^+, and Cs^+) and (ii) Ca^{+2}, Ba^{+2}.

Example 7.2: Soluble and Insoluble Ionic Compounds

Classify the following ionic compounds as soluble or insoluble:

a. LiF
b. NH_4Cl
c. $NaNO_3$
d. $Ca(CH_3COO)_2$
e. AgCl
f. Na_2SO_4
g. $CaSO_4$
h. LiOH
i. $Mg(OH)_2$
j. Na_2CO_3
k. $CaCO_3$
l. Na_2S
m. FeS
n. BaO

Solution:

a. LiF: soluble
b. NH_4Cl: soluble
c. $NaNO_3$: soluble
d. $Ca(CH_3COO)_2$: soluble
e. AgCl: insoluble
f. Na_2SO_4: soluble
g. $CaSO_4$: insoluble
h. LiOH: soluble
i. $Mg(OH)_2$: insoluble
j. Na_2CO_3: soluble
k. $CaCO_3$: insoluble
l. Na_2S: soluble
m. FeS: insoluble
n. BaO: soluble

7.3.3 Solubility versus Electrolyte Strength

A strong electrolyte may not necessarily be very soluble in water. However, those dissolved will be fully dissociated into ions.

7.4 PROPERTIES OF IONIC SOLUTIONS

7.4.1 Major Ions in Water

Waterbodies in our environment (i.e., rivers, lakes, groundwater, and oceans) contain many dissolved ions; and their concentrations are different. Table 7.1 shows a comparison of river water and seawater; the concentrations are in mM (10^{-3}M) and in mg/L (Source: Water Encyclopedia, 2023).

As shown, sodium and chloride ions have the highest concentrations in seawater, while bicarbonate is the dominant ion in river water. It should be noted that each water type should also contain many other dissolved ions in trace amounts.

7.4.2 Electrical Neutrality of Aqueous Solutions

All aqueous solutions should be electrically neutral; it means that the sum of positive charges carried by cations is the same as the sum of all negative charges carried by anions.

Example 7.3: Electrical Neutrality of Water

A laboratory analyzed a river water sample for water quality. The major ion concentrations are the same as those shown in Table 7.1.

a. Find the sums of the cation and anion concentrations, in mM.
b. Find the sums of the cation and anion concentrations, in mg/L.
c. Find the sums of the cation and anion concentrations, in meq/L.
d. What does the reported data say about the electrical neutrality of this water sample?

TABLE 7.1
Comparison of Major Ions of River Water and Seawater

	River Water		Seawater	
Ions	**(mM)**	**(mg/L)**	**(mM)**	**(mg/L)**
Na^+	0.23	5.3	468	10,800
K^+	0.03	1.2	10.2	400
Ca^{+2}	0.33	13.2	10.2	410
Mg^{+2}	0.15	3.6	53.2	1,300
Cl^-	0.16	5.7	545	19,300
$SO_4^=$	0.069	6.6	28.2	2,700
HCO_3^-	0.86	52.5	2.38	150

Solution:

a. The sum of the cation concentrations (in mM) = 0.23 + 0.03 + 0.33 + 0.15 = 0.74
The sum of the anion concentrations (in mM) = 0.16 + 0.069 + 0.86 = 1.089
b. The sum of the cation concentrations (in mg/L) = 5.3 + 1.2 + 13.2 + 3.6 = 23.3
The sum of the anion concentrations (in mg/L) = 5.7 + 6.6 + 52.5 = 64.8
c. The sum of the cation concentrations (in meq/L) = 0.23 + 0.03 + 0.33(2) + 0.15(2) = 1.22
The sum of the anion concentrations (in meq/L) = 0.16 + 0.069(2) + 0.86 = 1.16
d. From part (c), the sum of cation concentrations is 1.22 meq/L which is close to that of the anion concentration (1.16 meq/L).

Discussion:

1. In part (c), the concentration values of divalent ions (i.e., Ca^{+2}, Mg^{+2}, and $SO_4^{=}$) are multiplied by a factor of 2 because their equivalency is equal to 2.
2. Comparing the sum of charges of cations and that of anions is a quick way to evaluate the quality of the reported data.
3. The sum of concentrations of cations and that of anions, in mM or in mg/L, are quite different for this water sample; these sums should not be used to check the electrical neutrality of water.

7.4.3 Salinity of Aqueous Solutions

Salinity is an indicator of the dissolved salt content of a waterbody. It is a strong contributor to conductivity and has impacts on many aspects of chemistry of the waterbody and the biological processes occurring within. Salts can be toxic to freshwater plants and can make water unsafe for drinking, irrigation, and livestock watering. Salinity is one of the primary factors used to determine whether a given study site is a part of estuarine or coastal system (USEPA, 2023).

The salinity of water is often expressed as mass of salt in water in a unit mass of water. With that, one part per million (ppm) is equal to 1 mg/kg of water. Below is the classification system for saline water used by the U.S. Geological Survey (USGS, 2018):

- Freshwater - Less than 1,000 ppm
- Slightly saline water - From 1,000 to 3,000 ppm
- Moderately saline water - From 3,000 to 10,000 ppm
- Highly saline water - From 10,000 to 35,000 ppm

Ocean water contains about 35,000 ppm of salt. The salinity of seawater is often expressed in grams of salt per kilogram of water; that is parts per thousand (ppt). One ppt of salinity is also called one *practical salinity unit* (psu). In open ocean, the range of salinity is generally from 32 to 37 ppt (or psu).

7.4.4 Ionic Strength of Aqueous Solution

As mentioned in Section 5.2, the equilibrium constant (K_{eq}) for a general reaction equation (Eq. 5.6) can be expressed as in Eq. (5.16):

$$\text{Overall Reaction}: aA + bB \leftrightarrow cC + dD \tag{5.6}$$

$$K_{eq} = \frac{[C]^c[D]^d}{[A]^a[B]^b} \tag{5.16}$$

The reported K_{eq} values usually relate the molar concentrations of all the reaction participants (i.e., [*A*], [*B*], [*C*], and [*D*] here.) and are unit-less. The reason for them being dimensionless is because activities, not molar concentrations of the reaction participants, should be used in Eq. (5.16).

In chemical thermodynamics, *activity* (a) of a chemical species is a measure of the "effective concentration" of this species in a solution. Ions in a solution will interact with each other and with water molecules. Consequently, ions will behave as less concentrated as the measured concentrations. In other words, the activity of a chemical species should be smaller than its measured concentration. For an ideal gas/solution, there are no interactions among the species present; the activity is equal to the measured concentration.

Activity of a species in a solution depends on many factors, including temperature, pressure, and the composition of the solution. For a gas component in a gas mixture, its activity is its effective partial pressure, and it is usually referred to as its *fugacity*. The activity (a) and the measured concentration (C) in an aqueous solution are related by the *activity coefficient* (γ) as:

$$a = \gamma \times C \tag{7.4}$$

The activity coefficients for chemical species in an ideal solution are equal to one; while those in a non-ideal solution are less than unity.

The activity coefficients of ions in an aqueous solution depend on the ionic strength of the solution. The *ionic strength* of a solution (I) can be found from the molar concentrations of all the ionic species present as:

$$I = \frac{1}{2}\sum (Z_i^2 \times C_i) \tag{7.5}$$

where Z_i and C_i are the charge and the molar concentration of ionic species i.

Example 7.4: Ionic Strength of an Aqueous Solution

Find the ionic strength of 0.1 M magnesium chloride ($MgCl_2$) solution.

Solution:

a. For the 0.1 M $MgCl_2$ solution, $[Mg^{+2}] = 0.1$ M and $[Cl^-] = 0.2$ M.

b. Use Eq. (7.5) to find the ionic strength of the solution:

$$I = \frac{1}{2}\sum (Z_i^2 \times C_i) = \frac{1}{2}\left[\left(2^2 \times 0.1\right) + \left(1^2 \times 0.2\right)\right] = 0.3 \text{ M}$$

Discussion:

1. The ionic strength of 0.1 M $MgCl_2$ solution is equal to 0.3 M.
2. Although proton (H^+) and hydroxide ion (OH^-) are always present in aqueous solutions, their concentrations are usually smaller than those of other dissolved ions so that they are not included in the calculation of ionic strength.

Typical ionic strengths of natural waters are (Aqion, 2021):

- Surface water: 0.001 to.005 M
- Potable water and groundwater: 0.001 to 0.02 M
- Seawater: 0.7 M

Example 7.5: Ionic Strength of a River Water Sample

A laboratory analyzed a river water sample for water quality. The major ion concentrations are the same as those shown in Table 7.1. Estimate the ionic strength of this river water sample.

Solution:

a. As shown in Table 7.1, the concentration of Na^+, K^+, Ca^{+2}, Mg^{+2}, Cl^-, $SO_4^=$, and HCO_3^- are 0.23, 0.03, 0.33, 0.15 0.16, 0.069, and 0.86 mM, respectively.
b. Use Eq. (7.5) to find the ionic strength of the solution:

$$I = \frac{1}{2}\left\{\left[1^2 \times (0.23 + 0.03 + 0.16 + 0.86)\right] + \left[2^2 \times (0.33 + 0.15 + 0.069)\right]\right\} = 1.74 \text{ mM}$$

Discussion:

1. The solution took a "shortcut" by summing the concentrations of all the mono-valent ions and those of the divalent ions first in calculation of the ionic strength.
2. The calculated value, 1.74 mM (i.e., 0.00174 M), falls in the range of ionic strength of surface water (i.e., 0.001 to 0.005 M).

There are many empirical equations (e.g., Debye–Hückel equation) being developed to relate the activity coefficient of an ionic specifies in aquatic solutions to its charge and the ionic strength of the solution. However, concentrations of compounds of concern (COCs) in aqueous solutions are relatively low, and the ionic strength of the solution is also low in most environmental engineering applications. Consequently, the ionic interactions can be ignored so that the concentrations, instead of activities, can be used without incurring significant errors in most practical applications.

7.5 SOLUBILITY OF SLIGHTLY SOLUBLE IONIC COMPOUNDS IN WATER

When a slightly soluble ionic compound is added to water, some of it will get dissolved (and dissociated) and the rest will remain in its solid form, either suspended in the solution or settled at the bottom of the container/beaker. After a while, an equilibrium condition will be reached between the undissolved solid and the corresponding dissociated ions. It should be noted that even if an ionic compound is considered insoluble in water, it will still dissolve/dissociate to a limited extent.

7.5.1 Solubility Product Constant

For dissolution of an ionic compound, A, in water, the reaction equation can be written as:

$$aA_{(s)} \overset{H_2O}{\leftrightarrow} cC_{(aq)} + dD_{(aq)} \tag{7.6}$$

Following the discussion in Section 7.5.2, the equilibrium constant (K_{eq}), for the reaction equation above, should be written as:

$$K_{eq} = \frac{[C]^c [D]^d}{[A]^a} \tag{7.7}$$

Since the concentration of a pure solid remains constant and it is set to be unity in an equilibrium constant, Eq. (7.7) becomes:

$$K_{sp} = [C]^c [D]^d \tag{7.8}$$

The equilibrium constant for this case is often called the *solubility product constant* (K_{sp}). The solubility product constant provides insight into the equilibrium between the solid and the concentrations of dissociated ions in the aqueous solution. The more soluble a substance is, the larger the K_{sp} value would be. The values of K_{sp} are readily available in chemistry or chemical engineering handbooks as well as on the Internet. Table 7.2 tabulates the K_{sp} values of some ionic compounds. It should be noted that the reported values of solubility product constants are typically derived from laboratory experiments. Consequently, the reported K_{sp} values can be different from different sources, plausibly due to differences in the experimental settings.

Example 7.6: Solubility and Solubility Product Constant

Use the values of the solubility product constant in Table 7.2 to find the solubility of silver chloride (AgCl) and lead chloride ($PbCl_2$) in water at $T = 25°C$.

Solution:

a. The dissociation equation for AgCl can be written as:

$$AgCl_{(s)} \leftrightarrow Ag^+_{(aq)} + Cl^-_{(aq)} \tag{7.9}$$

TABLE 7.2
Solubility Product Constants of Selected Compounds (T = 25°C)

Compounds	K_{sp}
$AgCl$	1.8×10^{-10}
$PbCl_2$	1.7×10^{-5}
$CaCO_3$	3.4×10^{-9}
$CaSO_4$	7.1×10^{-5}
$BaSO_4$	1.1×10^{-10}
$Ca(OH)_2$	5.0×10^{-6}
$Mg(OH)_2$	2.0×10^{-11}
$Al(OH)_3$	1.9×10^{-33}
$Fe(OH)_3$	2.8×10^{-39}
PbS	8.0×10^{-28}

The solubility product constant can then be expressed as:

$$K_{sp} = \left[Ag^{+}\right]\left[Cl^{-}\right] = 1.8 \times 10^{-10}$$

Let the concentrations of silver ion and chloride ion be x, then

$$K_{sp} = 1.8 \times 10^{-10} = (x)(x)$$

Then, $[Ag^+] = [Cl^-] = x = \underline{1.34 \times 10^{-5}\ M}$.

b. Similarly, the dissociation equation for $PbCl_2$ can be written as:

$$PbCl_{2(s)} \leftrightarrow Pb^{+2}_{(aq)} + 2\ Cl^{-}_{(aq)} \qquad (7.10)$$

The solubility product constant can then be expressed as

$$K_{sp} = \left[Pb^{+2}\right]\left[Cl^{-}\right]^2 = 1.7 \times 10^{-5}$$

Let the concentration of lead ion be x (and that of chloride ion be $2x$), then

$$K_{sp} = 1.7 \times 10^{-5} = (x)(2x)^2 = 4x^3$$

Then, $[Pb^{+2}] = x = 1.62 \times 10^{-2}$ M and $[Cl^-] = 2x = 3.24 \times 10^{-2}$ M

Discussion:

1. One chloride ion will form from the dissociation of one AgCl and two chloride ions will form from the dissociation of one $PbCl_2$.
2. $PbCl_2$ is very soluble in water when compared to AgCl.

3. This example illustrates that the solubility of a compound can be found from its solubility product constant. On the other hand, if the solubility of a compound is known, the value of its solubility product constant can be derived accordingly.

Example 7.7: Solubility and Solubility Product Constant

The solubility product of magnesium hydroxide, $Mg(OH)_2$, is 2.0×10^{-11} M^3 at 20°C.

a. Write/express the equilibrium constant for the reaction shown below.
b. What is the hydroxyl ion concentration (i.e., $[OH^-]$) at pH = 9, in M?
c. Estimate the maximum soluble Mg^{+2} concentration in a water sample of pH = 9, in M.
d. Estimate the maximum soluble Mg^{+2} concentration in a water sample of pH = 11, in M.
e. Is magnesium hydroxide more soluble or less soluble at a higher pH?

$$\mathrm{Mg(OH)}_{2(s)} \rightarrow \mathrm{Mg}^{+2}_{(aq)} + 2\mathrm{OH}^-_{(aq)} \tag{7.11}$$

Solution:

a. The solubility product constant for $Mg(OH)_2$ can be expressed as:

$$K_{sp} = \left[\mathrm{Mg}^{+2}\right]\left[\mathrm{OH}^-\right]^2 = 2.0\times10^{-11}$$

b. At pH = 9.0, $[OH^-] = 10^{-5}$ M (more in the next chapter).
c. Let the concentration of magnesium ion be x, then

$$K_{sp} = 2.0\times10^{-11} = (x)\left(10^{-5}\right)^2$$

d. Then, $[Mg^{+2}] = x = $ 0.2 M
At pH = 11.0, $[OH^-] = 10^{-3}$ M (more in the next chapter).
Let the concentration of magnesium ion be x, then

$$K_{sp} = 2.0\times10^{-11} = (x)\left(10^{-3}\right)^2$$

e. Then, $[Mg^{+2}] = x = 2.0\times10^{-5}$ M
$Mg(OH)_2$ becomes 10,000 times less soluble when pH is raised from 9 to 11.

Discussion:

1. Metal oxides and metal hydroxides are less soluble at higher pHs.
2. One treatment alternative for the removal of heavy metals from an aqueous solution is to raise the solution pH to precipitate out the heavy metals.

7.5.2 Ionic Product versus Solubility Product Constant

Ionic product is the product of the concentrations of ions present in a solution by taking the stoichiometric coefficients into account. Comparing the value of the ionic product with that of the corresponding solubility product constant can tell if a precipitate will occur in the solution. There are three potential scenarios:

- When the ionic product $= K_{sp}$: The system is in equilibrium; the dissolution and the precipitation are occurring at the same rate.
- When the ionic product $> K_{sp}$: precipitation will occur. The solution is "*over-saturated*".
- When the ionic product $< K_{sp}$: precipitation will not occur. The solution is "*under-saturated*".

Example 7.8: Ionic Product and Solubility Product Constant

If a solution ($T = 25°C$) contains 2.0×10^{-2} M of Pb^{+2} and Cl^- ions. Will it form lead chloride ($PbCl_2$) precipitate?

Solution:

a. The dissociation equation for $PbCl_2$ can be written as:

$$PbCl_{2(s)} \leftrightarrow Pb^{+2}_{(aq)} + 2\ Cl^-_{(aq)}$$

b. The ionic product can then be found as:

$$\text{Ionic product} = \left[Pb^{+2}\right]\left[Cl^-\right]^2 = \left(2.0 \times 10^{-2}\right)\left(2.0 \times 10^{-2}\right)^2 = 8 \times 10^{-6}$$

c. Since the ionic product (8×10^{-6}) is smaller than K_{sp} (1.7×10^{-5}), shown in Table 7.2, the precipitation will not occur.

Discussion:

1. The solubility of $[Pb^{+2}] = 1.62 \times 10^{-2}$ M; it was determined in Example 7.6 by using the K_{sp} value of $PbCl_2$ and by adding $PbCl_2$ into pure water.
2. In this example, although $[Pb^{+2}] = 2.0 \times 10^{-2}$ M, which is larger than its solubility, $PbCl_2$ will not precipitate because the $[Cl^-]$ is too low.

7.5.3 Common Ion Effect

In real aqueous solutions, there are many other ionic species present differently from the ones of concern. Taking a saturated $PbCl_2$ solution as an example:

$$PbCl_{2(s)} \leftrightarrow Pb^{+2}_{(aq)} + 2\ Cl^-_{(aq)} \tag{7.10}$$

By adding sodium chloride (NaCl) into this solution, more chloride ions will appear in the solution because of the solubility of NaCl. From Le Chatelier's principle, the reaction equation (i.e., Eq. 7.10) will shift toward the left. Consequently, more Pb^{+2} will precipitate with Cl^- to form $PbCl_{2(s)}$; the solubility of $PbCl_2$ in this aqueous solution will decrease. Since both $PbCl_2$ and NaCl are dissolved in the same solution, the chloride ions are common to both salts. In this system/solution, the chloride ions are the *common ions*. The effect incurred by common ions is called the *common ion effect*.

Example 7.9: Common Ion Effect

A concentrated NaCl solution is added to a saturated solution, prepared by dissolving $PbCl_2$ into water ($T = 25°C$) as in Example 7.6. The chloride ions immediately after the NaCl addition become 0.1 M. What will be the lead ion concentration after a new equilibrium is achieved?

Solution:

a. Let the concentration of lead ion be x in equilibrium, then

$$K_{sp} = 1.7 \times 10^{-5} = (x)\left(10^{-1}\right)^2$$

b. Then, $[Pb^{+2}] = x = 1.7 \times 10^{-3}$ M

The new lead concentration (1.7×10^{-3} M) is approximately ten times less than its original solubility (1.62×10^{-2} M).

Discussion:

1. This example illustrates that the solubility of an ionic compound decreases due to the common ion effect.
2. It should be noted that a slightly soluble ionic compound may become more soluble in a solution containing "non-participating" ions. It is the opposite of the common ion effect, and it is called the *"diverse ion effect"* (or the *salt effect*). Its solubility increases as the ionic strength of the solution increases.

REFERENCES

Aqion (2021). "Activity and Ionic Strength", https://www.aqion.de/site/69, last modified, January 15, 2021.

USEPA (2023). "Indicators: Salinity", United States Environmental Protection Agency, https://www.epa.gov/national-aquatic-resource-surveys/indicators-salinity#, last updated on June 20, 2023.

USGS (2018). "Saline Water and Salinity", U.S. Geological Survey, U.S. Department of Interior, https://www.usgs.gov/special-topics/water-science-school/science/saline-water-and-salinity.

Water Encyclopedia (2023). "Ocean Chemical Processes", https://www.waterencyclopedia.com/Mi-Oc/Ocean-Chemical-Processes.html.

EXERCISE QUESTIONS

1. Classify the following compounds as (i) molecular or ionic compound and (ii) strong, weak, or non-electrolyte.
   a. Toluene ($C_6H_5CH_3$) and 1,1,1-trichloroethane ($C_2H_3Cl_3$)
   b. Ethanol (C_2H_5OH) and formaldehyde (HCHO)
   c. Formic acid (HCOOH) and hydrochloric acid (HCl)
   d. Sodium hydroxide (NaOH) and hydrobromic acid (HBr)

2. Classify the following ionic compounds as soluble or insoluble:
   a. NaF
   b. NH_4NO_3
   c. $Mg(NO_3)_2$
   d. $Mg(CH_3COO)_2$
   e. $CuCl_2$
   f. $BaSO_4$
   g. Ag_2SO_4
   h. KOH
   i. $Ba(OH)_2$
   j. Li_2CO_3
   k. $MgCO_3$
   l. K_2S
   m. Cu_2S
   n. BeO
3. A laboratory analyzed a seawater sample for water quality. The major ion concentrations are the same as those shown in Table 7.1.
   a. Find the sums of the cation and anion concentrations, in mM.
   b. Find the sums of the cation and anion concentrations, in mg/L.
   c. Find the sums of the cation and anion concentrations, in meq/L.
   d. What does the reported data say about the electrical neutrality of this water sample?
4. Find the ionic strength of 0.05 M calcium fluoride (CaF_2) solution.
5. A laboratory analyzed a seawater sample for water quality. The major ion concentrations are the same as those shown in Table 7.1. Estimate the ionic strength of this seawater sample.
6. Use the values of the solubility product constant in Table 7.2 to find the solubility of calcium carbonate ($CaCO_3$) and calcium sulfate ($CaSO_4$) at $T = 25°C$.
7. The solubility product of calcium hydroxide, $Ca(OH)_2$, is 5.0×10^{-6} M^3 at 20°C.
   a. Write/express the equilibrium constant for the reaction shown below.
   b. What is the hydroxyl ion concentration (i.e., $[OH^-]$) at pH = 9, in M?
   c. Estimate the maximum soluble Ca^{+2} concentration in a water sample of pH = 9, in M.
   d. Estimate the maximum soluble Ca^{+2} concentration in a water sample of pH = 11, in M.
   e. Is calcium hydroxide more soluble or less soluble at a higher pH?

$$Ca(OH)_{2(s)} \rightarrow Ca^{+2}_{(aq)} + 2OH^-_{(aq)}$$

8. A solution ($T = 25°C$) contains 1.0×10^{-5} M of Ag^+ and Cl^- ions. Will silver chloride (AgCl) precipitate?
9. A concentrated NaCl solution is added to the saturated solution prepared by dissolving AgCl into water ($T = 25°C$) as in Question #8. The chloride ions immediately after the NaCl addition become 0.1 M. What will be the silver ion concentration after a new equilibrium is achieved?

8 Acid-Base Equilibria in Aqueous Solutions

8.1 ACIDS AND BASES

8.1.1 DEFINITIONS OF ACIDS AND BASES

There are three classic definitions of acids and bases. The earliest is by S. Arrhenius in the 1880s. The *Arrhenius acid* is a substance that can dissociate (or ionize), when dissolved in water, to produce a hydrated hydrogen ion $\left(H^+_{(aq)}\right)$ and an anion. As shown in Eq. (8.1), hydrochloric acid (HCl) is an acid because of its release of H^+.

$$HCl_{(aq)} \rightarrow H^+_{(aq)} + Cl^-_{(aq)} \tag{8.1}$$

The *Arrhenius base* is a substance that can dissociate (or ionize), when dissolved in water, to produce a hydrated hydroxide ion $\left(OH^-_{(aq)}\right)$ and a cation. As shown in Eq. (8.2), sodium hydroxide (NaOH) is a base because of its release of OH^- (the hydroxyl ion or the hydroxide ion).

$$NaOH_{(s)} \rightarrow Na^+_{(aq)} + OH^-_{(aq)} \tag{8.2}$$

Both Eqs. (8.1) and (8.2) show the forward reactions only; it is because HCl is a strong acid and NaOH is a strong base. The dissociation (or ionization) of strong acids or strong bases in water is almost complete (more in later sections). $H^+_{(aq)}$ does not actually exist in water, and it forms a hydronium ion $\left(H_3O^+_{(aq)}\right)$ as:

$$H^+_{(aq)} + H_2O_{(l)} \rightarrow H_3O^+_{(aq)} \tag{8.3}$$

"$H^+_{(aq)}$" and "$H_3O^+_{(aq)}$" are often used interchangeably; and a hydrogen ion (H^+) is also often called as a *proton*.

If ammonia is added into water, the following reaction will occur:

$$NH_{3\,(aq)} + H_2O_{(l)} \leftrightarrow NH^+_{4\,(aq)} + OH^-_{(aq)} \tag{8.4}$$

As shown in Eq. (8.4), although NH_3 does not contain a hydroxyl group, its addition will increase the concentration of hydroxyl ions in water so that water will become more basic. J. Brønsted and T. Lowry in the 1920s introduced more inclusive definitions of acids and bases. A *Brønsted-Lowry acid* is a compound that donates a proton (H^+) to another compound; while a *Brønsted-Lowry base* is a compound that accepts a proton from another compound. In the reaction equation shown in Eq. (8.4), NH_3 receives a proton to become an ammonium ion $\left(NH_4^+\right)$, so it is a Brønsted-Lowry

DOI: 10.1201/9781003502661-8

base (i.e., a base). An acid-base reaction is the transfer of proton(s) from an acid (the *proton donor*) to a base (the *proton acceptor*).

The *Lewis theory*, introduced by G. N. Lewis in 1923, broadens the definitions of acids and bases even further. The Lewis theory revolves around electron transfer, instead of proton transfer. A *Lewis acid* is a compound that can attach itself to an unshared pair of electrons in another molecule (i.e., an electron pair acceptor); while a *Lewis base* is an electron pair donor. Lewis's definition is applicable beyond aqueous solutions. Since this book is focused on aqueous solutions, the two earlier acid/base definitions appear to be sufficient for the scope in this book.

8.1.2 Equilibrium Constants for Acids and Bases

The general reaction for ionization of a weak acid (i.e., only partially ionized/dissociated in water), HA, can be written as:

$$HA_{(aq)} \leftrightarrow H^+_{(aq)} + A^-_{(aq)} \tag{8.5}$$

In this reaction, HA and A^- are a *conjugate acid-base pair*, in which HA is the parent acid and A^- is its conjugate base. The equilibrium constant for this reaction is the *acid ionization constant* (K_a) as:

$$K_a = \frac{[H^+][A^-]}{[HA]} \tag{8.6}$$

Similarly, the general reaction for ionization of a weak base B in water can be written as:

$$B_{(aq)} + H_2O_{(l)} \leftrightarrow BH^+_{(aq)} + OH^-_{(aq)} \tag{8.7}$$

In this reaction, B and BH^+ are a conjugate acid-base pair, in which B is the parent base and BH^+ is its conjugate acid. The equilibrium constant for this reaction is the *base ionization constant* (K_b) as:

$$K_b = \frac{[BH^+][OH^-]}{[B]} \tag{8.8}$$

Pure water will ionize itself slightly into hydrogen and hydroxide ions as:

$$H_2O_{(l)} \leftrightarrow H^+_{(aq)} + OH^-_{(aq)} \tag{8.9}$$

The equilibrium constant for the water ionization is called the *ion-product of water* (K_w) as:

$$K_w = [H^+][OH^-] \tag{8.10}$$

TABLE 8.1
Values of the Ion-Product of Water versus Temperature

T (°C)	K_w	pH of Pure Water
10	0.293×10^{-14}	7.27
20	0.681×10^{-14}	7.08
25	1.008×10^{-14}	7.00
30	1.471×10^{-14}	6.92

As shown in Eq. (8.10), K_w is essentially the product of the hydrogen ion and hydroxyl ion concentrations. K_w values are temperature dependent [Note: K_w will also be affected by the presence of other ionic species in water]. The commonly-mentioned K_w value of 1.0×10^{-14} is for pure water at $T = 25°C$ only. The K_w values at some ambient temperatures are tabulated in Table 8.1.

Equation (8.9) also shows that water can dissociate into a proton (i.e., an acid) and a hydroxide ion (i.e., a base). Substances that can donate and accept protons are said to be *amphiprotic*. We talked about some oxides are amphoteric because they can act as acids and bases depending on the medium. The main difference is that amphiprotic refers to the ability to donate or accept protons; while amphoteric refers to the ability to act as an acid or a base. All amphiprotic species are amphoteric; however, not all amphoteric species are amphiprotic. For example, bicarbonate ion $\left(HCO_3^-\right)$ is amphiprotic because it can act as an acid to release a proton (see Eq. 8.11) and as a base to accept a proton (see Eq. 8.12).

$$HCO^-_{3(aq)} \rightarrow H^+_{(aq)} + CO^{=}_{3(aq)} \tag{8.11}$$

$$HCO^-_{3(aq)} + H^+_{(aq)} \rightarrow H_2CO_{3(aq)} \tag{8.12}$$

Bicarbonate ion is also amphoteric because it can work as an acid and as a base. As shown in Section 3.1, zinc oxide is amphoteric because it can react with an acid as a base (see Eq. 3.4) and with a base as an acid (see Eq. 3.5). However, it is not amphiprotic because it does not release or accept proton(s).

$$ZnO_{(s)} + 2HCl_{(l)} \rightarrow ZnCl_{2(s)} + H_2O_{(l)} \tag{3.4}$$

$$ZnO_{(s)} + 2OH^-_{(aq)} \rightarrow ZnO^{-2}_{2\,(aq)} + H_2O_{(l)} \tag{3.5}$$

8.1.3 Relationship between the Acid Ionization Constant and the Base Ionization Constant

As shown in Eq. (8.5), HA is the parent acid and A^- is its conjugate base. A reaction equation can be written with A^- as a base, then

$$A^-_{(aq)} + H_2O_{(l)} \leftrightarrow OH^-_{(aq)} + HA_{(aq)} \tag{8.13}$$

$$K_b = \frac{[OH^-][HA]}{[A^-]} \tag{8.14}$$

Multiplying Eq. (8.14) with Eq. (8.6), one can obtain:

$$K_a \times K_b = [H^+][OH^-] = K_w \tag{8.15}$$

Consequently, the acid ionization constant (K_a) of an acid and the base ionization constant (K_b) of its conjugate base has the following relationship:

$$K_a \times K_b = K_w \tag{8.16}$$

8.1.4 Monoprotic versus Polyprotic

A *monoprotic acid* is an acid that can give only one proton, and a *monoprotic base* can only accept one proton. For example, hydrochloric acid is a monoprotic acid and sodium hydroxide is a monoprotic base as shown below.

$$HCl_{(aq)} \leftrightarrow H^+_{(aq)} + Cl^-_{(aq)} \tag{8.17}$$

$$NaOH_{(aq)} \leftrightarrow OH^-_{(aq)} + Na^+_{(aq)} \tag{8.18}$$

A *polyprotic acid* is an acid that can give more than one proton, and a *polyprotic base* can accept more than one proton. For example, sulfuric acid is a polyprotic acid and magnesium hydroxide is a polyprotic base as shown below.

$$H_2SO_{4(l)} \rightarrow H^+_{(aq)} + HSO^-_{4\ (aq)} \tag{8.19}$$

$$HSO^-_{4\ (aq)} \leftrightarrow H^+_{(aq)} + SO^=_{4\ (aq)} \tag{8.20}$$

$$Mg(OH)_{2(aq)} \leftrightarrow 2\ OH^-_{(aq)} + Mg^{+2}_{(aq)} \tag{8.21}$$

From Eqs. (8.19) and (8.20), we can also say that bisulfate ion $\left(HSO^-_4\right)$ is a monoprotic acid as well as a monoprotic base, because it can give or accept one proton; while sulfate ion $\left(SO^=_4\right)$ is a polyprotic base because it can accept two protons. Sulfuric acid (H_2SO_4) is a <u>di</u>protic acid; while phosphoric acid (H_3PO_4) is a <u>tri</u>protic acid.

Polyprotic acids and bases have multiple acid or base dissociation constants, and they are named as K_{a1}, K_{a2}, K_{a3} or K_{b1}, K_{b2}, and K_{b3}.

8.2 pH

pH is an expression of the hydrogen ion concentration in a solution, and it is defined as:

$$pH = -\log_{10}[H^+];\ \text{or}\ [H^+] = 10^{-pH} \tag{8.22}$$

TABLE 8.2
Concentrations of H^+ and OH^- in water at different pHs (T=25°C)

pH	$[H^+]$ (From definition of pH)	$[OH^-]$ (Calculated from $K_w = [H^+][OH^-] = 10^{-14}$)
0	1.0 M	10^{-14} M
1	10^{-1} M	10^{-13} M
4	10^{-4} M	10^{-10} M
7	10^{-7} M	10^{-7} M
10	10^{-10} M	10^{-4} M
14	10^{-14} M	1.0 M

Similarly, pOH is defined as:

$$\text{pOH} = -\log_{10}\left[\text{OH}^-\right];\ \text{or}\ \left[\text{OH}^-\right] = 10^{-\text{pOH}} \tag{8.23}$$

The prefix "p" in pH or pOH is a math operator, "$-\log_{10}$". By applying this math operator to both sides of Eq. (8.10), then:

$$\text{pH} + \text{pOH} = \text{p}K_w \tag{8.24}$$

$$\text{pH} + \text{pOH} = \text{p}K_w = 14.0\ @T = 25\ °\text{C} \tag{8.25}$$

The "*p*" is also commonly used as a prefix for an equilibrium constant (e.g., pK_w in Eq. 8.24). Thus, Eq. (8.16) can be rewritten as:

$$\text{p}K_a + \text{p}K_b = \text{p}K_w \tag{8.26}$$

Because pH (and pOH) is a logarithmic function, one unit change in pH value implies a ten-time change in $[H^+]$. Table 8.2 tabulates concentrations of $[H^+]$ and $[OH^-]$ in water versus pH at T=25°C.

At T=25°C and pH=7, $[H^+] = [OH^-] = 10^{-7}$ M. Since these two concentrations are equal, we say the water is neutral at pH=7. When $[H^+] > [OH^-]$, we say the water is acidic; vice versa, the water is basic when $[OH^-] > [H^+]$. It should be noted that the neutral pH will not be 7.0 if $T \neq 25$°C. In addition, the pH values can be outside the range of 0–14.

Example 8.1: pH of Pure Water at Different Temperatures

Estimate the pH of pure water @ T=20°C.

Solution:

a. From Table 8.1, K_w of pure water=0.681 ×10^{-14} @ T=20°C.
b. For pure water, $[H^+] = [OH^-]$.
c. Thus, $K_w = 0.681 \times 10^{-14} = [H^+][OH^-] = [H^+]^2 \rightarrow [H^+] = 8.25 \times 10^{-8}$M

d. From the definition of pH (Eq. 8.22),

$$pH = -\log_{10}\left[H^+\right] = -\log_{10}\left(8.25\times10^{-8}\right) = 7.08$$

Discussion:

1. Since pure water should be neutral, the neutral pH of pure water at 20°C = 7.08.
2. As shown in Table 8.1, the neutral pH of pure water decreases as temperature increases, with a value of 7.0 at 25°C.
3. To be precise, it is incorrect to say the neutral pH of a water sample = 7.0 because the $pH_{neutral}$ depends on the K_w value of that water sample. It is also incorrect to say a water sample is acidic because its pH <7.0. A water sample is acidic, only if its $[H^+] > [OH^-]$.

Example 8.2: pH of Water after Dilution

The pH of a water sample is 3.0. If the water sample is diluted with pure water of the same volume, estimate the pH of the diluted water sample.

Solution:

a. From the definition of pH (Eq. 8.22), the proton concentration of the initial solution:

$$\left[H^+\right] = 10^{-pH} = 10^{-3}M$$

b. After the equal-volume dilution, the final proton concentration will be equal to half of its initial concentration; that is 5×10^{-4} M. Then use Eq. (8.22) again to find the new pH:

$$pH = -\log_{10}\left[H^+\right] = -\log_{10}\left(5\times10^{-4}\right) = 3.30$$

Discussion:

1. As expected, the pH of the solution would increase after dilution because the solution becomes less acidic.
2. The $[H^+]$ of the dilution water should be around 10^{-7} M, which is much smaller than the $[H^+]$ of the initial solution, so it is negligible in finding the $[H^+]$ of the final solution.

8.3 STRONG AND WEAK ACIDS

Not all the acids and bases will ionize/dissociate to the same extent; in other words, they do not have equal strength. The strength of acids and bases depends on their ability to dissociate in water. A *strong acid* (e.g., hydrochloric acid) will dissociate almost completely and release H^+ ions when added to water; similarly, a strong base (e.g., sodium hydroxide) will dissociate almost completely and release OH^- ions into the solution.

TABLE 8.3
Chemical Formula and Names in Aqueous Solutions of Common Strong Acids

Formula	Name
HBr	Hydrobromic acid
HCl	Hydrochloric acid
HI	Hydroiodic acid
HNO_3	Nitric acid
HCl_4	Perchloric acid
$HClO_3$	Chloric acid
H_2SO_4	Sulfuric acid

8.3.1 Common Strong Acids

Table 8.3 tabulates seven common strong acids with their chemical formula and names in aqueous solutions.

All seven strong acids shown in Table 8.3 are inorganic acids. HBr, HCl, and HI are *binary acids* because each has a hydrogen bonded to another chemical element, usually a non-metal. The other four (i.e., HNO_3, $HClO_4$, $HClO_3$, and H_2SO_4) are ternary acids (or known as oxoacid). An *oxoacid* (or *oxyacid*) is an acid that has elements of hydrogen and oxygen, along with another element, usually non-metal. It is actually a compound composed of hydrogen ions (H^+) bonded to a polyatomic anion (e.g., NO_3^-, ClO_4^-, ClO_3^-, and, $SO_4^=$ here).

The central element of oxoacids may have several oxidation states (see more on oxidation state in Chapter 9), the names of these oxoacids are related to their oxidation states. The general rules for nomenclatures of oxoacids are:

The central atom has only two oxidation states:

- the oxoacid with the higher oxidation state is indicated by the suffix "-ic".
- the one with the lower oxidation state is indicated by the suffix "-ous".
- For example, the oxidation stats of sulfur (S) in sulfuric acid (H_2SO_4) and sulfurous acid (H_2SO_3) are +6 and +4, respectively. Similarly, the oxidation states of phosphorus (P) in phosphoric acid (H_3PO_4) and phosphorus acid (H_3PO_3) are +5 and +3, respectively.

The central atom has more than two oxidation states:

- the oxoacid with the highest oxidation state is indicated by the prefix "per-".
- the one with the lowest oxidation state is indicated by the prefix "hypo-".
- The names of oxoacids of chlorine are shown in Table 8.4, as an example.

TABLE 8.4
Names of Oxoacids and Oxoanions of Chlorine

Formula	Oxidation State	Name of Acid	Name of Oxoanion
$HClO_4$	+7	Perchloric acid	Perchlorate
$HClO_3$	+5	Chloric acid	Chlorate
$HClO_2$	+3	Chlorous acid	Chlorite
$HClO$	+1	Hypochlorous acid	Hypochlorite

Oxoanions, derived from "-ic" acids, have the ending of "-ate"; and those from "-ous" acids have the ending of "-ite" (see Table 8.4). In addition, the prefix "bi-" is used to indicate the presence of a single hydrogen atom. The common ones include bicarbonate $\left(HCO_3^-\right)$, bisulfate $\left(HSO_4^-\right)$, and bisulfite $\left(HSO_3^-\right)$ ions.

The K_{a1} values of strong acids are large; for example, the reported K_{a1} values of HCl, HNO_3, and H_2SO_4 are approximately 1.0×10^6, 20, and 1,000, respectively. It implies that these strong acids will dissociate almost completely in an aqueous solution under normal pHs.

8.3.2 Common Weak Acids

Table 8.5 tabulates common weak acids with their chemical formula and names in aqueous solutions as well as their acid ionization constants at $T = 25°C$. Values of acid ionization constants can be readily found in handbooks of chemistry or chemical engineering as well as on the internet; however, the reported values are not always the same.

TABLE 8.5
Chemical Formula and Names in Aqueous Solutions of Common Weak Acids with Their K_a Values ($T = 25°C$)

Formula	Name	K_{a1}	K_{a2}	K_{a3}
HF	Hydrofluoric acid	7.1×10^{-4}		
HCN	Hydrocyanic acid	4.9×10^{-10}		
CH_3COOH	Acetic acid	1.8×10^{-5}		
C_6H_5COOH	Benzoic acid	6.5×10^{-5}		
HNO_2	Nitrous acid	4.5×10^{-4}		
HClO	Hypochlorous acid	3.1×10^{-8}		
$HClO_2$	Chlorous acid	1.1×10^{-2}		
H_2SO_3	Sulfurous acid	1.7×10^{-2}	6.3×10^{-8}	
H_2CO_3	Carbonic acid	4.5×10^{-7}	4.7×10^{-11}	
H_3PO_4	Phosphoric acid	7.0×10^{-3}	6.2×10^{-8}	4.8×10^{-13}

8.3.3 Speciation and Degree of Ionization/Dissociation

Chlorine gas used to be the principal disinfectant used in water and wastewater treatment. When chlorine gas is added to water, it will be hydrolyzed into hypochlorous acid (HOCl) as:

$$Cl_{2(g)} + H_2O_{(l)} \rightarrow HOCl_{(aq)} + H^+_{(aq)} + Cl^-_{(aq)} \tag{8.27}$$

Hypochlorous acid is a weak acid, and it will ionize/dissociate into proton and hypochlorite ion (OCl^-) in water as:

$$HOCl_{(aq)} \rightarrow H^+_{(aq)} + OCl^-_{(aq)} \tag{8.28}$$

Both hypochlorous acid and hypochlorite ion have the disinfecting power, and they are called *free chlorine* (versus combined chlorine) in water/wastewater disinfection. The free chlorine concentration in water is the sum of the concentrations of these two species as:

$$\text{Free Chlorine} = [HOCl] + [OCl^-] \tag{8.29}$$

Although both free chlorine species have disinfecting power, the strength of hypochlorous acid is much stronger. With that, the effectiveness of disinfection would be better if the water pH is lower so that the hypochlorous acid will be the dominant species (see Example 8.3 below).

The *percent ionization/dissociation* of a weak acid or a weak base is:

$$\text{Percent Ionization} = \frac{[\text{Concentration of Ionized Weak Acid or Base}]}{[\text{Initial Concentration of the Acid or Base}]} \times 100 \tag{8.30}$$

Example 8.3: Speciation and Percent Ionization of Weak Acid

For the reaction equation shown below ($T = 25°C$):

$$HOCl_{(aq)} \rightarrow H^+_{(aq)} + OCl^-_{(aq)} \qquad pK_a = 7.50$$

a. At what pH, the concentrations of HOCl and OCl^- are equal?
b. What would be the ratio of $[HOCl]/[OCl^-]$ when pH = 8.5?
c. What would be the ratio of $[HOCl]/[OCl^-]$ when pH = 6.5?
d. What would be the percent ionization of HOCl at pH = 6.5, 7.5, and 8.5?

Solution:

a. The equilibrium constant expression of this reaction equation is:

$$K_a = 10^{-7.5} = \frac{[H^+][OCl^-]}{[HOCl]}$$

When $[HOCl] = [OCl^-]$, $[H^+] = K_a = 10^{-7.5}$

Thus, pH = 7.5.

b. At pH = 8.5, $[H^+] = 10^{-8.5}$, then

$$K_a = 10^{-7.5} = \frac{(10^{-8.5})[OCl^-]}{[HOCl]}$$

Thus, $[HOCl]/[OCl^-] = 0.1$

c. Taking an approach similar to that in part (b), at pH = 6.5 $[HOCl]/[OCl^-] = 10.0$

d. The percent ionization can be found as the ratio of $[HOCl]/\{[OCl^-] + [HOCl]\}$. Thus, the percentages of ionization are 9%, 50%, and 91% at pH = 6.5, 7.5, and 8,5, respectively.

Discussion:

1. A pK_a value, instead of a K_a value, is used in this example to facilitate the illustration.
2. When pH = pK_a, the percent ionization is equal to 50%.
3. As illustrated, the [HOCl] is ten times higher than that of $[OCl^-]$ at pH = 6.5 (the percent ionization = 9%). The relationship is reversed at pH = 8.5. Consequently, disinfection will be more effective at lower pHs with the same free chlorine dose.

8.3.4 pH of a Weak Acid Solution

With a given K_a value, the pH of a weak acid solution can be calculated. The common approach is to set up an ICE table (where I = initial concentrations, C = changes of concentrations, and E = concentrations at equilibrium), in which the change is assigned with a variable x, or a multiple of x. The equilibrium concentrations are then inserted to the equilibrium constant to solve for x; the obtained value can then be used to find other equilibrium concentrations (see Example 8.4 below).

Example 8.4: Calculate the pH of a Weak Acid Solution

Vinegar is roughly 3 to 9% acetic acid (CH_3COOH) by volume. A vinegar has an acetic acid concentration of 60 g/L.

a. Illustrate the molar acetic acid concentration of this vinegar is 1.0 M.
b. Estimate the pH of this vinegar.

$$CH_3COOH_{(l)} \leftrightarrow CH_3COO^-_{(aq)} + H^+_{(aq)} \qquad K_a = 1.8 \times 10^{-5} \text{M} @ 25°\text{C}$$

Solution:

a. MW of CH_3COOH = 60, and $[CH_3COOH]_{\text{initial}}$ = 60 g/L ÷ 60 g/mole = 1 mole/L = 1.0 M

b. Set up an ICE table, with the change of the acetic acid concentration as $-x$:

	$[CH_3COOH]$	$[CH_3COO^-]$	$[H^+]$
Initial	1.0	0	0
Change	$-x$	x	x
Equilibrium	$1.0-x$	x	x

$$K_a = 1.8 \times 10^{-5} = \frac{[CH_3COO^-][H^+]}{[CH_3COOH]} = \frac{(x)(x)}{(1-x)}$$

c. Assuming x is much smaller than 1 so that $(1-x) \approx 1.0$, then $x = 4.24 \times 10^{-3}$ ($x \ll 1.0$, so the assumption is valid).

d. At equilibrium, $[H^+] = x = 4.24 \times 10^{-3}$ M → pH = 2.37.

Discussion:

1. The approach is straightforward.
2. The pH of the solution of a strong acid (e.g., HCl) with a concentration of 1.0 M would be very close to 0 (i.e., pH = $-\log[H^+] = -\log(1.0) = 0$).

8.4 STRONG BASES AND WEAK BASES

8.4.1 Definitions of Strong and Weak Bases

Similar to strong and weak acids, bases can be considered as strong or weak, depending on the extent of ionization in an aqueous solution. A *strong base* is a base that ionizes/dissociates completely in an aqueous solution. The most common strong bases are hydroxides of alkali metals and alkaline earth metals (see Table 8.6). Some of these oxides are not as strong because they are not very soluble in water. However, the portion that dissolved will dissociate completely and release hydroxyl ions.

A *weak base* is a base that ionizes only partially in an aqueous solution. Ammonia

TABLE 8.6
Chemical Formula and Names in Aqueous Solutions of Common Strong Bases

Formula	Name
LiOH	Lithium hydroxide
NaOH	Sodium hydroxide
KOH	Potassium hydroxide
$Ca(OH)_2$	Calcium hydroxide
$Sr(OH)_2$	Strontium hydroxide
$Ba(OH)_2$	Barium hydroxide

(NH_3) is one of the most commonly-mentioned weak bases in an aqueous solution;

TABLE 8.7
Chemical Formula and Names in Aqueous Solutions of Common Weak Bases with Their K_b Values ($T = 25°C$)

Formula	Name	K_b
Ammonia	NH_3	1.8×10^{-5}
CH_3NH_2	Methyl amine	5.0×10^{-4}
C_5H_5N	Pyridine	1.6×10^{-9}
$CO_3^{=}$	Carbonate ion	2.1×10^{-4}
CH_3COO^-	Acetate ion	5.6×10^{-10}

when dissolved, it could accept a proton from water to form an ammonium ion (NH_4^+). NH_4^+ is the conjugate acid of the base (i.e., NH_3) which releases a hydroxide ion as:

$$NH_{3\,(aq)} + H_2O_{(l)} \leftrightarrow NH_{4\,(aq)}^+ + OH_{(aq)}^- \tag{8.4}$$

The base ionization constant (K_b), which is the equilibrium constant of a base, of ammonia can be written as:

$$K_b = \frac{[NH_4^+][OH^-]}{[NH_3]} \tag{8.31}$$

Table 8.7 tabulates the chemical formula and names of some weak bases in aqueous solutions with their K_b values.

The numerical value of K_b is an indicator of the strength of a base. The larger the K_b value, the stronger the base. It is also interesting to note that the carbonate ion (conjugate base of bicarbonate ion) and acetate ion (the conjugate base of the acetic acid) appear in Table 8.7. The conjugate base of a weaker acid would be a stronger base itself (see Example 8.5 below).

Example 8.5: Relationship between an Acid and Its Conjugate Base

a. Use the data in Table 8.5 to write the expression for the acid ionization constant of bicarbonate.
b. Use the data in Table 8.7 to write the expression for the base ionization constant of carbonate.
c. What is the relationship between these two ionization constants?

Solution:

a. From Table 8.5, the K_{a2} value of carbonic acid (H_2CO_3) is equal to 4.7×10^{-11}.

$$HCO_{3(aq)}^- \rightarrow H_{(aq)}^+ + CO_{3(aq)}^=$$

Then from Eq. (8.6),

$$K_{a2} = \frac{[H^+][CO_3^=]}{[HCO_3^-]} = 4.7\times10^{-11}$$

b. From Table 8.7, the K_b value of carbonate ion is equal to 2.1×10^{-4}.

$$CO^=_{3(aq)} + H_2O_{(l)} \rightarrow HCO^-_{3(aq)} + OH^-_{(aq)}$$

Then from Eq. (8.6),

$$K_b = \frac{[HCO_3^-][OH^-]}{[CO_3^=]} = 2.1\times10^{-4}$$

c. Multiplying the two ionization constants, $K_{a2}\times K_b = K_w = 1.0\times10^{-14}$.

Discussion:

1. The relationship between two ionization constants follows that described in Eq. (8.16).
2. The K_a value of acetic acid is larger than that of bicarbonate ion (Table 8.5), and it implies that acetic acid is a stronger acid. The K_b value of acetate ion is smaller than that of carbonate ion (Table 8.7), and it implies that carbonate ion is a stronger base than acetate ion. In other words, the conjugate base of a weaker acid will be a stronger base.

8.4.2 pH of a Weak Base Solution

With a given K_b value, the pH of a weak base solution can also be calculated, using the same approach as that for a weak acid solution by setting up an ICE table (where I=initial concentrations, C=changes of concentrations, and E=concentrations at equilibrium), in which the change is assigned with a variable x, or a multiple of x. The equilibrium concentrations are then inserted to the equilibrium constant to solve for x; the obtained value can then be used to find other equilibrium concentrations (see Example 8.6 below).

Example 8.6: Calculate the pH of a Weak Base Solution

Estimate the pH of 1.0 M ammonia solution at 25°C.

$$NH_{3\,(aq)} + H_2O_{(l)} \leftrightarrow NH^+_{4\,(aq)} + OH^-_{(aq)} \quad K_b = 1.8\times10^{-5}\,M@25°C$$

Solution:

Set up an ICE table, with the change of ammonia concentration as –x:

	$[NH_3]$	$[NH_4^+]$	$[OH^-]$
Initial	1.0	0	0
Change	$-x$	x	x
Equilibrium	$1.0-x$	x	x

$$K_a = 1.8\times10^{-5} = \frac{[NH_4^+][OH^-]}{[NH_3]} = \frac{(x)(x)}{(1-x)}$$

a. Assuming x is much smaller than 1 so that $(1-x)\approx1.0$, then $x=4.24\times10^{-3}$ ($x \ll 1.0$, so the assumption is valid).

b. At equilibrium, $[OH^-]=x=4.24\times10^{-3}$ M → pOH = 2.37 → pH = 11.43.

Discussion:

1. The approach is straightforward.
2. It is interesting to note that K_a of acetic acid (Example 8.4) and K_b of ammonia (this example) have the same numeric value.

8.5 NEUTRALIZATION BETWEEN ACIDS AND BASES

A *neutralization reaction* is a reaction in which an acid and a base react in an aqueous solution. The final products will be water and salt. The *salt* is an ionic product composed of an anion from the acid and a cation from the base. For example, the reaction equation between sulfuric acid and sodium hydroxide can be written as

$$H_2SO_{4(aq)} + 2\ NaOH_{(aq)} \rightarrow Na_2SO_{4(aq)} + H_2O_{(l)} \tag{8.32}$$

In this case, sodium sulfate is very soluble in water, and it should be present in the ionic forms of Na^+ and $SO_4^=$. Example 8.7 illustrates calculations for neutralization of a strong acid and a strong base.

Example 8.7: Neutralization of an Acid and a Base

98 mg of sulfuric acid (H_2SO_4) is added to ten liters of water at 25°C.
[Note: Sulfuric acid is a strong acid; $H_2SO_4 \rightarrow 2\ H^+ + SO_4^=$]

a. Moles of H^+ in this solution = __________ g-moles.
b. Molar concentration of H^+ in this solution = __________ M
c. pH of this solution = __________

Concentrated sodium hydroxide (NaOH) is to be added to neutralize the acidic solution above.
[Note: Sodium hydroxide is a strong base; $NaOH \rightarrow Na^+ + OH^-$]

d. Moles of OH^- needed for neutralization = __________ g-moles
e. Moles of NaOH needed for neutralization = __________ g-moles
f. Mass of NaOH needed for the neutralization = __________ mg

Solution:

a. MW of $H_2SO_4 = 98$
   Moles of $H_2SO_4 = (98 \times 10^{-3} g) \div 98$ g/mole $= 10^{-3}$ mole
   Moles of $H^+ = (10^{-3}) \times 2 = 2.0 \times 10^{-3}$ moles

b. $[H^+] = (2.0 \times 10^{-3}$ moles$) \div 10$ L $= 2.0 \times 10^{-4}$ M

c. pH of the solution $= -\log(2.0 \times 10^{-4}) = 3.7$

d. Moles of OH^- needed = moles of H^+ present $= 2 \times 10^{-3}$ moles

e. Moles of NaOH needed = Moles of OH^- needed $= 2 \times 10^{-3}$ moles

f. MW of NaOH = 40
   Mass of NaOH $= (2 \times 10^{-3}$ moles$)(40{,}000$ mg/mole$) = 80$ mg.

Discussion:
If the stoichiometric amount of 80 mg NaOH is added to neutralize the solution and the reaction is complete, the final solution should be neutral.

8.6 BUFFER SOLUTION

8.6.1 Definition and Types

A *buffer* is a solution that can resist pH change upon an addition of acidic or basic substance. It is a mixture of a weak acid and a salt containing its conjugate base; or a weak base and a salt containing its conjugate acid. Upon the addition of a small amount of acid or base, it can neutralize it to maintain the pH of the solution without an abrupt change in pH.

There are acidic and basic buffers. An *acidic buffer solution* has an acidic pH, and it is a mixture of a weak acid (e.g., acetic acid, CH_3COOH) and a salt containing its conjugate base (e.g., sodium acetate, CH_3COONa). An aqueous solution of an equal concentration of these two in the solution has a pH of 4.74 (i.e., the pK_a value of acetic acid). On the other hand, a *basic buffer solution* has a basic pH, and it is a mixture of a weak base (e.g., NH_3) and a salt containing its conjugate acid (e.g., ammonium chloride, NH_4Cl). An aqueous solution of an equal concentration of these two in the solution has a pH of 9.26 (= $14.0 - pK_b$ of ammonia).

8.6.2 Henderson-Hasselbalch Approximation

The *Henderson–Hasselbach approximation* is a method to approximate pH (or pOH) of a buffer solution.

For an acidic buffer solution:

$$HA_{(aq)} \leftrightarrow H^+_{(aq)} + A^-_{(aq)} \tag{8.5}$$

$$pH_{\text{acidic buffer solution}} \approx pK_a + \log_{10}\frac{[A^-]}{[HA]} \tag{8.33}$$

For a basic buffer solution:

$$B_{(aq)} + H_2O_{(l)} \leftrightarrow BH^+_{(aq)} + OH^-_{(aq)} \tag{8.7}$$

$$pOH_{\text{basic buffer solution}} \approx pK_b + \log_{10}\frac{[BH^+]}{[B]} \tag{8.34}$$

Example 8.8: pH of an Acidic Buffer

a. What is the pH of a buffer containing 0.5 M of acetic acid and sodium acetate?
b. What is the pH of a buffer containing 0.5 M of acetic acid and 0.7 M of sodium acetate?

Solution:

a. From Table 8.5, K_a of acetic acid $= 1.8 \times 10^{-5} \rightarrow pK_a = 4.74$
   Use Eq. (8.33),

$$pH_{\text{acidic buffer solution}} \approx pK_a + \log_{10}\frac{[A^-]}{[HA]} = 4.74 + \log_{10}\frac{(0.5)}{(0.5)} = 4.74$$

b. Use Eq. (8.33) again,

$$pH_{\text{acidic buffer solution}} \approx pK_a + \log_{10}\frac{[A^-]}{[HA]} = 4.74 + \log_{10}\frac{(0.7)}{(0.5)} = 4.89$$

Example 8.9: pH of a Basic Buffer

a. What is the pH of a buffer containing 0.5 M of ammonia (NH_3) and ammonium chloride?
b. What is the pH of a buffer containing 0.5 M of ammonia and 0.7 M of ammonium chloride?

Solution:

a. From Table 8.7, K_a of ammonia $= 1.8 \times 10^{-5} \rightarrow pK_b = 4.74$
   Use Eq. (8.34),

$$pOH_{\text{basic buffer solution}} \approx pK_b + \log_{10}\frac{[BH^+]}{[B]} = 4.74 + \log_{10}\frac{(0.5)}{(0.5)} = 4.74$$

$$pH = 14 - pOH = 9.26.$$

b. Use Eq. (8.34) again,

$$pOH_{\text{basic buffer solution}} \approx pK_b + \log_{10}\frac{[BH^+]}{[B]} = 4.74 + \log_{10}\frac{(0.7)}{(0.5)} = 4.89$$

$$pH = 14 - pOH = 9.11.$$

8.7 ALKALINITY AND CARBONATE/BICARBONATE BUFFER SYSTEM

8.7.1 ALKALINITY

Alkalinity is an important water quality parameter, and it is a measure of the capacity of water to neutralize acids. Alkalinity comes from ions in water that can incorporate protons into them so that the protons are not available to work as a free acid to lower the solution pH. Thus, alkalinity is also known as the "*buffering capacity*" of water. A sufficient level of alkalinity is important in ambient water bodies; with the acid-neutralizing capacity, a waterbody can resist pH change which is critical to aquatic life, for example.

Many ionic species can neutralize acids; for example, a bicarbonate ion $\left(HCO_3^-\right)$ and a carbonate ion $\left(CO_3^=\right)$ can neutralize one proton and two protons, respectively, to form a carbonic acid molecule $\left(H_2CO_3\right)$. In the meantime, a phosphate ion $\left(PO_4^{-3}\right)$ can neutralize three protons to form a phosphoric acid molecule (H_3PO_4). *Total alkalinity* of a water sample is determined by measuring the amount of acid needed to bring the sample pH down to 4.2. At this pH, all the alkaline compounds in the sample are completely "used up", and the results are typically reported as milligrams per liter of calcium carbonate (mg/L as $CaCO_3$) – more about this unit in Chapter 11.

Many other species (e.g., hydroxyl, borate, and silicate ions) in water can also neutralize acids. However, alkalinity is often referred to as concentrations of carbonate and bicarbonate ions. This is because they typically present in much higher concentrations in ambient water bodies than other species with buffering capacity. One definition of alkalinity (in *M*) is:

$$\text{Alkalinity (in } M) = 2\left[CO_3^=\right]+\left[HCO_3^-\right]+\left[OH^-\right]-\left[H^+\right] \tag{8.35}$$

In Eq. (8.35), a multiple of 2 is added to the concentration of carbonate ion; it is because each carbonate ion can neutralize two protons. In addition, a negative sign is added to the concentration of protons because it adds acidity to water.

8.7.2 BICARBONATE/CARBONATE BUFFER SYSTEM

As mentioned, the bicarbonate/carbonate buffer system is the most significant buffer in our natural water bodies; it regulates the pH of the water bodies. The following equations illustrate how the system works.

$$CO_{2(g)} \xrightarrow{H_2O} CO_{2(aq)} \tag{8.36}$$

$$CO_{2(aq)} + H_2O_{(l)} \rightarrow H_2CO_{3(aq)} \tag{8.37}$$

$$H_2CO_{3(aq)} \leftrightarrow H^+_{(aq)} + HCO^-_{3\,(aq)} \quad pK_{a1} \approx 6.3 \tag{8.38}$$

$$HCO^-_{3\,(aq)} \leftrightarrow H^+_{(aq)} + CO^=_{3\,(aq)} \quad pK_{a2} \approx 10.3 \tag{8.39}$$

$$Ca^{+2}_{(aq)} + CO^=_{3(aq)} \leftrightarrow CaCO_{3(s)} \tag{8.40}$$

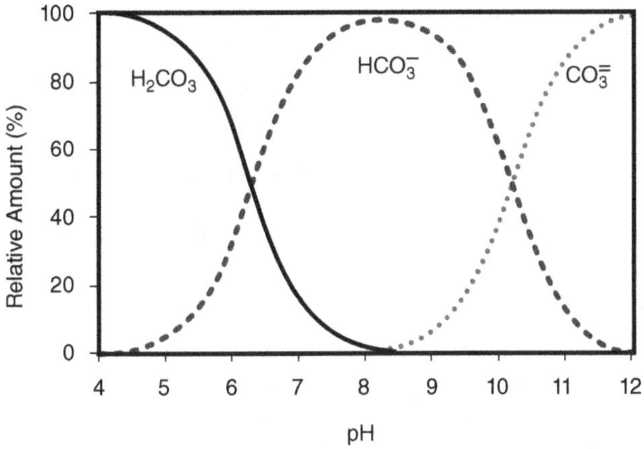

FIGURE 8.1 Relative concentrations of carbonate species vs. pH.

If water exposed to the atmosphere, some carbon dioxide in the atmosphere ($CO_{2(g)}$) will get dissolved into water to form $CO_{2(aq)}$ (Eq. 8.36). It will then react with water to form carbonic acid (H_2CO_3) (Eq. 8.37). Equations (8.38) and (8.39) are the reaction equations among three carbonate species (i.e., H_2CO_3, HCO_3^-, and $CO_3^=$). Minerals within the system can also contribute to the buffering capacity. For example, when acid enters an aquifer containing calcite (i.e., a mineral consisting largely of $CaCO_3$), the calcite would dissolve and release carbonate ions (see Eq. 8.40). Within the normal pH range, bicarbonate ion will be the dominate carbonate species (see Figure 8.1 and the examples below).

Example 8.11: Speciation of Carbonate Species in Water

As discussed, the alkalinity of natural water is mainly due to the presence of bicarbonate and carbonate ions. For a natural water system, the total molar concentration of carbonate species (i.e., the sum of H_2CO_3, HCO_3^-, and CO_3^{-2}) is 0.1 M. Find the concentrations of each individual carbonate species at 25°C, when (a) pH = 6.3, (b) pH = 10.3, and (c) pH = 7.3 (Assume $pK_{a1} = 6.3$ and $pK_{a2} = 10.3$ for this water).

Solution:

From Eqs. (8.38) and (8.39), the two acid association equilibrium constants can be written as

$$K_{a1} = 10^{-6.3} = \frac{[H^+][HCO_3^-]}{[H_2CO_3]} \tag{8.41}$$

$$K_{a2} = 10^{-10.3} = \frac{[H^+][CO_3^=]}{[HCO_3^-]} \tag{8.42}$$

a. At pH = 6.3 (i.e., $[H^+] = 10^{-6.3}$ M), $[H_2CO_3] = [HCO_3^-]$ from Eq. (8.41) and $[CO_3^=] = 10^{-4} \times [HCO_3^-]$ from Eq. (8.42).

Total carbonate concentration = 0.1 M = $[H_2CO_3] + [HCO_3^-] + [CO_3^=] = 2.0001 \times [HCO_3^-]$

Thus, $[H_2CO_3] = [HCO_3^-] \approx 0.05$ M; and $[CO_3^=] \approx 5.0 \times 10^{-6}$ M

b. At pH = 10.3 (i.e., $[H^+] = 10^{-10.3}$ M), $[CO_3^=] = [HCO_3^-]$ from Eq. (8.42) and $[H_2CO_3] = 10^{-4} \times [HCO_3^-]$ from Eq. (8.41).

Total carbonate concentration = 0.1 M = $[H_2CO_3] + [HCO_3^-] + [CO_3^=] = 2.0001 \times [HCO_3^-]$

Thus, $[CO_3^=] = [HCO_3^-] \approx 0.05$ M; and $[H_2CO_3] \approx 5.0 \times 10^{-6}$ M

c. At pH = 7.3 (i.e., $[H^+] = 10^{-7.3}$ M), $[H_2CO_3] = 0.1 \times [HCO_3^-]$ from Eq. (8.41) and $[CO_3^=] = 0.001 \times [HCO_3^-]$ from Eq. (8.42).

Total carbonate concentration = 0.1 M = $[H_2CO_3] + [HCO_3^-] + [CO_3^=] = 1.101\,[HCO_3^-]$

Thus $[HCO_3^-] \approx 0.091$ M; $[H_2CO_3] \approx 0.009$ M; and $[CO_3^=] \approx 9.0 \times 10^{-5}$ M

Discussion:

1. When pH = pK_{a1}, $[H_2CO_3] = [HCO_3^-]$; when pH = pK_{a2}, $[CO_3^=] = [HCO_3^-]$.
2. $[H_2CO_3]$ is the dominant carbonate species when pH < 6.3. $[CO_3^=]$ is the dominant species when pH > 10.3. $[HCO_3^-]$ is the dominant species, when 6.3 < pH < 10.3.
3. The more exact values of K_{a1} and K_{a2} of carbonic acid can be found in Table 8.5.

Example 8.12: Alkalinity of Water

Estimate the alkalinity of the water in Example 8.11 when (a) pH = 6.3, (b) pH = 10.3, and (c) pH = 7.3.

Solution:

From Eq. (8.35),

$$\text{Alkalinity (in M)} = 2\,[CO_3^=] + [HCO_3^-] + [OH^-] - [H^+]$$

a. At pH = 6.3,

$$\text{Alkalinity (in M)} = (2)(5.0 \times 10^{-6}) + (0.05) + 10^{-7.7} - 10^{-6.3} = 0.05 \text{ M}$$

b. At pH = 10.3,

$$\text{Alkalinity (in M)} = (2)(0.05) + (0.05) + 10^{-3.7} - 10^{-10.3} = 0.15 \text{ M}$$

c. At pH = 7.3,

$$\text{Alkalinity (in M)} = (2)(9.0 \times 10^{-5}) + (0.091) + 10^{-6.7} - 10^{-7.3} = 0.091 \text{ M}$$

Discussion:

1. The contributions of $[OH^-]$ and $[H^+]$ to the final alkalinity values in all three pHs are insignificant. Consequently, they are often neglected in determining the alkalinity values.
2. With the same total carbonate concentration of 0.1 M, the alkalinity values are 0.05, 0.091, and 0.15 M at pH = 6.3, 7.3, and 10.3, respectively.

EXERCISE QUESTIONS

1. Name the following two oxoacids and two oxyanions:
   a. HNO_3 and HNO_2
   b. NO_3^- and NO_2^-
2. a. Estimate the pH of pure water @ $T = 10°C$.
   b. Compare your value with the value given in Table 8.1.
   c. Compare your value with the neutral pH found in Example 8.1 for $T = 20°C$.
3. The pH of a water sample is 10.0.
   a. If the water sample is diluted with pure water of the same volume, estimate the pH of the diluted water sample.
   b. If the water sample is diluted with pure water of 19 times of its original volume, estimate the pH of the diluted water sample.
4. For the dissociation of phenol into phenolate shown below ($T = 25°C$)

$$C_6H_5OH_{(aq)} \rightarrow H^+_{(aq)} + C_6H_5O^- \quad pK_a = 10.0$$

   a. At what pH, the concentrations of C_6H_5OH and $C_6H_5O^-$ are equal?
   b. What would be the ratio of $[C_6H_5OH]/[C_6H_5O^-]$ when pH = 12.0?
   c. What would be the ratio of $[C_6H_5OH]/[C_6H_5O^-]$ when pH = 7.0?
   d. What would be the percent ionization of C_6H_5OH at pH = 7.0, 10.0, and 12.0?
5. The pH of Coca Cola Classic is ~2.5. Phosphoric acid (H_3PO_4) is often used in soda to give it a tangy flavor and to act as a preservative.
   a. Find the K_{a1} value of phosphoric acid at 25°C from Table 8.5.
   b. Write the equilibrium constant expression for K_{a1} of phosphoric acid.
   c. Estimate the pH of the solution when the molar ratio of $\left[H_2PO_4^{-1}\right]/\left[H_3PO_4\right] = 1$.
   d. Determine the molar ratio of $\left[H_2PO_4^{-1}\right]/\left[H_3PO_4\right]$ at pH = 3.15.
   e. Determine the percentage of phosphoric acid in its <u>unionized</u> form at pH = 3.15.
6. Use the K_b value of NH_3 in Table 8.7 to estimate the K_a value of its conjugate acid, NH_4^+, at 25°C [Note: Review Example 8.5, if needed]
7. a. Estimate the pH of 2.0 M ammonia solution at 25°C.
   b. Compare your value with the pH of 1.0 M ammonia solution obtained in Example 8.6.

8. 74 mg of calcium hydroxide ($Ca(OH)_2$) is added to 2 L of water at 20°C. [Note: Calcium hydroxide is a strong base; $Ca(OH)_2 \rightarrow Ca^{+2} + 2\ OH^-$]
   a. Moles of OH^- in this solution = ____________ g-moles.
   b. Molar concentration of OH^- in this solution = ____________ M
   c. pOH of this solution = ____________
   d. pH of this solution = ____________

   Concentrated hydrochloric acid (HCl) is to be added to neutralize the caustic solution above. [Note: Hydrochloric acid is a strong acid; $HCl \rightarrow H^+ + Cl^-$]
   e. Moles of H^+ needed for neutralization = ____________ g-moles
   f. Moles of HCl needed for neutralization = ____________ g-moles
   g. Mass of HCl needed for the neutralization = ____________ mg
9. Sodium benzoate (C_6H_5COONa) is the sodium salt of benzoic acid (C_6H_5COOH).
   a. What is the pH of a buffer containing 0.4 M of benzoic acid and 0.4 M sodium benzoate?
   b. What is the pH of a buffer containing 0.4 M of benzoic acid and 0.8 M sodium benzoate?
10. Methylammonium chloride (CH_3NH_3Cl) is an ammonium salt composed of methyl amine (CH_3NH_2) and hydrogen chloride.
    a. What is the pH of a buffer containing 0.4 M of methyl amine and 0.4 M methylammonium chloride?
    b. What is the pH of a buffer containing 0.4 M of methyl amine and 0.8 M methylammonium chloride??
11. Carbonate buffer system is the most important one for natural water systems.

$$H_2CO_{3(aq)} \leftrightarrow HCO^{-1}_{3(aq)} + H^+_{(aq)} \qquad pK_1 = 6.35@25°C$$

$$HCO^{-1}_{3(aq)} \leftrightarrow CO^{-2}_{3(aq)} + H^+_{(aq)} \qquad pK_2 = 10.35@25°C$$

    At pH = 8.35 & T = 25°C,
    a. Estimate the concentration ratio of $\left[HCO_3^-\right]/\left[H_2CO_3\right]$.
    b. Estimate the concentration ratio of $\left[HCO_3^-\right]/\left[CO_3^{-2}\right]$.
    c. Which is the dominant carbonate species at this pH (i.e., pH = 8.35)?
    d. At a lower pH, will $[H_2CO_3]$ increase or decrease?
12. If the total molar concentration of carbonate species of the water in Question #11 is equal to 0.1 M (pH = 8.35 & T = 25°C), answer the following questions.
    a. What are the molar concentrations of each individual carbonate species?
    b. What is the alkalinity of this water sample (in M)?

9 Oxidation-Reduction Reactions in Aqueous Solutions

9.1 INTRODUCTION

An *oxidation-reduction* (or "*redox*") reaction is a reaction that involves the transfer of electrons between chemical species (i.e., atoms, ions, or molecules) that participate in the reaction.

The following reaction equation shows an oxidation-reaction reaction:

$$Ni_{(s)} + Cu^{+2}_{(aq)} \rightarrow Ni^{+2}_{(aq)} + Cu_{(s)} \tag{9.1}$$

To understand the electron transfers between two metals (i.e., nickel and copper), the overall reaction can be considered as the sum of two "half-reactions" as:

$$Ni_{(s)} \rightarrow Ni^{+2}_{(aq)} + 2e^{-} \tag{9.2}$$

$$Cu^{+2}_{(aq)} + 2e^{-} \rightarrow Cu_{(s)} \tag{9.3}$$

Equation (9.2) shows that each nickel atom lost two electrons (e^-) to become a nickel ion (Ni^{+2}). An atom that loses electrons is "*oxidized*". The species being oxidized in a redox reaction is called the *reducing agent* of the redox reaction. Equation (9.2) is the *half-reaction for the oxidation reaction*. On the other hand, as shown in Eq. (9.3), a cupric ion (Cu^{+2}) gains two electrons to become a copper atom. The cupric ion is reduced, and it is the *oxidizing agent* in the redox reaction. Eq. 9.3 is the *half-reaction for the reduction reaction.*

9.2 OXIDATION STATE/NUMBER

Knowing the change of the oxidation state/number of an atom in a chemical reaction will help us understand what happened to it (or its role) in that reaction. An *oxidation state/number* of an atom can be considered as the number of electrons that it gained or lost (when compared to its neutral and uncombined form).

The oxidation state of an atom can be assigned using the following guidelines:

1. An atom of a free element has an oxidation state of 0. For example, Na atom in Na metal, Cl atom in Cl_2 gas, H atom in H_2 gas, and O atom in O_2 gas, all have an oxidation state of zero.

DOI: 10.1201/9781003502661-9

2. A monatomic ion has an oxidation state equals to its charge. For example, the oxidation number of Na in Na^+ is +1, that of Cl in Cl^- is −1, that of H in H^+ is +1, and that of Ca in Ca^{+2} is +2.
3. When combined with other elements, alkali metals (Group 1) always have an oxidation state of +1.
4. When combined with other elements, alkaline earth metals (Group 2) always have an oxidation state of +2.
5. Hydrogen has an oxidation state of +1 in most compounds, except when combined with metals. For example, the oxidation state number of H in metal hydrides (e.g., NaH and MgH_2) is −1, instead of +1.
6. Oxygen has an oxidation state of +2 in most compounds, except in peroxide and superoxide. Peroxides (e.g., hydrogen peroxide (H_2O_2) and sodium peroxide (Na_2O_2)) are a class of chemical compounds in which two oxygen atoms are linked together by a single covalent bond; the oxidation state of O in a peroxide is −1. Superoxides (e.g., sodium superoxide (NaO_2) and potassium superoxide (KO_2) are a class of chemical compounds that contains superoxide ion $\left(O_2^-\right)$; the oxidation number of O in a superoxide is −1/2.
7. Fluorine (F) has an oxidation state of −1 in all compounds.
8. Other halogens (Cl, Br, and I) also have an oxidation state of −1, except when combined with oxygen.
9. The sum of the oxidation states of all atoms in a neutral compound is equal to zero. For example, the oxidation states of H_2O, NaCl, $HClO_4$, H_2SO_4, and H_3PO_4 are all equal to zero.
10. The sum of the oxidation states of all atoms in a polyatomic ion is equal to the charge of the ion. For example, the oxidation numbers of ClO_4^{-1}, $SO_4^=$, and PO_4^{-3} are −1, −2, and −3, respectively.

In summary, in a redox reaction, the species that lost electrons (the oxidation state increased) is said to be oxidized and it works as the reducing agent. On the other hand, the species that gained electrons (the oxidation state decreased) is said to be reduced and it is the oxidizing agent.

Example 9.1: Oxidation State/Number

Find the oxidation number/state of the specific atom in different compounds.

a. Oxidation numbers of C in CH_4, C, CO, and CO_2.
b. Oxidation numbers of C in H_2CO_3, HCO_3^-, and $CO_3^=$.
c. Oxidation numbers of Cl in Cl_2, OCl^-, $HClO_3$, and $HClO_4$.
d. Oxidation states of N in NH_3, NH_4^+, NO_2^-, and NO_3^-.
e. Oxidation states of S in H_2S, S_8, SO_2, and SO_4^{-2}.
f. Oxidation states of Fe in $FeCl_2$, $Fe(OH)_3$, and Fe_2O_3.
g. Oxidation states of Cr in dichromate ion $\left(Cr_2O_7^=\right)$ and chromate ion $\left(CrO_4^=\right)$.

Solution:

a. Oxidation numbers of C in CH_4, C, CO, and CO_2 are −4, 0, +2, and +4, respectively.

b. Oxidation numbers of C in H_2CO_3, HCO_3^-, and $CO_3^=$ are all +6.

c. Oxidation numbers of Cl in Cl_2, OCl^-, $HClO_3$, $HClO_4$ are 0, +1, +5, and +7, respectively.

d. Oxidation states of N in NH_3, NH_4^+, NO_2^-, and NO_3^- are −3, −3, +3, and +5, respectively.

e. Oxidation state of S in H_2S, S_8, SO_2, and SO_4^{-2} are −2, 0, +4, and +6, respectively.

f. Oxidation state of Fe in $FeCl_2$, $Fe(OH)_3$, and Fe_2O_3 are +2, +3, and +3, respectively.

g. Oxidation state of Cr in dichromate ion and chromate ion $\left(CrO_4^=\right)$ are both +6.

Discussion:

1. In part (a), the oxidation state of C in methane (CH_4) is −4 which is much smaller than that of CO_2 (+4). Methane is considered more reduced than CO_2. The reduced compounds are often formed under anaerobic conditions and possess larger energies. Combustion of methane gas is a redox reduction, in which methane is oxidized to CO_2 and water, while oxygen gas is reduced to water.
2. In part (b), carbon atoms in all three carbonate species have the same oxidation state. It implies that the dissociation of an acid is not a redox reaction.
3. In part (c), the central atom (i.e., Cl) of three oxoacids have different oxidation states (also see Table 8.4).
4. In part (d), all four inorganic nitrogen species are commonly present in water/wastewater. NH_3 (and NH_4^+) are more reduced, and they are the dominant inorganic nitrogen species found in anaerobic digesters and/or at the bottom of deep lakes/reservoirs.
5. The oxidation state of iron in ferrous chloride is +2, while that in ferric oxide is +3.
6. The terms "oxidation state" and "oxidation number" are used interchangeably here.

9.3 OXIDATION-REDUCTION POTENTIAL

9.3.1 Standard Reduction Potential

Oxidation-reduction potential (ORP) or *redox potential* is a measure of the tendency of a chemical species to acquire or to lose electrons [Note: It is not the likelihood for a specific oxidation or reduction reaction to occur]. The *Standard reduction potential* $\left(E^o_{red}\right)$ for a half reaction is measured at $T = 25°C$ and $P = 1$ atm, with 1 M concentrations; and it is defined/measured related to the *standard hydrogen electrode* (SHE) which is used as the reference electrode with an assigned potential of 0.00 Volt (V).

Table 9.1 tabulates standard reduction potential $\left(E^{o}_{red}\right)$ of some common half reactions [Note: The values can be readily found in handbooks of chemistry and chemical engineering as well as on the Internet. However, the values from different sources may not always be the same]. The table can be viewed as a ranking of substances according to their oxidizing or reducing power. The E^{o}_{red} of SHE lies about halfway down the list, with a value of 0.00 V. The larger the standard reduction potential, the substance would have a stronger oxidizing power, while the stronger reducing agent would have a more negative reduction potential. As shown, the strength of a substance as an oxidizing agent will increase from $Li^{+}_{(aq)}$ (E^{o}_{red}= −3.04 V; the bottom row) to $F_{2(g)}$ (E^{o}_{red}= 2.87 V; the top row), with fluorine gas being the strongest oxidizing agent and $Li_{(s)}$ is the strongest reducing agent.

TABLE 9.1
Standard Reduction Potentials of Some Half Reactions

Reduction Half-Reaction	Reduction Potential (V)
$F_{2(g)} + 2e^- \rightarrow 2F^-_{(aq)}$	2.87
$H_2O_{2(aq)} + 2H^+_{(aq)} + 2e^- \rightarrow 2H_2O_{(l)}$	1.76
$MnO^-_{4(aq)} + 8H^+_{(aq)} + 5e^- \rightarrow Mn^{+2}_{(aq)} + 4H_2O_{(l)}$	1.51
$Cl_{2(g)} + 2e^- \rightarrow 2Cl^-_{(aq)}$	1.36
$Cr_2O^{-2}_{7(aq)} + 14H^+_{(aq)} + 6e^- \rightarrow 2Cr^{+3}_{(aq)} + 7H_2O_{(l)}$	1.33
$O_{2(g)} + 4H^+_{(aq)} + 4e^- \rightarrow 2H_2O_{(l)}$	1.23
$Br_{2(g)} + 2e^- \rightarrow 2Br^-_{(aq)}$	1.09
$NO^-_3 + 4H^+_{(aq)} + 3e^- \rightarrow NO_{(g)} + 2H_2O_{(l)}$	0.94
$Ag^+_{(aq)} + e^- \rightarrow Ag_{(s)}$	0.80
$Fe^{+3}_{(aq)} + e^- \rightarrow Fe^{+2}_{(aq)}$	0.77
$O_{2(g)} + 2H^+_{(aq)} + 4e^- \rightarrow 2H_2O_{2(aq)}$	0.70
$MnO^-_{4(aq)} + 2H_2O_{(l)} + 3e^- \rightarrow MnO_{2(s)} + 4OH^-_{(aq)}$	0.60
$I_{2(s)} + 2e^- \rightarrow 2I^-_{(aq)}$	0.54
$O_{2(g)} + 2H_2O_{(l)} + 4e^- \rightarrow 4OH^-_{(aq)}$	0.40
$Cu^{+2}_{(aq)} + 2e^- \rightarrow Cu_{(s)}$	0.34
$AgCl_{(s)} + e^- \rightarrow Ag_{(s)} + Cl^-_{(aq)}$	0.22
$Cu^{+2}_{(aq)} + e^- \rightarrow Cu^+_{(aq)}$	0.16
$2H^+_{(aq)} + 2e^- \rightarrow H_{2(g)}$	0.00
$Ni^{+2}_{(aq)} + 2e^- \rightarrow Ni_{(s)}$	−0.26
$Fe^{+2}_{(aq)} + 2e^- \rightarrow Fe_{(s)}$	−0.44
$Zn^{+2}_{(aq)} + 2e^- \rightarrow Zn_{(s)}$	−0.76
$2H_2O_{(l)} + 2e^- \rightarrow H_{2(g)} + 2OH^-_{(aq)}$	−0.83
$Al^{+3}_{(aq)} + 3e^- \rightarrow Al_{(s)}$	−1.68
$Na^+_{(aq)} + e^- \rightarrow Na_{(s)}$	−2.71
$Li^+_{(aq)} + e^- \rightarrow Li_{(s)}$	−3.04

These "standard reduction" potentials $\left(E^{o}_{red}\right)$ are often just referred as the "*redox*" *potentials* (E^{o}). There are standard "oxidation" potentials $\left(E^{o}_{ox}\right)$, which indicate the tendency of a chemical species to be oxidized. The value for standard oxidation potential has the exact value, but is an opposite sign from that of the standard reduction potential of the same chemical species. Thus,

$$E^{o}_{oxidation} = -\ E^{o}_{reduction} \tag{9.2}$$

9.3.2 Common Oxidizing Agent

Strong oxidizing agents are typically compounds with elements in higher possible oxidation states or with high electronegativity; and they gain electrons in the redox reaction (i.e., the *electron acceptor*). Common ones include (i) non-metallic elements that accept electrons to form anions such as halogens (i.e., F_2, Cl_2, Br_2, I_2) and O_2, (ii) ions with an element with a large oxidation number (e.g., MnO_4^-, $Cr_2O_7^{-2}$, and NO_3^-), (iii) metal cations that could accept electrons to form neutral molecules (e.g., Ag^+, Cu^{+2}, and Fe^{+3}), and (iv) strong acids (e.g., HNO_3 and H_2SO_4).

ChemTalk (2023) has a list of favorite oxidizing agents that are commonly used in chemistry labs and experiments:

1. Hydrogen peroxide (H_2O_2)
2. Potassium dichromate ($K_2Cr_2O_7$)
3. Sodium or calcium hypochlorite (NaOCl or $Ca(OCl)_2$)
4. Nitric acid (HNO_3)
5. Oxygen (O_2) or air
6. Ozone (O_3)
7. Potassium perchlorate ($KClO_4$)
8. Potassium chlorate ($KClO_3$)
9. Potassium permanganate ($KMnO_4$)
10. Ammonium or sodium persulfate ($(NH_4)_2S_2O_8$ or $Na_2S_2O_8$).

9.3.3 Common Reducing Agent

Strong reducing agents are typically compounds with elements in lower possible oxidation states or with low electronegativity; and they lose electrons in the redox reaction (i.e., the *electron donor*). Common ones include (i) alkali metals (e.g., Li and Na), (ii) metals (e.g., Al, Fe, Ni, and Zn), (iii) metal hydrides (e.g., LiH, NaH, $LiAlH_4$, and CaH_2), and (iv) H_2.

ChemTalk (2023) has a list of favorite reducing agents that are commonly used in chemistry labs and experiments:

1. Ascorbic acid ($C_6H_8O_6$)
2. Glucose ($C_6H_{12}O_6$)
3. Zinc metal (Zn)
4. Oxalic acid ($C_2H_2O_4$)

5. Sodium sulfite (Na_2SO_3)
6. Sodium bisulfite ($NaHSO_3$)
7. Tin(II) chloride ($SnCl_2$)
8. Sodium thiosulfate ($Na_2S_2O_3$)
9. Lithium aluminum hydride ($LiAlH_4$)
10. Magnesium metal (Mg)

It should be noted that some elements and compounds can be oxidizing or reducing agents, depending on the species they work with in a specific reaction. As shown in Table 9.1, the reduction potentials of Cu^{+2}, Ni^{+2}, and Zn^{+2} are in the order of Cu^{+2} (+0.34V) > Ni^{+2} (−0.26V) > Zn^{+2} (−0.76V). Nickel ion serves as the reducing agent when works with copper ion; but will serve as the oxidizing agent when works with zinc ion.

9.4 ELECTROCHEMICAL REACTION

Electrochemistry is a branch of chemistry that deals with the conversion between chemical energy and electrical energy. Redox reactions, which involve transfer of electrons, are behind all electrochemical processes.

9.4.1 Electrochemical Cells

An *electrochemical cell* is a device that can generate electrical energy from the chemical reactions occurring within it, or it can use the electrical energy supplied to it to facilitate chemical reactions to occur within it. An electrochemical cell generally consists of a cathode, an anode, a solution (in which the redox reactions can occur), a conductor (for electron transfer) and a salt bridge for ions to move through. As mentioned, chemical reactions would be spontaneous, if the change of Gibbs free energy (ΔG) is negative (while the reaction will move in the opposite direction, if $\Delta G > 0$). The thermodynamic driving force for redox reactions in an electrical cell is the *cell potential* (E_{cell}); or called the *electromotive force* (emf). The cell potential under standard conditions can be found as:

$$E^{o}_{cell} = E^{o}_{red}(\text{cathode}) - E^{o}_{red}(\text{anode}) \tag{9.3}$$

If the cell potential is positive, a cell would operate in the spontaneous direction, and it is called a *galvanic cell* (or *voltaic cell*). In a spontaneous electrochemical reaction, electrons are transferred and energy would be released. A reduction reaction will occur in the *cathode* of the electrochemical cell, and it is called the positive electrode while the electrons move into the electrode. On the other hand, an oxidation reaction occurs in the *anode*; and it is called the negative electrodes while electrons move out of the anode.

Take the reaction between hydrogen gas and oxygen gas to form water as an example [Note: It is a spontaneous reaction in the direction of the arrow]. The overall reaction can be viewed as the sum of two half reactions:

$$\text{Overall Reaction: } 2H_{2(g)} + O_{2(g)} \rightarrow 2H_2O_{(l)} \quad (9.4)$$

$$\text{Cathode: } O_{2(g)} + 4H^+_{(aq)} + 4e^- \rightarrow 2H_2O_{(l)} \;\; E^o_{\text{red}} = 1.23 \text{ V} \quad (9.5)$$

$$\text{Anode: } H_{2(g)} \rightarrow 2H^+_{(aq)} + 2e^- \;\; E^o_{\text{red}} = 0.00 \text{ V} \quad (9.6)$$

As shown, a reduction occurs at the cathode, in which O_2 is reduced and electrons are added into the half reaction to form water. The values of standard reduction potentials of these two half reactions are obtained from Table 8.1. Using Eq. (9.3), the cell potential can be found as 1.23 V (= 1.23 – 0.00), which is positive, indicating this electrochemical reaction is spontaneous. It should be noted that voltage for the reaction at the anode needs not to be doubled, even the reduction half-reaction needs to be doubled to balance the overall redox equation.

Example 9.2: Cell Potential of a Galvanic Cell

A galvanic cell is based on the following reduction half-reactions at 25°C, where $[Ag^+] = 1.0$ M and $[Ni^{2+}] = 1.0$ M:

- $Ag^+_{(aq)} + e^- \rightarrow Ag_{(s)} \quad E^o_{\text{red}} = +0.80$ V
- $Ni^{+2}_{(aq)} + 2e^- \rightarrow Ni_{(s)} \quad E^o_{\text{red}} = -0.28$ V

a. What is the overall reaction?
b. What is the standard cell potential?

Solution:

a. In a galvanic cell, the reaction is spontaneous and E^o_{cell} should be > 0. With that, the reduction of silver ion would occur at the cathode. The overall reaction would be:

$$2Ag^+_{(aq)} + Ni_{(s)} \rightarrow 2Ag_{(s)} + Ni^{+2}_{(aq)} \quad (9.7)$$

b. Use Eq. (9.3) to find the standard cell potential:

$$E^o_{\text{cell}} = E^o_{\text{red}}(\text{cathode}) - E^o_{\text{red}}(\text{anode}) = 0.80 - (-0.28) = 1.08 \text{ V}$$

Discussion:

1. Even a stoichiometric coefficient of 2 is needed (for Ag and Ag^+) in Eq. (9.7) to balance the charges, a multiplier of two is not needed to calculate the standard cell potential in part (b). The reason is that E^o is an intensive property.
2. Since both the concentrations of Ag^+ and Ni^{+2} are 1.0 M and $T = 25$°C (i.e., the standard states), the calculated cell potential is the standard cell potential.

9.4.2 Relationship between Cell Potential and Gibbs Free Energy

The standard cell potential is the electrochemical version of the Gibbs free energy, and it is related to Gibbs free energy (ΔG^o) as

$$\Delta G^o = -nFE^o \tag{9.8}$$

where "n" = the moles of electrons involved in the electrochemical reaction and F = the Faraday constant, which is 96,485 Coulombs/mole (C/mole). As shown, ΔG^o and E^o are opposite in sign; therefore, an electrochemical reaction will be spontaneous, if $\Delta G^o < 0$ (i.e., $E^o > 0$).

The Coulomb (C) is the SI unit of electrical charge, which is equal to the elemental charges delivered by a constant current of one ampere in one second; and it is equal to 6.241×10^{18} elemental charge. The *elemental charge*, usually denoted by a symbol *e*, is the electrical charge carried by a single electron. In other words, one elemental charge is equal to a charge of 1.602×10^{-19}C (= $1 \div 6.241 \times 10^{18}$). A *Faraday unit of charge* (96,585 C) is equal to 6.022×10^{23} elemental charges (= 96,585 C $\times 6.241 \times 10^{18}$ elemental charge/C); that is essentially one mole of elemental charges. The relationships among charge (in C), current (in Ampere), voltage (in V), energy (in Joule), and time (in second) are:

$$1 \text{ Coulomb } (\text{C}) = 1 \text{ Ampere } (\text{A}) \times 1 \text{ second} \tag{9.9}$$

$$1 \text{ Joule } (\text{J}) = 1 \text{ Coulomb } (\text{C}) \times 1 \text{ Volt } (\text{V}) \tag{9.10}$$

Example 9.3: Cell Potential and Gibbs Free Energy

For the galvanic cell described in Eqs. (9.4)–(9.6):

$$\text{Overall Reaction: } 2\text{H}_{2(g)} + \text{O}_{2(g)} \rightarrow 2\text{H}_2\text{O}_{(l)} \tag{9.4}$$

$$\text{Cathode: } \text{O}_{2(g)} + 4\text{H}^+_{(aq)} + 4e^- \rightarrow 2\text{H}_2\text{O}_{(l)} \qquad E^o_{\text{red}} = 1.23 \text{ V} \tag{9.5}$$

$$\text{Anode: } \text{H}_{2(g)} \rightarrow 2\text{H}^+_{(aq)} + 2e^- \quad \rightarrow E^o_{\text{red}} = 0.00 \text{ V} \tag{9.6}$$

a. Find the ΔG^o value of the overall reaction.
b. Find the cell potential of this galvanic cell.
c. Is the relationship in Eq. (9.8) applicable for this case?

Solution:

a. The ΔG^o_f values of $H_{2(g)}$, $O_{2(g)}$, and $H_2O_{(g)}$ are 0.0, 0.0, and −120.4 kJ/K/mole, respectively (Table 5.1)

Use Eq. (5.22) to find the change of Gibbs free energy of this reaction (P = 1 atm and T = 25 °C):

$$\Delta G^o_{\text{rxn}} = [(2)(-120.4)] - [(2)(0.0) + (1)(0.0)] = -240.8 \text{ kJ/mole}$$

b. Use Eq. (9.3) to find the cell potential:

$$E^{o}_{cell} = E^{o}_{red}(\text{cathode}) - E^{o}_{red}(\text{anode}) = 1.23 - (0.00) = 1.23 \text{ V}$$

c. $-nFE^{o} = -(2)(96{,}585)(1.23) = 2.38 \times 10^{5}\ \text{C}\cdot\text{V} = 2.38 \times 10^{5}\ \text{J} \approx 241\ \text{kJ}$

Discussion:

This example illustrates the validity of Eq. (9.8). The calculated values of $-nFE^{o}$ and ΔG^{o} are essentially the same.

9.4.3 The Nernst Equation

Since the electrochemical reactions do not always occur under the standard conditions (i.e., concentrations are not equal to 1.0 M, $T \neq 25°C$, and/or $P \neq 1$ atm), more generally,

$$\Delta G = -nFE \tag{9.11}$$

As shown in Eq. (5.27), the relationship between ΔG^{o} (the change of free-energy under standard conditions) and ΔG (the change of free-energy under any other conditions) is

$$\Delta G = \Delta G^{o} + RT(lnQ) \tag{5.27}$$

where R = ideal gas constant and Q = reaction quotient [Note: $\Delta G = \Delta G^{o}$, when $Q = 1$].

Inserting Eqs. (9.8) and (9.11) into Eq. (5.27), the following equation can be readily derived:

$$E = E^{o} - \left(\frac{RT}{nF}\right) lnQ \tag{9.12}$$

Inserting the value of the Faraday constant and using "$\log_{10}$" instead of natural log, then @$T = 25°C$,

$$E = E^{o} - \left(\frac{0.0592}{n}\right) log_{10}Q \tag{9.13}$$

Eq. (9.12) (or Eq. 9.13) is called the *Nernst equation*, which can be used to calculate the cell potential of a non-standard galvanic cell.

Example 9.4: Use of the Nernst Equation

(Continuation of Example 9.2) Determine the cell potential of a galvanic cell based on the following reduction half-reactions at 25°C, where $[Ag^{+}] = 0.05$ M and $[Ni^{2+}] = 0.5$ M:

- $Ag^{+}_{(aq)} + e^{-} \rightarrow Ag_{(s)}$ $\quad E^{o}_{red} = +0.80$ V
- $Ni^{+2}_{(aq)} + 2e^{-} \rightarrow Ni_{(s)}$ $\quad E^{o}_{red} = -0.28$ V

Solution:

a. The overall reaction would be:

$$2Ag^{+}_{(aq)} + Ni_{(s)} \rightarrow 2Ag_{(s)} + Ni^{+2}_{(aq)} \tag{9.7}$$

b. The standard cell potential = 1.08 V (from Example 9.2).

c. Use Eq. (9.13) to find its cell potential:

$$E = E^{o} - \left(\frac{0.0592}{n}\right) Log_{10}Q = 1.08 - \left(\frac{0.0592}{2}\right) \log_{10}\left(\frac{0.5}{0.05}\right) = 1.05 \text{ V}$$

Discussion:

1. Since two moles of electrons involved in the reaction, $n=2$.
2. The value of $Q=10$ in this example, and it causes a change of 59.2 mV (= 0.0592 V) per mole of electrons transferred.

At equilibrium, $E=0$ and the equilibrium constant is equal to the reaction quotient, thus:

$$0 = E^{o} - \left(\frac{RT}{nF}\right) lnK \tag{9.14}$$

$$K = Exp\left[nFE^{o}/RT\right] \tag{9.15}$$

Example 9.5: Standard Cell Potential and Equilibrium Constant

(Continuation of Example 9.2) Determine the equilibrium constant of the redox reaction in a galvanic cell based on the following reduction half-reactions at 25°C.

- $Ag^{+}_{(aq)} + e^{-} \rightarrow Ag_{(s)} \quad E^{o}_{red} = +0.80 \text{ V}$
- $Ni^{+2}_{(aq)} + 2e^{-} \rightarrow Ni_{(s)} \quad E^{o}_{red} = -0.28 \text{ V}$

Solution:

a. The overall reaction would be:

$$2Ag^{+}_{(aq)} + Ni_{(s)} \rightarrow 2Ag_{(s)} + Ni^{+2}_{(aq)} \tag{9.7}$$

b. The standard cell potential = 1.08 V (from Example 9.2).

c. Use Eq. (9.15) to find the equilibrium constant:

$$K = \left[Ni^{+2}_{(aq)}\right] / \left[Ag^{+}_{(aq)}\right] = \text{Exp}\left[nFE^{o}/RT\right] = \text{Exp}\left[\frac{(2 \text{ mole})(96{,}585 \text{ C})(1.08 \text{ V})}{\left(8.3145 \text{ J} \cdot \frac{\text{mole}}{\text{K}}\right)(298\text{K})}\right] = e^{84.2}$$

Discussion:

1. At equilibrium, the cell potential is equal to zero; it means the cell does not create voltage.
2. The equilibrium constant value is huge; and it implies that when the value of *Q* is smaller than this *K* value, the reaction proceeds to the right and creates the voltage.

9.4.4 Pourbaix Diagrams

A *Pourbaix diagram* is a plot of the redox potentials of electrochemical reactions, between a metal and its various oxidized species, versus pH. To plot a Pourbaix diagram, the relevant Nernst equations are used. As the Nernst equations are derived entirely from thermodynamics, the diagram can be used to determine which species is thermodynamically stable at a given set of the redox potential and pH.

Figure 9.1 is the Pourbaix diagram for 1.0 M iron solutions. The vertical axis, *pe* (or pE), represents the reducing agent (e^-), analogous to the pH scale which represents the concentration of acid (H^+). Values of *pe* are obtained by dividing the standard cell potential (E^o) with 0.059. The small *pe* values represent a reducing environment (see H_2 at the bottom portion in the plot); the large *pe* values represent an oxidizing environment (see O_2 on the top portion in the plot). The other observation is that the region of ferric ion (Fe^{+3}) is one top of the region of ferrous ion (Fe^{+2}); similarly, the region of ferric hydroxide ($Fe(OH)_3$) is on top of the region of ferrous hydroxide ($Fe(OH)_2$) [Note: Fe^{+3} is more oxidized than Fe^{+2}]. In addition, iron species

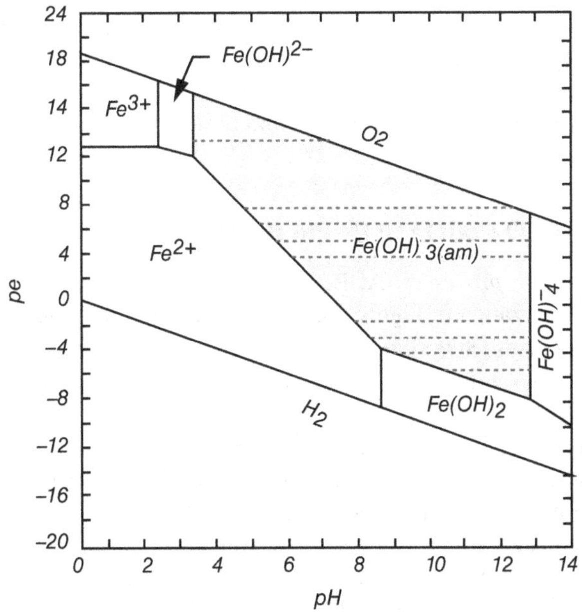

FIGURE 9.1 The Pourbaix diagram for 1.0 M iron solutions. (USEPA, 1991.)

are in the dissolved ionic forms at lower pHs; while they are in a solid form at relatively higher pHs.

9.4.5 Fuel Cell

A *fuel cell* consists of a galvanic (voltaic) cell that is continuously supplied with fuel (e.g., H_2) and an oxidant (e.g., O_2) and continuously generates electricity out of chemical reactions. The galvanic cell, shown in Eqs. (9.4)–(9.6), is one of the fuel cells commonly used. The standard cell potential is 1.23 V, and the actual voltage in operation is smaller. Individual fuel cells are connected in series to generate a higher voltage to charge the battery pack to power electrical buses as an example. The source of hydrogen can come from methane. However, the use of fuel cells in transport is often limited by lack of facilities for hydrogen gas storage and distribution.

9.4.6 Electrolytic Cell versus Galvanic Cell

An *electrolytic cell* is the opposite of a galvanic cell. As mentioned, the electrochemical reaction in a galvanic cell is spontaneous. For example, the reaction of hydrogen gas and oxygen gas to form water (see Eq. 9.4) is spontaneous. On the other hand, in an electrolytic cell, the energy needed for a chemical reaction is from the electrical energy applied to the cell.

An electrolytic cell can be defined as an electrochemical device that uses the applied electrical energy to facilitate a non-spontaneous redox reaction. The device is often used for the electrolysis of certain compounds; for example, it can be used to electrolyze water to produce hydrogen gas and oxygen gas (the backward reaction of Eq. 9.4). Other applications of electrolytic cells include extraction of aluminum from bauxite (i.e., a sedimentary rock with a relatively high aluminum content) and electroplating (i.e., a process of forming a thin protective layer of a specific metal on the surface of another metal) (Table 9.2).

9.5 ADVANCED OXIDATION PROCESSES

Advanced oxidation processes (AOPs) can be defined as treatment processes that involve in-situ generation of highly potent chemical oxidants (e.g., hydroxyl radical (·OH), sulfate radicals $\left(SO_4^- \cdot\right)$, in sufficient quantities to degrade refractory organic compounds, trace organic contaminants, and/or certain inorganic pollutants in water, wastewater, and air.

In chemistry, a *free radical* is an atom, ion, or molecule that has an unpaired electron. A free radical is typically very reactive toward other substances. A *free radical process* typically goes through three stages in a chain reaction: (i) initiation (i.e., formation of radicals), (ii) propagation (attacking other substances to generate more free radicals in different forms), and (iii) termination (reacting with other radicals to become inactive). In common AOPs, hydroxyl radicals (·OH) are produced with one or more primary oxidants as initiators (e.g., O_3 and H_2O_2) and an energy source

TABLE 9.2
Reduction Potentials of Some Common Oxidizing Agents

Oxidizing Agents	Reduction Potential (V)	Relative Oxidation Power ($Cl_2 = 1.00$)
Fluorine	2.87	2.25
Hydroxyl radical	2.80	2.05
Atomic oxygen	2.42	1.78
Ozone	2.07	1.52
Hydrogen peroxide	1.77	1.30
Perhydroxyl radical	1.70	1.25
Permanganate	1.68	1.23
Hypochlorous acid	1.49	1.10
Chlorine	1.36	1.00
Bromine	1.07	0.79
Iodine	0.54	0.40

Source: Reif (1998).

(e.g., UV light). Catalysts (e.g., titanium oxide (TiO_2)) can be added to enhance the reaction.

A few reaction mechanisms for some common AOPs are shown below:

- UV/Hydrogen Peroxide

$$H_2O_2 \xrightarrow{hv} 2 \cdot OH \tag{9.16}$$

- UV/Ozone

$$O_3 + H_2O \xrightarrow{hv} H_2O_2 + O_2 \xrightarrow{hv} 2 \cdot OH + O_2 \tag{9.17}$$

- H_2O_2/Ozone (without UV)

$$2O_3 + H_2O_2 \rightarrow 2 \cdot OH + 3O_2 \tag{9.18}$$

- Fenton agent (ferrous ion + hydrogen peroxide)

$$Fe^{+2} + H_2O_2 \rightarrow Fe^{+3} + \cdot OH + OH^- \tag{9.19}$$

These AOPs are especially applicable to contaminants that are reactive with hydroxyl radicals; for example, benzene (C_6H_6), toluene ($C_6H_5CH_3$), ethylbenzene ($C_6H_5C_2H_5$), xylenes ($C_6H_4(CH_3)_2$, phenol (C_6H_5OH)) and those with double bonds (e.g., vinyl chloride (C_2H_3Cl), dichloroethylene ($C_2H_2Cl_2$), trichloroethylene (C_2HCl_3), and perchloroethylene (C_2Cl_4).

REFERENCES

ChemTalk (2023). "Common Oxidizing Agents & Reducing Agents", https://chemistrytalk.org/oxidizing-reducing-agents/, last updated November 20, 2023.

Reif, F. (1998). Drinking Water Improvement in the Americas with Mixed Oxidant Gases Generated On-site for Disinfection (NOGGOD)", *Bull. Pan Am. Health Org.*, V. 22, No. 4, p. 394–415.

USEPA (1991). "*Site Characterization for Subsurface Remediation*", EPA/625/4-91/026, United States Environmental Protection Agency.

EXERCISE QUESTIONS

1. Find the oxidation number/state of the specific atom in different compounds.
   a. Oxidation numbers of C in CH_4, C_2H_4, C_2H_2, HCHO, and CH_3OH.
   b. Oxidation numbers of C in CO, CO_2, and H_2CO_3.
   c. Oxidation numbers of Cl in Cl_2, NaCl, $CaCl_2$, NaClO, and $NaClO_4$.
   d. Oxidation states of N in NH_3, NH_4Cl, N_2O, and HNO_3.
   e. Oxidation states of S in FeS, H_2S, SO_3, $NaHSO_4$ and H_2SO_4.
   f. Oxidation states of Al in $AlCl_3$, Al_2O_3, and Al^{+3}.
   g. Oxidation states of Mn in manganate ion $\left(MnO_4^{=}\right)$ and permanganate ion $\left(MnO_4^{-}\right)$.
2. A galvanic cell is based on the following reduction half-reactions at 25°C, where $[Zn^{+2}] = 1.0\,M$ and $[Cu^{2+}] = 1.0$ M:
   - $Zn^{+2}_{(aq)} + 2e^- \rightarrow Zn_{(s)} \quad E^o_{red} = -0.76$ V
   - $Cu^{+2}_{(aq)} + 2e^- \rightarrow Cu_{(s)} \quad E^o_{red} = +0.34$ V

   a. What is the overall reaction?
   b. What is the standard cell potential?
3. A galvanic cell is based on the following reduction half-reactions at 25°C, where $[Zn^{+2}] = 0.02\,M$ and $[Cu^{2+}] = 0.10$ M:
   - $Zn^{+2}_{(aq)} + 2e^- \rightarrow Zn_{(s)} \quad E^o_{red} = -0.76$ V
   - $Cu^{+2}_{(aq)} + 2e^- \rightarrow Cu_{(s)} \quad E^o_{red} = +0.34$ V

   a. What is the overall reaction?
   b. What is the cell potential?
4. A galvanic cell is based on the following reduction half-reactions at 25°C, where $[Zn^{+2}] = 1.0\,M$ and $[Cu^{2+}] = 1.0$ M:
   - $Zn^{+2}_{(aq)} + 2e^- \rightarrow Zn_{(s)} \quad E^o_{red} = -0.76$ V
   - $Cu^{+2}_{(aq)} + 2e^- \rightarrow Cu_{(s)} \quad E^o_{red} = +0.34$ V

   What is the equilibrium constant of the redox reaction in this galvanic cell?

10 Fundamentals of Chemical Reaction Kinetics

10.1 INTRODUCTION

From thermodynamics, one can tell if a chemical reaction can occur. *Chemical kinetics* is essentially the study of how fast a chemical reaction would proceed. In a typical chemical reaction, one reactant (or more than one) reacts under a specific condition to form product(s). The concentrations of the reactants will decrease with time, while those of the products will increase (until the reaction reached an equilibrium). In equilibrium, the concentrations of the reactants and the products will stay constant.

The *reaction rate* describes the speed of this chemical reaction. The rate of the reaction depends on the concentrations of the reactants, the reaction conditions (e.g., mixing, the temperature and pressure of the system), and the presence of other substances (e.g., catalysts and inhibitors).

10.2 RATES OF CHEMICAL REACTIONS

10.2.1 Factors Affecting Reaction Rates

Chemical reactions occur at different rates. Factors affecting the rate of a chemical reaction include the following:

- ***Concentration of the reactants.*** Contact/collision among the reactants is the necessary first step for a chemical reaction to occur. Larger concentrations of the reactants would increase the possibility of collisions so that the reaction rate would be faster. For chemical reactions involving gaseous species, larger partial pressures of those species (i.e., higher concentrations) will increase the reaction rate through the increased number of collisions.
- ***Physical state of the reactants.*** For a homogenous solution, the possibility of collisions among reactants would be larger, when compared to a solution not homogenized. For a heterogeneous solution containing solids that participate in the reaction (e.g., a solid catalyst), solids with larger surface areas would provide chances for more contacts.
- ***Temperature of the reaction.*** At higher temperatures, kinetic energy of molecules would be larger to incur more collisions and to provide more energy to overcome the reaction barrier (more later in this chapter). Consequently, the reaction rate would be faster under elevated temperatures (e.g., combustion occurs rapidly at elevated temperatures).

DOI: 10.1201/9781003502661-10

However, elevated temperatures may have an inhibitory effect on some biochemical reactions and biological processes.

- ***Presence of a catalyst/inhibitor.*** A catalyst may lower the activation energy of a chemical reaction, which will increase the rate of reaction (more later in this chapter). On the other hand, some substances can be an inhibitor to a chemical reaction which would retard the reaction.

10.2.2 Reaction Rates and Stoichiometry

For an irreversible chemical reaction,

$$aA + bB \rightarrow cC + dD \tag{10.1}$$

A and B are the reactants and C and D are the products, while a, b, c, and d are the stoichiometric constants. The reaction rate equation is typically expressed in terms of concentrations. The reaction rate at a particular time (i.e., the *instantaneous rate*) can be expressed as

$$\text{Instantaneous Reaction Rate} = -\frac{1}{a}\frac{\Delta[A]}{\Delta t} = -\frac{1}{b}\frac{\Delta[B]}{\Delta t} = \frac{1}{c}\frac{\Delta[C]}{\Delta t} = \frac{1}{d}\frac{\Delta[D]}{\Delta t} \tag{10.2}$$

The "brackets" around A, B, C, and D indicate that they are the concentrations of these species (or expressed as C_A, C_B, C_C, and C_D). $\Delta[A]/\Delta t$ represents the rate of change of the concentration of reactant A within the time interval (Δt). The negative sign indicates that the concentrations of reactants A and B are decreasing (i.e., the reactants are disappearing); while the concentrations of products C and D are increasing (with a positive sign). The rate of concentration changes among the reactants and the products are related to the stoichiometric constants as shown in Eq. (10.2).

Considering the following chemical reaction with specific stoichiometric coefficients,

$$2A + 2B \rightarrow C + 3D \tag{10.3}$$

the instantaneous reaction rate would be as:

$$\text{Instantaneous Reaction Rate} = -\frac{1}{2}\frac{\Delta[A]}{\Delta t} = -\frac{1}{2}\frac{\Delta[B]}{\Delta t} = \frac{1}{1}\frac{\Delta[C]}{\Delta t} = \frac{1}{3}\frac{\Delta[D]}{\Delta t} \tag{10.4}$$

For this case, the concentrations of reactants A and B will decrease at the same rate, while the concentration of product C will increase at half of the decreasing rate of A or B; and the concentration of product D will increase at three times the rate of product C.

10.3 RATE EQUATIONS OF CHEMICAL REACTIONS

The *rate equation* (or *rate law*) is used to show the relationship between the reaction rate and the concentration(s) of the reactant(s). The rate equation for the reaction in Eq. (10.1) can be written as

$$\text{Reaction Rate} = k[A]^m[B]^n \tag{10.5}$$

where k is the rate constant, and m and n are the *orders of the reaction* (or the *reaction order*). The reaction orders are not necessarily positive integers; they can be zero, negative, or even fractional. The sum of m and n is called the *overall reaction order.* It should be noted that the reaction rate constant k is independent of the concentrations of the reactants and the reaction order. It also has nothing to do with the stoichiometric constants [Note: m and n could be different from the stoichiometric constants]. However, the value of a rate constant usually depends on the reaction conditions, such as system temperature and pressure. The reaction orders are determined experimentally. Some common scenarios for the rate equations are shown below:

$$Reaction\ Rate = k \tag{10.6a}$$

$$Reaction\ Rate = k[A] \tag{10.6b}$$

$$Reaction\ Rate = k[A][B] \tag{10.6c}$$

$$Reaction\ Rate = k[A]^2 \tag{10.6d}$$

$$Reaction\ Rate = k[A]^2[B] \tag{10.6e}$$

The rate equation of Eq. (10.6a) is for a *zeroth-order reaction*; in which the reaction rate is independent of the concentrations of both reactants A and B. It implies that the reactants are present in abundance. Eq. (10.6b) is for a *first-order reaction*, in which the reaction rate depends only on the concentration of reactant A. It implies that the reactant B is present in abundance. Eq. (10.6c) is for a *second-order reaction* (i.e., first order with reactant A and first order with reactant B), in which the reaction rate depends on concentrations of both reactants A and B. Eq. (10.6d) is also for a second-order reaction; however, its rate depends only on the concentrations of reactant A to a second-order. Eq. (10.6e) is for a third-order reaction; its reaction rate depends on both reactants, but is different from Eq. (10.6c). For all the reaction types, the larger the rate constant, the reaction rate will be faster.

10.3.1 Differential and Integral Rate Equations

Rate equations shown in Eq. (10.6) can be expressed in the form of differential equations. Equations (10.7)–(10.9) show the rate equations for zeroth, first, and second-order reactions with respect to reactant A, respectively:

$$\gamma_A = \frac{dC_A}{dt} = -k\ \left(0^{\text{th}}\ \text{order rxn}\right) \tag{10.7}$$

$$\gamma_A = \frac{dC_A}{dt} = -kC_A\ \left(1^{\text{st}}\ \text{order rxn}\right) \tag{10.8}$$

$$\gamma_A = \frac{dC_A}{dt} = -k(C_A)^2 \left(2^{\text{nd}} \text{ order rxn}\right) \tag{10.9}$$

where γ_A = the rate of conversion of reactant A, and C_A = the concentration of reactant A at time t. Values of γ_A come with a unit of "concentration/time".

Zeroth-order reaction. As shown in Eq. (10.7), the rate of change of C_A is constant; and it is independent C_A. The unit of the rate constant (k) is "concentration over time". The rate equation (Eq. 10.7) can be integrated to:

$$C_A = C_{A0} - kt \left(0^{\text{th}} \text{ order rxn}\right) \tag{10.10}$$

where C_{A0} = the initial concentration of reactant A. The reaction order and the value of the rate constant are usually determined from fitting the experimental data (i.e., concentration versus time) using the integrated rate equation. Figure 10.1 is a plot of C_A versus time for a zeroth-order reaction. As shown, the concentration of reactant A decreases linearly.

First-order reaction. As shown in Eq. (10.8), the rate of change of the concentration of reactant A (C_A) is proportional to C_A. In other words, the larger the reactant concentration, the faster the reaction rate; thus, the instantaneous rate is the highest at the start of the reaction. The first-order kinetics is applicable in many environmental engineering applications. The unit of k is "1/time". The rate equation (Eq. 10.8) can be integrated to:

$$C_A/C_{A0} = e^{-kt}; \text{ or } \ln\left(C_A/C_{A0}\right) = -kt \left(1^{\text{st}} \text{ order rxn}\right) \tag{10.11}$$

Figure 10.2 is a plot of the C_A versus time for a first-order reaction. As shown, the concentration of reactant A decreases exponentially.

For a first-order reaction, a plot of "*ln*(concentration)" versus time will be a straight line with a slope of $-k$; and it can be used to determine the value of the rate constant (see Figure 10.3).

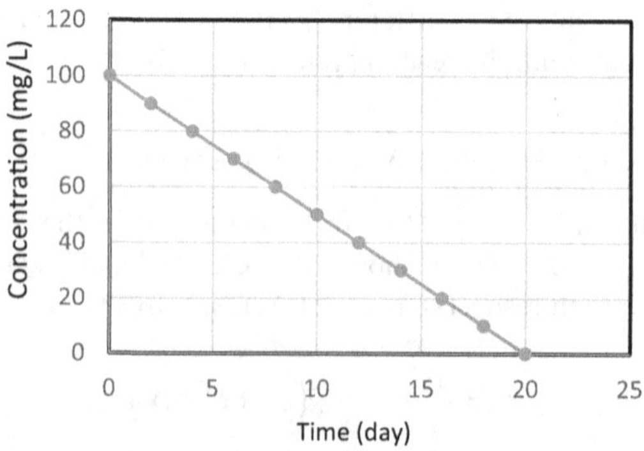

FIGURE 10.1 Concentration versus time for a zeroth-order reaction (C_{A0} = 100 mg/L and k = 5 mg/L/d).

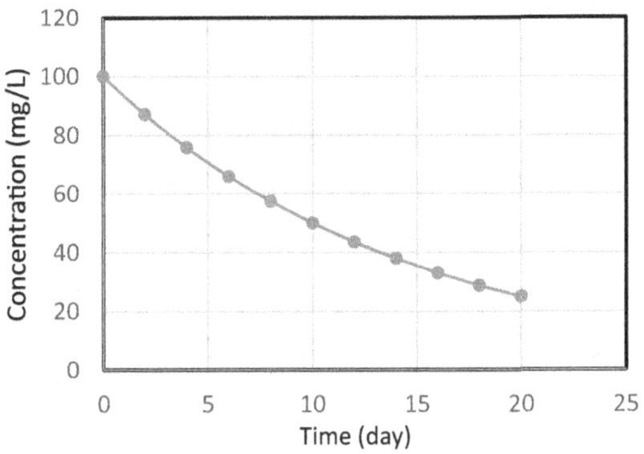

FIGURE 10.2 Concentration versus time for a first-order reaction (C_{A0} = 100 mg/L and k = 0.0693/d).

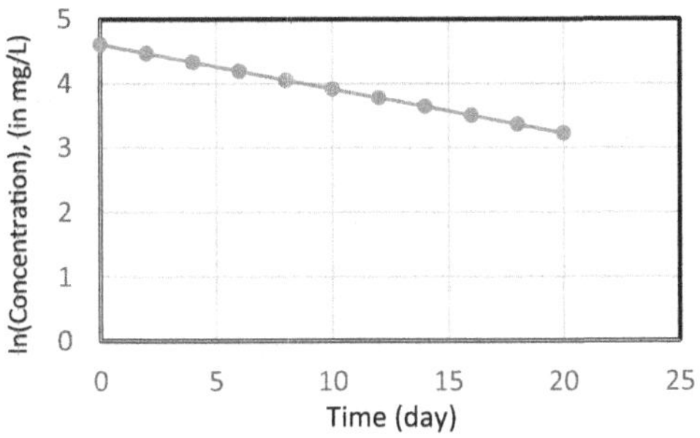

FIGURE 10.3 Concentration versus time for a first-order reaction (C_{A0} = 100 mg/L and k = 0.0693/d).

Second-order reaction. As shown in Eq. (10.9), the rate of change of C_A is proportional to the square of C_A. This is the simplest second-order reaction because it only involves one reactant. For this reaction, the larger the reactant concentration, the reaction rate will also be faster; thus, the instantaneous rate is the highest at the start of the reaction (similar to the first-order reactions). The unit of k is "1/[(concentration)(time)]". The rate equation (Eq. 10.9) can be integrated to:

$$\frac{1}{C_A} = \frac{1}{C_{A0}} + kt\ \left(2^{\text{nd}} \text{ order rxn}\right) \tag{10.12}$$

Figure 10.4 is a plot of the C_A versus time for a second-order reaction. As shown, the concentration of reactant A decreases continuously [Note: but not exponentially].

For a second-order reaction, a plot of "1/(concentration)" versus time will be a straight line with a slope of k; and it can be used to determine the value of the rate constant (see Figure 10.5).

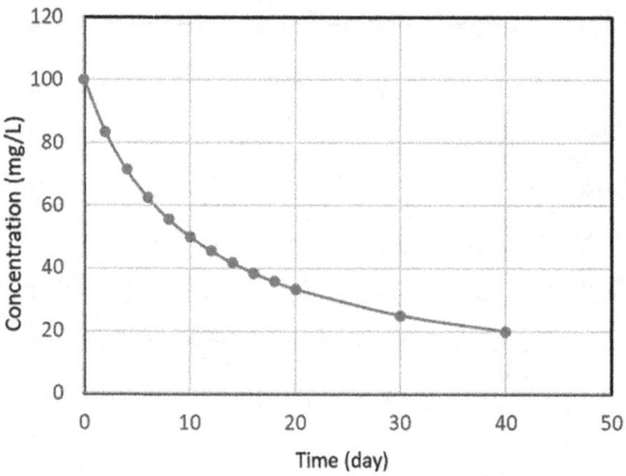

FIGURE 10.4 Concentration versus time for a second-order reaction (C_{A0} = 100 mg/L and k = 0.001 ((mg/L)·d)$^{-1}$).

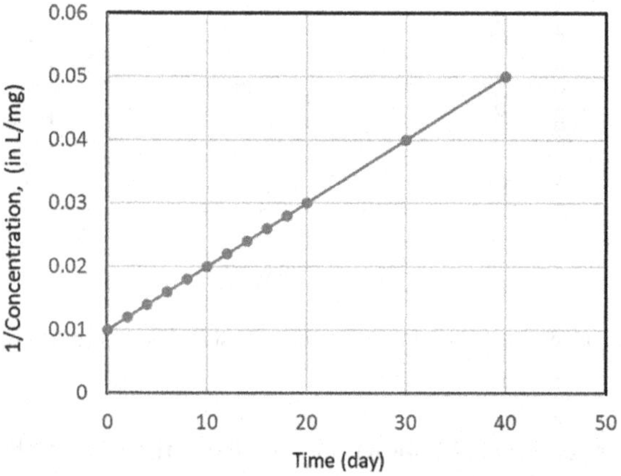

FIGURE 10.5 "1/Concentration" versus time for a second-order reaction (C_{A0} = 100 mg/L and k = 0.001 ((mg/L)·d)$^{-1}$).

Example 10.1: Estimate the Rate Constant from Two Known Concentration Values

Experimental data were used to analyze three chemical reactions. From the relationships of concentration and time, it was found that these three reactions are 0th-, 1st-, and second-order reactions, respectively. If the initial concentrations of all three reactions are 100 mg/L and all dropped to 50 mg/L at the end of 10 days, estimate the rate constants of these three reactions.

Solution:

Strategy: Insert the given concentrations into the integrated rate equations to find the values of the corresponding rate constants.

a. Zeroth-order reaction. Insert the initial concentration and the concentration at day 10 into Eq. (10.10) to obtain k:

$$50 = 100 - k(10)$$

So, $k = 5$ mg/L/d.

b. First-order reaction. Insert the initial concentration and the concentration at day 10 into Eq. (10.11) to obtain k:

$$\left(\frac{50}{100}\right) = e^{-k(10)}$$

So, $k = 0.0693$/d.

c. Second-order reaction. Insert the initial concentration and the concentration at day 10 into Eq. (10.12) to obtain k:

$$\frac{1}{50} = \frac{1}{100} + k(10)$$

So, $k = 0.001/(\text{mg/L·d})$.

Discussion:

1. Units of the rate constant are different if the orders of the reactions are different.
2. Numeric values of the rate constants of these three reactions are quite different.
3. Compare the values with the slopes of the straight line in Figures 10.1, 10.3, and 10.5.

Example 10.2: Rates of Concentration Changes for Different Reaction Orders

(Continuation from Example 10.1) With the rate constants derived from Example 10.1, estimate the additional time needed for the concentration to drop from 50 to 25 mg/L for each of these three reactions.

Solution:

Strategy: Insert the known data pair (50 and 25 mg/L) and the value of the rate constant into each integrated rate equation to find the time needed.

a. Zeroth-order reaction. Insert the data into Eq. (10.10) to obtain t:

$$25 = 50 - (5)(t)$$

So, $t = 5$ days.

b. First-order reaction. Insert the data into Eq. (10.11) to obtain t:

$$\left(\frac{25}{50}\right) = e^{-0.0693(t)}$$

So, $t = 10$ days.

c. Second-order reaction. Insert the data into Eq. (10.12) to obtain t:

$$\frac{1}{25} = \frac{1}{50} + 0.001(t)$$

So, $t = 20$ days.

Discussion:

1. The initial concentration (C_0) used is 50 mg/L and the time t found is the time needed to drop the concentration from 50 to 25 mg/L. If 100 mg/L were used, the obtained t values would be the time needed from day 0.
2. For all three reactions, it takes 10 days to drop the concentration from 100 to 50 mg/L (i.e., a 50% decrease). However, it takes 5, 10, and 20 days for the reactions of 0th, 1st, and second order to drop the concentration another 50%, respectively. The differences come from the fact that the rate of reaction (γ) is independent of C, proportional to C, and to C^2 for the reactions of zeroth, first, and second order, respectively.
3. Compare the calculated values with those shown in Figures 10.1–10.5.

Example 10.3: Reaction Rates are Additive

An accidental gasoline spill occurred at a site 20 days ago. The initial total petroleum hydrocarbon (TPH) concentration in soil was determined to be 2,000 mg/kg. The decrease in TPH concentration could be attributed to natural biodegradation and volatilization. Assuming both removal mechanisms are first-order, and the rate constants for natural biodegradation (k_b) and volatilization (k_v) are 0.003 and 0.006/d, respectively. Estimate how long will it take for the concentration to drop below the permissible level of 100 mg/kg, due to these two natural attenuation processes.

Solution:

Strategy: Two removal mechanisms (i.e., biodegradation and volatilization) are occurring simultaneously, and both are first-order. Consequently, these two mechanisms are additive; and they can be represented by one single equation with a combined rate constant.

$$\frac{dC}{dt} = -k_bC - k_vC = -(k_b + k_v)C = -kC \tag{10.13}$$

a. The reaction rate from both removal mechanisms = 0.003 + 0.006 = 0.009/d.

b. For the concentration to drop below 100 mg/kg, it will take (from Eq. 10.11):

$$\left(\frac{100}{2,000}\right) = e^{-0.009(t)}$$

$t = 332.8$ days.

Discussion:

1. The rate constants are additive, only if the reactions have the same rate law.
2. Since the rate constant of volatilization is twice that of biodegradation, the removal of TPH mass by volatilization should be twice of that by natural biodegradation.

10.3.2 Half-Life

Half-life ($t_{1/2}$) can be defined as the time for the concentration of a compound to drop to half of its initial value. Example 10.2 illustrates that the half-lives of compounds following the 0th- and second-order reactions depend on the initial concentration. However, the half-life of compounds following the first-order reaction is a constant (i.e., independent of the initial concentration). Many reactions occurring in the environment were found to be of first-order. Half-life is often used to describe the fate of a compound in the environment. The $t_{1/2}$ of a compound experiencing a first-order reaction can be found from Eq. (10.11) by substituting C_A with one half of C_{A0} (i.e., $C_A = 0.5C_{A0}$):

$$t_{1/2} = \frac{\ln 2}{k} = \frac{0.693}{k} \tag{10.14}$$

As shown in Eq. (10.14), the half-life and the rate constant are inversely proportional for first-order reactions. In other words, if the value of $t_{1/2}$ is given, the value of k can be readily found from Eq. (10.14); and vice versa.

On some occasions, the decay rate is expressed as T_{90} instead of $t_{1/2}$ (e.g., natural decay of coliform in surface water bodies). T_{90} is the time required for 90% of the compound converted (or its concentration dropped to 10% of the initial value). Value of T_{90} of a compound experiencing a first-order reaction can be found from Eq. (10.11) by substituting C_A with one tenth of C_{A0} (i.e., $C_A = 0.1C_{A0}$):

$$T_{90} = \frac{\ln 10}{k} = \frac{2.30}{k} \qquad \text{[Eq. 10.15]}$$

Example 10.4: Half-life Calculation

The half-life of 1,1,1-trichloroethane (1,1,1-TCA) in the subsurface was determined to be 180 days. Assume that all the removal mechanisms are first-order, determine (1) the rate constant and (2) the time needed to drop the concentration down to 25% and 10% of the initial concentration, respectively.

Solution:

a. The rate constant can be easily determined from Eq. (10.14) as:

$$t_{1/2} = 180 = \frac{\ln 2}{k} = \frac{0.693}{k}$$

Thus, $k = 0.00385$/d.

b. It takes two half-lives ($t_{1/2}$) to drop the concentration from the initial concentration to 25% (or another $t_{1/2}$ from 50% to 25%), thus $t = 2(180) = 360$ days.

c. Use Eq. (10.11) to determine the time needed to drop the concentration down to 10% of the initial value (i.e., $C = 0.1C_0$):

$$\frac{C}{C_0} = \frac{1}{10} = e^{-(0.00385)(t)}$$

Thus, $t = 598$ days.

Discussion:

1. The approach is only applicable to first-order reactions.
2. It will take three $t_{1/2}$ (i.e., 540 days) to drop the concentration to 12.5% of the original value.
3. Part (c) can also be solved using the equation for T_{90} (i.e., Eq. 10.15).

Example 10.5: Half-life Calculation

Methyl mercury (CH_3Hg^+) is an organic form of mercury. It is toxic and bioaccumulative. If the metabolic process for expelling it from the human body is a first-order reaction and the average excretion rate is 2% of the total body burden per day, determine the half-life of this compound in the body and how long it will take to drop the concentration in the body by 90%.

Solution:

a. As given, $k = 0.02$/day, the half-life can be readily found from Eq. (10.14):

$$t_{1/2} = \frac{\ln 2}{k} = \frac{0.693}{0.02} = 34.5 \text{ days}$$

b. Time to have a 90% reduction in the original concentration can be found using Eq. (10.15):

$$T_{90} = \frac{2.30}{k} = \frac{2.30}{0.02} = 115 \text{ days}$$

Discussion:

1. The approach is only applicable to first-order reactions.
2. The 90% reduction/removal is often called one-log reduction/removal.
3. Part (b) can also be solved using Eq. (10.11).

10.4 REACTION MECHANISMS

Equation (10.1) shows the overall reaction between reactants A and B and final products C and D in a single/simple step. The reaction may go through several "elementary" steps in a microscopic scale. Different intermediate compounds (or called "*intermediates*") may be formed in an elementary step and consumed in a subsequent step or before the completion of the overall reaction. Elementary steps usually do not move at the same speed; some faster/smaller than others. The slowest elementary step is called "*rate-determining*" step which controls the reaction rate of the overall reaction. The *reaction mechanism* describes the sequence and details of all the elementary steps.

For example, tetrachloroethylene (PCE, C_2Cl_4) is a common groundwater pollutant. If in-situ degradation is occurring, PCE could be stepwise degraded in several intermediate steps anaerobically: C_2Cl_4 → trichloroethylene (TCE, C_2HCl_3) → dichloroethylene (DCE, $C_2H_2Cl_2$) → vinyl chloride (C_2H_3Cl). The vinyl chloride can then be aerobically degraded to complete the PCE demineralization to produce chloride ions, carbon dioxide, and water. During this process, the concentration profiles of the intermediates (i.e., C_2HCl_3, $C_2H_2Cl_2$, and C_2H_3Cl) could show different peaks sequentially.

10.5 EFFECTS OF TEMPERATURE ON REACTION RATES

10.5.1 Collision Theory

Collision theory is used to predict the chemical reaction rates. The following statements are commonly used to describe the collision theory:

- A chemical reaction can occur only after the reacting species (atoms or molecules) come together or have collisions.
- Not all collisions will result in a chemical reaction – a collision will produce a chemical reaction only if the collided species possesses a minimum amount of internal energy (i.e., activation energy) to convert the reactants into an "activated complex".
- In addition, the colliding species must be oriented favorably at collision to facilitate the necessary rearrangement of atoms and electrons to form the activated complex.

Temperature is the most important parameter in controlling the reaction rates. Increased temperature means (1) increasing average velocity of reacting species and, consequently, the number of collisions between the species; and (2) increasing kinetic energy of the reacting species, and, thus, increasing the percentage of collisions with sufficient energy to form the activated complex.

10.5.2 Activation Energy

Activation Energy (E_a) is the minimum amount of energy that must be provided to the reactants to initiate a chemical reaction. It is considered as the energy barrier that needs to be overcome. At the top of the energy barrier, the atoms of the reactants formed an "*activated complex*", or they are in the "*transition state*" (see Figure 10.4). The difference between the energy level of the activated complex and that of the reactants is the *activation energy*, and it always has a positive value.

The difference between the sum of the energies of the products and that of the reactants is the *heat of reaction*. If the difference is negative (i.e., the products

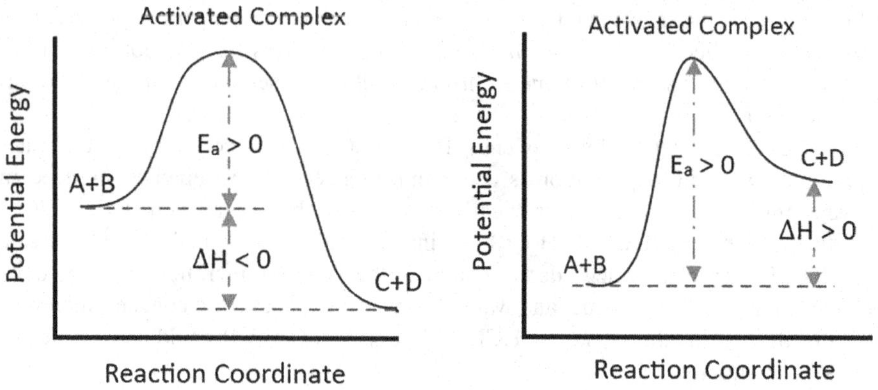

FIGURE 10.6 Activation energy and heat of reaction.

possess a smaller total energy than the reactants) as shown in Figure 10.6, it is an "*exothermic reaction*" (i.e., $\Delta H_{\text{reaction}} < 0$); energy will be released to the environment). Oppositely, it is an "*endothermic reaction* ($\Delta H_{\text{reaction}} > 0$)".

Activation energy of a chemical reaction is supplied, in most cases, by thermal energy which can come from intermolecular collisions or thermal excitation of bond-stretching vibrations.

Temperature of a substance is directly proportional to the average kinetic energy of its particles. Since the mass of these particles is constant, so that the particles must move faster as the temperature rises. However, the particles are not moving at the same speed. The *Maxwell-Boltzmann distribution* describes the distribution of the moving speeds of these particles in a gas sample. The distribution is often illustrated graphically, with particle speed (or the kinetic energy) on the *x*-axis and the relative number of particles on the *y*-axis (Figure 10.5). The area under a curve or a part of the curve is proportional to the number of molecules represented (Figure 10.7).

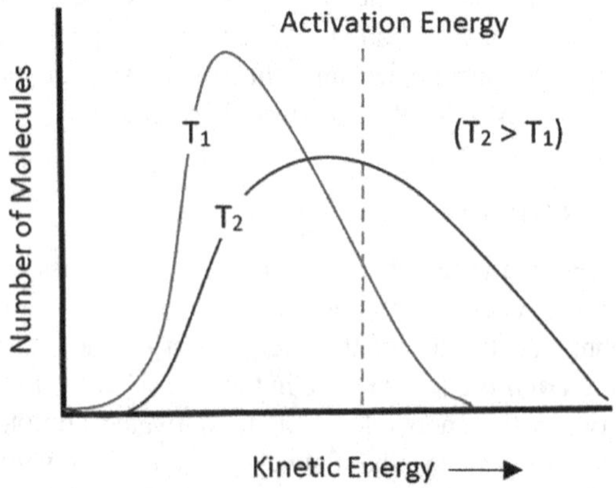

FIGURE 10.7 Kinetic energy as a function of temperature.

10.5.3 The Arrhenius Equation

A lower activation energy means a faster reaction (a larger reaction rate) because the barrier is smaller. Activation energy is independent of temperature. However, molecules would possess larger kinetic energies at elevated temperatures, a reaction will be faster at higher temperatures. The "*Arrhenius Equation*" relates the reaction rate constant (k) with the activation energy (E_a) and the absolute temperature (T) as:

$$k = Ae^{-E_a/RT} \tag{10.16}$$

where R is the universal gas constant and A is the frequency factor which is related to the frequency of collisions and the orientations of molecules at collisions (some orientations are more favorable than others).

The Arrhenius Equation can also be used to find the reaction rates at different temperatures as

$$\ln\frac{k_2}{k_1} = -\frac{E_a}{R}\left(\frac{1}{T_2} - \frac{1}{T_1}\right) \tag{10.17}$$

In addition, the pressure of the system can also be an important factor affecting the rate of reaction, especially for the gaseous reactions. At higher pressures, the concentrations of gaseous compounds would be higher so that the rates of most reactions would be greater. Consequently, the rates of most of reactions are faster at elevated temperatures and pressures. Using fossil fuel combustion as an example, coal is not combustible under ambient conditions. Starting fluid would be needed initially to provide energy to overcome the barrier (i.e., the activation energy). After that, the reaction will get started and proceed rapidly at higher temperatures. Fuel combustion is an exothermic reaction in which heat is released to its surroundings.

Example 10.6: Relationships among Activation Energy, Reaction Rate, and Temperatures

The rate of a chemical reaction doubles when the temperature changes from 20°C to 30°C (an increase of 10°C).

a. Estimate the activation energy of the reaction, assuming it is independent of temperature.
b. Compared to the reaction rate @20°C, how much faster if the reaction occurs at 520°C?

Solution:

a. The ratio of two rate constants and two temperatures are used in Eq. (10.17) as:

$$\ln\frac{k_2}{k_1} = \ln 2 = -\frac{E_a}{R}\left(\frac{1}{T_2} - \frac{1}{T_1}\right) = -\frac{E_a}{8.314}\left(\frac{1}{303} - \frac{1}{293}\right)$$

$E_a = 51{,}162$ J/mole

b. The ratio of the reaction rates at 520°C:

$$\ln\frac{k_2}{k_1} = -\frac{E_a}{R}\left(\frac{1}{T_2}-\frac{1}{T_1}\right) = -\frac{51{,}162}{8.314}\left(\frac{1}{793}-\frac{1}{293}\right)$$

$k_{520}/k_{20} = 5.64 \times 10^5$

Discussion:

1. $R = 8.314$ J/mole-K
2. Temperatures used in the equations need to be absolute temperatures.
3. The reaction rate is much faster at elevated temperatures. In this case, the reaction rate @520°C is 5.64×10^5 times faster than that at 20°C.

10.5.4 Catalysts

A *catalyst* is a substance that participates in a chemical reaction and accelerates the reaction rate; however, it does not undergo a permanent chemical change at the end of the reaction. Compound E in Eq. (10.18) would be a catalyst for the reaction shown in Eq. (10.1) if its participation accelerates the reaction rate:

$$aA + bB + E \rightarrow cC + dD + E \tag{10.18}$$

Catalysts generally provide a different reaction mechanism so that the activation energy becomes lower (and the reaction rate becomes faster) – See Figure 10.8. Some catalysts can increase the value of the frequency factor (A). When a catalyst is present

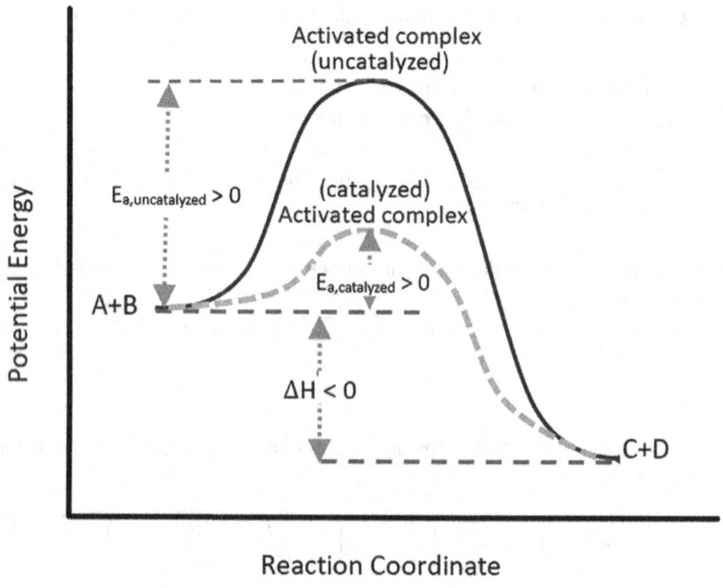

FIGURE 10.8 Activation energy for catalyzed and uncatalyzed reactions.

in the same phase as the reactants, they are "*homogenous catalysts*". In photochemical smog, ozone is formed when sunlight and heat cause reactions between nitrogen dioxide (NO_2) and volatile organic compounds (VOCs). In this case, VOCs work as homogeneous catalysts. Catalytic converters in automobiles typically use precious metals (such as platinum and palladium), as the catalyst, on a supporting substrate to convert the pollutants in engine exhausts to less toxic ones. It is "*heterogeneous catalysis*" because these catalysts are not in the gaseous phase and the initial step of reaction is adsorption of gaseous pollutants onto the solid surface. Enzymes are the catalysts that participate in biological reactions, and they are usually proteins.

Example 10.7: Effect of a Catalyst in a Chemical Reaction

(Continuation of Example 10.6) If the activation energy of the reaction in Example 10.6 (51,162 J/mole) is reduced by 20% due to employing a catalyst in the reaction. How much faster the reaction would be at 520°C?

Solution:

a. Assuming the frequency factor stays the same, the ratio of two rate constants for two reactions of different activation energies (at same *T*) can be found from using Eq. (10.16) as:

$$\frac{k_2}{k_1} = Ae^{-(E_{a2}/RT)} \div Ae^{-(E_{a1}/RT)} = \exp\left[(E_{a1} - E_{a2})/RT\right] \tag{10.18}$$

b. The ratio of the reaction rates at 520°C:

$$\frac{k_2}{k_1} = exp\left[\frac{(E_{a1} - E_{a2})}{RT}\right] = exp\left[\frac{(20\%)(51{,}162)}{(8.314)(793)}\right] = 4.72$$

Discussion:

The reaction rate at 520°C is 4.72 times of the original value with a 20% decrease in activation energy.

EXERCISE QUESTIONS

1. Experimental data were used to analyze three chemical reactions. From the relationships of concentration and time, it was found that these three reactions are of zeroth, first, and second-order, respectively. If the initial concentrations are the same (80 mg/L) and all dropped to 40 mg/L at the end of 10 minutes, estimate the rate constants of these three reactions
2. a. With the rate constants derived from Question #1, estimate the time for the concentrations to drop from 40 to 20 mg/L for each of these three reactions
   b. Compared the times needed for concentrations to drop 50% in Questions #1 and part (a) for all three reactions.
3. An accidental gasoline spill occurred at a site 20 days ago. The initial TPH concentration at a specific location is 1,000 mg/kg. The decrease in TPH

concentration could be attributed to natural biodegradation and volatilization. Assuming both removal mechanisms are first-order, and the rate constants for natural biodegradation (k_b) and volatilization (k_v) are 0.002 and 0.005/d, respectively.

   a. Estimate how long will it take for the concentration to drop to 500 mg/kg, due to these two natural attenuation processes.
   b. Estimate how long will it take for the concentration to drop to permissible 100 mg/kg, due to these two natural attenuation processes.

4. The half-life of perchloroethylene (C_2Cl_4) in subsurface was determined to be 100 days. Assume that all the removal mechanisms are first-order, determine (1) the rate constant and (2) the time needed to drop the concentration down to 50%, 25% and 10% of the initial concentration, respectively.
5. Methyl mercury (CH_3Hg^+) is an organic form of mercury. It is toxic and bio-accumulative. The metabolic process for expelling it from the human body is a first-order reaction and the average excretion rate is 1% of the total body burden per day.
   a. Determine the half-life of this compound in the body and how long it will take to drop the concentration in the body by 50%, 75%, and 90%, respectively.
   b. Compare your results with that found in Example 10.5.
6. The rate of a chemical reaction doubles when the temperature changes from 10°C to 25°C (an increase of 15°C).
   a. Estimate the activation energy of the reaction, assuming it is independent of temperature.
   b. Compared to the reaction rate at 10°C, how much faster if the reaction occurs at 600°C?
7. a. If the activation energy of the reaction in Question #6 is reduced by 20% due to employing a catalyst in the reaction. How much faster the reaction would be at 600°C?
   b. If the activation energy of the reaction in Question #6 is reduced by 30% due to employing a catalyst in the reaction. How much faster the reaction would be at 600°C?

11 Types and Design of Chemical Reactors

11.1 INTRODUCTION

As mentioned in Chapter 4, the main objective of a treatment process is to reduce the concentration/toxicity of contaminants in an environmental medium (e.g., water, soil, and air). A treatment process can be physical, chemical, biological, or thermal. It is often necessary to assemble a treatment process train, which consists of several unit processes in series, to meet the treatment goal. For each unit process, physical device(s) is needed to hold the medium (i.e., water, wastewater, soil, or air) to be treated for a sufficient time; appropriate operating conditions (e.g., temperature, pressure, pH, chemical doses, right types of microorganism, dissolved oxygen concentrations, etc.) need to be applied so that the treatment goal can be reached. These physical devices can be considered as reactors in which the desired treatment processes take place.

11.2 MASS BALANCE CONCEPT

Mass balance (or *material balance*) concept serves as a basis for designing environmental engineering systems (reactors). The mass balance concept is nothing but conservation of mass. Matter can neither be created nor destroyed (a nuclear process is one of the few exceptions), but it can be changed in form. The fundamental approach to analysis is to explore the changes occurring in a reactor by applying the mass-balance concept. The following is the general form of the mass balance equation:

$$\begin{bmatrix}\text{Rate of mass}\\ \text{ACCUMULATED}\end{bmatrix} = \begin{bmatrix}\text{Rate of mass}\\ \text{IN}\end{bmatrix} - \begin{bmatrix}\text{Rate of mass}\\ \text{OUT}\end{bmatrix} \pm \begin{bmatrix}\text{Rate of mass}\\ \text{GENERATED or}\\ \text{DESTROYED}\end{bmatrix} \quad (11.1)$$

This general mass balance equation can also be expressed as:

$$V\frac{dC}{dt} = \sum Q_{in}C_{in} - \sum Q_{out}C_{out} \pm (V \times \gamma) \quad (11.2)$$

where V = the volume of the system (the reactor), C = the compound concentration inside the reactor, Q_{in} = the influent flow rate, C_{in} = the influent concentration,

DOI: 10.1201/9781003502661-11

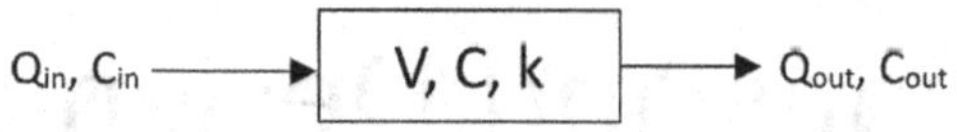

FIGURE 11.1 Mass balance around a chemical reactor.

Q_{out} = the effluent flow rate, C_{out} = the effluent concentration, γ = the reaction rate, and t = time. Figure 11.1 illustrates the mass balance around a reactor (where k = reaction rate constant). The subsequent sections will demonstrate the role of the reaction in the mass balance equation and how it affects the reactor design.

Performing a mass balance on an environmental engineering system is similar to balancing a checking account. The rate of mass accumulated (or depleted) in a reactor can be viewed as the rate that money is accumulated in (or withdrawn from) the checking account. How fast the balance would change depends on how much and how often the money is deposited and withdrawn (i.e., the rate of mass input and output), interest incurred (i.e., the rate of mass generated), and bank charges for service and ATM fees (i.e., the rate of mass destroyed).

In using the mass balance concept to analyze an environmental engineering system, we usually start with drawing a process flow diagram and employing the following procedure:

Step 1: Draw system boundaries or boxes around the unit processes/operations or flow junctions to facilitate calculations.
Step 2: Place known flow rates and concentrations of all streams, sizes, and types of reactors, as well as operating conditions such as temperature and pressure on the diagram.
Step 3: Calculate and convert all known mass inputs, outputs, and accumulation/disappearance to having the same units and place them on the diagram.
Step 4: Mark the unknown (or the ones to be found) inputs, outputs, and accumulation/disappearance on the diagram.
Step 5: Perform the necessary analyses/calculations using the procedure described in this chapter.

A few special cases or reasonable assumptions would simplify the general mass balance equation (i.e., Eq. 11.1) and make the analysis easier. Three common ones are:

1. <u>No Reactions Occurring</u>: If the system has no chemical reactions occurring (e.g., a physical process such as sedimentation), there will be no increases or decreases of compound mass due to reactions. The mass balance equation would become:

$$\begin{bmatrix} \text{Rate of mass} \\ \text{ACCUMULATED} \end{bmatrix} = \begin{bmatrix} \text{Rate of mass} \\ \text{IN} \end{bmatrix} - \begin{bmatrix} \text{Rate of mass} \\ \text{OUT} \end{bmatrix} \quad (11.3a)$$

$$V\frac{dC}{dt} = \sum Q_{in}C_{in} - \sum Q_{out}C_{out} \quad (11.3b)$$

2. Batch Reactor: For a batch reactor, there is no input into and output out of the reactor (more details later). The mass balance equation can be simplified into:

$$\left[\begin{array}{c}\text{Rate of mass}\\ \text{ACCUMULATED}\end{array}\right] = \pm\left[\begin{array}{c}\text{Rate of mass}\\ \text{GENERATED or}\\ \text{DESTROYED}\end{array}\right] \quad (11.4a)$$

$$V\frac{dC}{dt} = \pm(V\times\gamma) \quad (11.4b)$$

3. Steady-State Conditions: To maintain the stability of treatment processes, treatment systems are usually operated under steady-state conditions after the start-up period. A *steady-state condition* basically means that flow and concentrations at any locations within the treatment process train are not changing with time. Although the influent concentration and the influent flow rate typically fluctuate, engineers may want to incorporate devices such as equalization tanks to dampen the fluctuations. This is especially true for treatment processes that are sensitive to fluctuations in mass loading (e.g., biological processes).

   For a reactor under a steady-state condition, although reactions are occurring, the rate of mass accumulation in the reactor would be zero (because of no change in concentration). Consequently, the term on the left-hand side of Eq. (11.1) becomes zero. The mass balance equation can then be reduced to:

$$0 = \left[\begin{array}{c}\text{Rate of mass}\\ \text{IN}\end{array}\right] - \left[\begin{array}{c}\text{Rate of mass}\\ \text{OUT}\end{array}\right] \pm \left[\begin{array}{c}\text{Rate of mass}\\ \text{GENERATED or}\\ \text{DESTROYED}\end{array}\right] \quad (11.5a)$$

$$0 = \sum Q_{\text{in}}C_{\text{in}} - \sum Q_{\text{out}}C_{\text{out}} \pm (V\times\gamma) \quad (11.5b)$$

Assumption of steady state is frequently used in the analysis of flow reactors. It should be noted that a batch reactor is not operated under steady state (i.e., un-steady state) because the concentration in the reactor is changing; and it is not a flow reactor because there is no flow in and out of the reactor when it is in operation.

Example 11.1: Mass Balance Equation - Air Dilution (No Chemical Reaction Occurring)

A glass bottle containing 600 mL of methylene chloride (CH_2Cl_2, specific gravity = 1.335) was accidentally left uncapped in a poorly ventilated room (5 m × 6 m × 4 m) over a weekend (T = 20°C). On the following Monday, it was found that one-third of methylene chloride had volatilized. An exhaust fan ($Q = 200$ ft^3/min) was turned on to vent the fouled air out of the room.

a. What is the original methylene chloride concentration before ventilation?
b. How long will it take to reduce the concentration down below the Occupational Health and Safety Association's (OSHA's) short-term exposure limit (STEL) of 125 ppmV?

Strategy: This is the first of the three special cases (i.e., no reactions occurring) of the general mass balance equation; thus, Eq. (11.3b) applies:

$$V\frac{dC}{dt} = \sum Q_{in}C_{in} - \sum Q_{out}C_{out}$$

The equation can be further simplified with the following assumptions:

1. The air leaving the laboratory is only through the exhaust fan and the air ventilation rate is equal to the rate of air entering the laboratory ($Q_{in} = O_{out} = Q$)
2. The air entering the room does not contain methylene chloride ($C_{in} = 0$)
3. The air in the room is fully mixed, thus the concentration of methylene chloride in the room is uniform and is the same as that of the air vented by the fan ($C = C_{out}$).

$$V\frac{dC}{dt} = -Q_{out}C_{out} = -QC \qquad (11.6)$$

It is a first-order differential equation, and it can be integrated with $C = C_0$ at $t = 0$:

$$\frac{C}{C_0} = e^{-\left(\frac{Q}{V}\right)t}; \text{or } C = C_0\, e^{-\left(\frac{Q}{V}\right)t} \qquad (11.7)$$

Solution:

a. Mass of methylene chloride volatilized = (liquid volume) × (density)

= [(1/3)(600 mL)](1.335 g/mL) = 267 g = 2.67 × 10^5 mg

Vapor concentration in mass/volume = (mass) ÷ (volume)

= (2.67 × 10^5 mg)/[(5m)(6m)(4m)] = 2,225 mg/m^3

MW of methylene chloride = 85. At $T = 20°C$ and $P = 1$ atm, from Eq. (6.12),

1 ppmV of methylene chloride = (85/24.05) mg/m^3 = 3.53 mg/m^3

Initial methylene chloride concentration = 2,225 mg/m^3 ÷ 3.53 mg/m^3/ppmV = <u>630 ppmV</u>

b. The system flow rate (Q) = ventilation rate

$$= 200\ ft^3/min = (200\ ft^3/min) \div (35.3\ ft^3/m^3) = 5.66\ m^3/min$$

The initial concentration, $C_0 = 630$ ppmV and $C_{final} = 125$ ppmV; use Eq. (11.7):

$$125 = (630)\, e^{-\left(\frac{5.66}{5\times6\times4}\right)t}$$

Thus, t = 34.3 minutes.

Discussion:

1. The actual time required would be longer than 34.3 minutes, because the assumption of completely mixed inside the room may not be valid. In addition, if the ambient air contains some methylene chloride, the cleanup time would be longer.
2. The initial methylene concentration exceeds the STEL.

11.3 TYPES OF REACTORS

Reactors are typically classified based on their flow characteristics as well as the mixing conditions within the reactors. Reactors may be operated either in a batch or in a continuous flow mode. For a *batch reactor*, the reactor is charged with the reactants, the content is well mixed and left to react. At the end of a specified time duration, the resulted mixture (i.e., the products) is discharged. A batch reactor is an unsteady-state reactor because the composition of the reactor content changes with time (see Figure 11.2). The capital cost of a batch reactor is usually less than that of a continuous flow reactor, but it is very labor-intensive, and the operating cost is higher. It is usually limited to small installations and to the cases when the raw materials are expensive (the operating conditions would be carefully monitored and adjusted, if necessary).

In a *continuous flow reactor*, the feed to the reactor and the discharge from it are continuous. In most of the cases, the flow reactors are operated under *steady-state conditions* in which the influent flow rate, its composition, the reaction condition in the reactor, and the effluent flow rate and its concentration are constant with respect to time. Frequently, the reaction kinetics is studied in a laboratory by using a batch reactor. The obtained reaction rate constant is then applied to the design of a continuous flow reactor. In general, there are two ideal types of flow reactors, they are: *continuous flow stirred tank reactor* (CFSTR) and *plug flow reactor* (PFR). They are classified mainly by the mixing conditions within the reactors (see Figure 11.2).

A CFSTR consists of a stirred tank that has feed stream(s) of the reactants and discharge stream(s) of the reacted materials. The CFSTR is usually round, square, or slightly rectangular in a plan view; and it is mandatory to have sufficient mixing. The stirring/mixing of a CFSTR is extremely important. It is assumed that the fluid in a CFSTR is perfectly mixed (i.e., the content is uniform throughout the entire

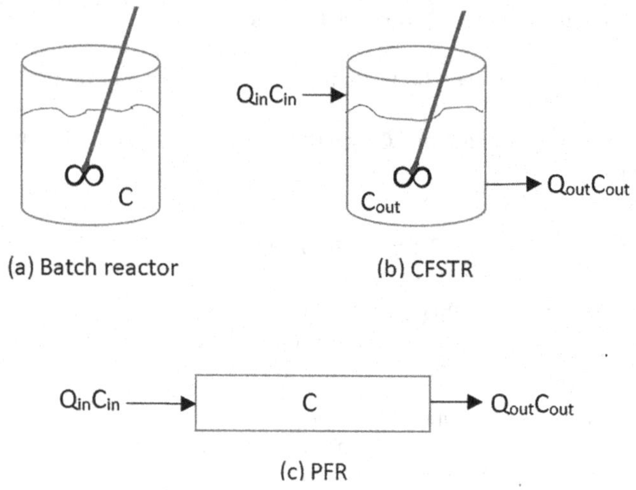

FIGURE 11.2 Types of reactors: (a) batch reactor, (b) CFSTR, and (c) PFR.

reactor volume). As a result of mixing, the composition of the discharge stream(s) is the same as that of the reactor content. Therefore, it is also called a *completely stirred tank reactor* (CSTR) or *completely mixed flow reactor* (CMF). Under the steady-state conditions, the concentration of the effluent and that at any location within the reactor are the same and should not change with time.

A PFR ideally has the geometric shape of a long tube or tank, and the flow is continuous, and the fluid particles pass through the reactor sequentially. The reactants enter at the upstream end of the reactor, and the products leave at the downstream end. Ideally, there is no induced mixing between elements of fluid along the direction of flow. The fluid particles that enter the reactor first will leave first. The composition of the reacting fluid in the PFR changes in the direction of flow. For the case of COC removal or destruction, the concentration will be the largest at the entrance and dropped continuously to the effluent value at the exit condition. Under the steady-state conditions, the effluent concentration and concentration at any location within the PFR reactor should not change with time.

It should be noted that CFSTRs and PFRs are "ideal" reactors. The continuous flow reactors in the real world behave somewhere between these ideal cases. The ideal CFSTRs are more resistant to shock loadings because the influent would be mixed with the reactor content immediately. They are a better choice if the process is sensitive to shock loadings (e.g., biological processes). On the other hand, the ideal PFRs provide the same residence time for all the components in the influent flow. They are a better choice for chlorine contact tanks in which a minimum contact time between pathogens and disinfectants is needed [Note: The residence times of the influent parcels in an ideal CFSTR vary from extremely short to extremely long].

11.3.1 Batch Reactors

Let us consider a batch reactor with a first-order reaction. Inserting Eq. (10.8) into Eq. (11.4b), the mass balance equation becomes:

$$V\frac{dC}{dt} = V \times \gamma = V(-kC); \text{ thus } \frac{dC}{dt} = -kC \tag{11.8}$$

It is a first-order differential equation and can be integrated with $C = C_i$ @ $t = 0$; then the concentration at any time t (C_t) can be found as:

$$\frac{C_t}{C_i} = e^{-kt}; \text{ or } C_t = C_i \times e^{-kt} \tag{11.9}$$

Table 11.1 tabulates the design equations for batch reactors in which zeroth-, first-, and second-order reactions take place.

Example 11.2: Batch Reactor (Find the Required Residence Time with a Known Rate Constant)

A batch reactor is to be designed to treat wastewater containing 20 mg/L of benzene (C_6H_6). The required removal is 90%.

a. The rate constant is 0.5 hr^{-1}, what is the required reaction time in this batch reactor?
b. What is the required reaction time if the desired final concentration is 1.0 mg/L?

Strategy: Although the order of the reaction is not mentioned in the problem statement, it is a first-order reaction because the unit of k is 1/time.

Solution:

a. For a 90% reduction (η = 90%), $C_f = C_i(1 - \eta)$ = (20)(1 − 90%) = 2.0 mg/L.

Insert the known values into Eq. (11.9),

$$\frac{C_t}{C_i} = e^{-kt} = \frac{2.0}{20} = e^{-(0.5)t}$$

t = 4.6 hours

TABLE 11.1
Design Equations for Batch Reactors

Order of Reaction	Design Equation	
0	$C_t = C_i - kt$	(11.10)
1	$C_t = C_i \times e^{-kt}$	same as (11.9)
2	$C_t = \dfrac{C_i}{1+(kt)C_i}$	(11.11)

b. To achieve a final concentration of 1.0 mg/L, use Eq. (11.9) again:

$$\frac{C_t}{C_i} = e^{-kt} = \frac{1.0}{20} = e^{-(0.5)t}$$

$$t = 6.0 \text{ hours}$$

Example 11.3: Batch Reactor (Determine the Required Residence Time with an Unknown Rate Constant)

A batch reactor was installed to treat water contaminated with benzene. A trial run was conducted with an initial benzene concentration of 25 mg/L. After 10 hours of batch-wise operation, the benzene concentration dropped to 5.0 mg/L. However, it is required to reduce the concentration down to 1.0 mg/L. Determine the required residence time to achieve the final concentration of 1.0 mg/L.

Strategy: It requires a two-step approach to solve problem. The first is to determine the rate constant using the given information. Then, use this obtained k value to estimate the residence time for other conversions. The given information did not tell us the order of the reaction. We assume it is a first-order reaction, and it should be confirmed with additional test data.

Solution:

a. Insert the known values into Eq. (11.9), to find the value of the rate constant (k):

$$\frac{C_t}{C_i} = e^{-kt} = \frac{5.0}{25} = e^{-k(10)}$$

$$k = 0.161 \text{ hr}^{-1}$$

b. The time required to achieve a concentration of 1.0 mg/L by using Eq. (11.9) again:

$$\frac{C_t}{C_i} = e^{-kt} = \frac{1.0}{25} = e^{-0.161t}$$

$$t = 20.0 \text{ hours}$$

Discussion:

It is assumed that the reaction is a first-order reaction. One should check the validity of this assumption, for example, by running the pilot experiment longer to collect more data. For example, if the run is extended to 20 hours and the final concentration is close to 1.0 mg/L, the assumption of first-order kinetics should be relatively valid.

Example 11.4: Determine the Rate Constant from Batch Experiments

An in-vessel bioreactor is designed to remediate soil impacted by cresol (C_7H_8O). A bench-scale batch reactor was set up to determine the order and the rate constant of the reaction. The following concentrations of cresol in the batch reactor at various times were determined as:

Time (hours)	Cresol Concentration (mg/kg)
0	400
0.5	260
1	190
2	90
5	37

Use these data to determine the reaction order and the value of the rate constant.

Strategy: To determine the reaction order, a trial-and-error approach is often taken. From Table 11.1, if it is a zeroth-order reaction, the plot of concentration versus time should be a straight line. On the other hand, the plot of ln(*C*) versus time should be a straight line if it is first-order kinetics. If the reaction is of second order, the plot of (1/*C*) versus time will be a straight line, with a slope of *k*. The value of *k* is then obtained from the slope of the straight line (see Chapter 10).

Solution:

Many reactions of environmental concern are first-order reactions. Assuming the reaction is first-order, plot the concentration-time data on a semi-log scale (Figure 11.3).

A straight line fits the data very well, so the assumption of the first-order kinetics is valid (see Chapter 10). The slope of the straight line is 0.28/hr. It should be noted that Eq. (11.9) is an exponential equation; and the plot is linear in the figure of $\log_{10}C$ versus *t*. Consequently, the value of *k* to be used in Eq. (11.9) should be (the slope from the semi-$\log_{10}$ plot) × 2.303 (which is the natural log of 10).

$$k = (0.32)(2.303) = 0.74/\text{hr}$$

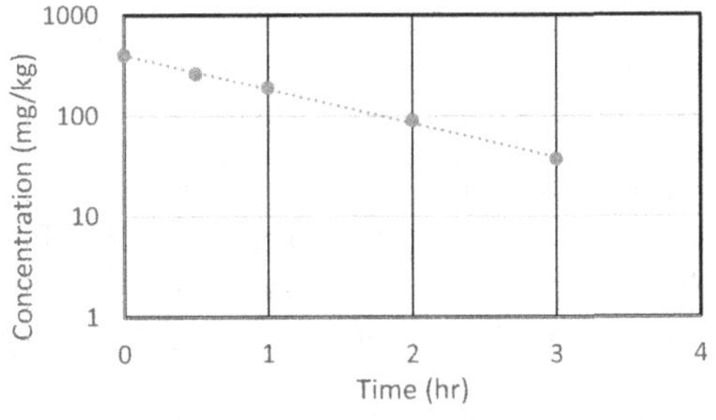

FIGURE 11.3 Concentration versus Time (for a first-order reaction).

Discussion:

1. A plot of ln(*C*) versus *t* should be made, instead of $\log_{10}(C)$ versus *t*.
2. Using the obtained rate constant to estimate the concentration at another time *t* can serve as a check. For example, the concentration at $t = 2$ hours can be calculated:

$$\frac{C_t}{C_i} = e^{-kt} = \frac{C_t}{400} = e^{-(0.74)(2)}$$

$$\rightarrow C_t = 91.0 \text{ mg/L}$$

The value (91.0 mg/kg) is comparable to the experimental data (90 mg/kg).

Example 11.5: Batch Reactor with Second-Order Kinetics

A batch reactor is to be designed to treat water contaminated with 20 mg/L of benzene. The required reduction is 90%. The rate constant is $0.5[(\text{mg/L})(\text{hr})]^{-1}$. What is the reaction residence time in the batch reactor?

Solution:

Strategy: Although the order of the reaction is not mentioned in the problem statement, it is a second-order reaction because the unit of *k* is $[(\text{mg/L})(\text{hr})]^{-1}$.

a. For a 90% reduction ($\eta = 90\%$), $C_f = (20)(1 - 90\%) = 2.0$ mg/L.

b. Insert the known values into Eq. (11.11):

$$2.0 = \frac{20}{1 + (0.5t)(20)}$$

$$\rightarrow t = 0.9 \text{ hour}$$

Discussion:

The only difference between the reactors in Examples 11.2 and 11.5 is the reaction kinetics. With the same numerical value of the reaction rate constants, the required residence time to achieve the same conversation rate is much shorter for the reaction with the second-order kinetics (0. 9 hour) than that of the first-order kinetics (4.6 hours).

11.3.2 Continuous Flow Stirred Tank Reactors

Let us now consider a steady-state CFSTR with a first-order reaction. As mentioned, by definition, the concentration in the effluent from a CFSTR is the same as that in the reactor; and the concentration in the reactor is uniform and constant. Under steady-state conditions, the flow rate is constant (i.e., $Q_{\text{in}} = Q_{\text{out}}$). Inserting Eq. (10.8) into Eq. (11.5b), the mass balance equation becomes:

$$0 = QC_{\text{in}} - QC_{\text{out}} + (V)(-kC_{\text{reactor}}) = QC_{\text{in}} - QC_{\text{out}} + (V)(-kC_{\text{out}}) \quad (11.12)$$

TABLE 11.2
Design Equations for CFSTRs

Order of Reaction	Design Equation	
0	$C_{out} = C_{in} - k\tau$	(11.14)
1	$\frac{C_{out}}{C_{in}} = \frac{1}{1+k\tau}$	same as (11.13)
2	$\frac{C_{out}}{C_{in}} = \frac{1}{1+(k\tau)C_{out}}$	(11.15)

With a simple mathematical manipulation, Eq. (11.12) can be rearranged as:

$$\frac{C_{out}}{C_{in}} = \frac{1}{1+k\left(\frac{v}{Q}\right)} = \frac{1}{1+k\tau} \tag{11.13}$$

where τ = residence time (or hydraulic detention time) = (tank volume)/(flow rate) = V/Q. Table 11.2 tabulates the design equations for CFSTRs in which zeroth-, first-, and second-order reactions take place.

Example 11.6: A Reactor with First-Order Kinetics (CFSTR)

A biological reactor is used to treat wastewater containing 1,500 mg/L of glucose ($C_6H_{12}O_6$). The required effluent glucose concentration is 50 mg/L. From a bench-scale study, the rate equation was found to be:

$$\gamma = -0.25C \text{ in mg/L/min}$$

The content in the reactor is fully mixed. Assuming the reactor behaves as a CFSTR, determine the required residence time.

Solution:

Strategy: It is a first-order reaction, and the reaction rate constant = 0.25/min.

Inserting the known values into 11.13,

$$\frac{C_{out}}{C_{in}} = \frac{50}{1{,}500} = \frac{1}{1+(0.25\tau)}$$

$$\rightarrow \tau = 116 \text{ minutes}$$

Example 11.7: A Reactor with Second-Order Kinetics (CFSTR)

A biological reactor is used to treat wastewater containing 1,500 mg/L of glucose ($C_6H_{12}O_6$). The required effluent glucose concentration is 50 mg/L. From a bench-scale study, the rate equation was found to be:

$$\gamma = -0.25C^2 \text{ in mg/L/hr}$$

The content in the reactor is fully mixed. Assuming the reactor behaves as a CFSTR, determine the required residence time.

Solution:

Strategy: It is a second-order reaction, and the reaction rate constant is equal to 0.25/[(mg/L)(hr)].

Inserting the known values into 11.15,

$$\frac{C_{\text{out}}}{C_{\text{in}}} = \frac{50}{1{,}500} = \frac{1}{1+(0.25\tau)(50)}$$

$$\rightarrow \tau = 2.32 \text{ hours } (= 139.2 \text{ minutes})$$

Discussion:

1. The rate constants in Examples 11.6 and 11.7 have the same numerical value of 0.25. Please pay attention to the difference in their units.
2. The calculated values of residence time should have units corresponding to those of their rate constants. For example, the residence time in Example 11.6 is in minutes (i.e., 116 minutes), while that in Example 11.7 is in hours (i.e., 2.32 hours).

11.3.3 Plug-Flow Reactors (PFRs)

Let us now consider a steady-state PFR with a first-order reaction. As mentioned, by definition, there is no longitudinal mixing within the PFR. The concentration in the reactor (C_{reactor}) decreases from C_{in} at the inlet to C_{out} at the exit. Under the steady-state condition, the flow rate is constant (i.e., $Q_{\text{in}} = Q_{\text{out}}$). Inserting Eq. (10.8) into Eq. (11.5b), the mass balance equation becomes:

$$0 = QC_{\text{in}} - QC_{\text{out}} + (V)(-kC_{\text{reactor}}) \qquad (11.16)$$

In this case, C_{reactor} is a variable. The equation can be solved by considering an infinitesimal section of the reactor and integrating the equation. The solution can be found as:

$$\frac{C_{\text{out}}}{C_{\text{in}}} = e^{-k\left(\frac{V}{Q}\right)} = e^{-k\tau} \qquad (11.17)$$

TABLE 11.3
Design Equations for PFRs

Order of Reaction	Design Equation	
0	$C_{out} = C_{in} - k\tau$	(11.18)
1	$\frac{C_{out}}{C_{in}} = e^{-k\tau}$	same as (11.17)
2	$\frac{C_{out}}{C_{in}} = \frac{1}{1+(k\tau)C_{in}}$	(11.19)

Table 11.3 tabulates the design equations for PFRs in which zeroth-, first-, and second-order reactions take place.

When comparing the design equations for PFRs (Table 11.3) with those for CFSTRs (Table 11.2), the following remarks can be made:

1. Zeroth-Order Reactions: The design equations are identical for both reactor types. It means that the conversion rate is independent of the reactor types, provided all the other conditions are the same.
2. First-Order Reactions: For the first-order reactions, the ratio of the effluent and the influent concentrations is inversely proportional to the residence time for CFSTRs; while it is exponentially proportional to the negative value of the residence time for PFRs. In other words, the effluent concentration from PFRs decreases more rapidly with an increase in the residence time than that from CFSTRs, if the rate constants are the same. In addition, if the residence time and the rate constant are the same, the effluent concentration from a PFR would be smaller than that from a CFSTR.
3. Second-Order Reactions: The design equations for the second-order reactions appear to be similar in format for PFRs and CFSTRs; the only difference is the C_{out} in Eq. (11.15) is replaced by C_{in} in Eq. (11.19). It implies that the effluent concentration from a PFR would be smaller than that from a CFSTR, for the same C_{in}, k, and τ.

It is also interesting to note that the design equations for PFRs (Table 11.3) are similar to those for the batch reactors (Table 11.1), except the residence time (τ) in Table 11.3, while time t (a variable) is in Table 11.1. Be noted that a batch reactor is an unsteady-state reactor, in which the concentration in the reactor varies with time t; while a PFR is typically operated in a steady-state mode and C_{out} is the effluent concentration.

Example 11.8: A Reactor with First-Order Kinetics (PFR)

A biological reactor is used to treat wastewater containing 1,500 mg/L of glucose ($C_6H_{12}O_6$). The required effluent glucose concentration is 50 mg/L. From a bench-scale study, the rate equation was found to be:

$$\gamma = -0.25C \text{ in mg/L/min}$$

Assuming the reactor behaves as a PFR, determine the required residence time.

Solution:

Strategy: It is a first-order reaction, and the reaction rate constant is equal to 0.25/min.

Inserting the known values into (11.17),

$$\frac{C_{out}}{C_{in}} = e^{-k\tau} = \frac{50}{1{,}500} = e^{-0.25\tau}$$

$$\rightarrow \tau = 13.6 \text{ minutes}$$

Discussion:

1. For the same inlet concentration and reaction rate constant, the required residence time to achieve a specified final concentration is 13.6 minutes for a PFR which is much shorter than that for a CFSTR, 116 minutes (Example 11.6).
2. For the first-order kinetics, the reaction rate is proportional to the concentration inside the reactor (i.e., $\gamma = kC_{reactor}$). The higher the reactor concentration, the faster the reaction rate. For CFSTRs, the reactor concentration is equal to the effluent concentration (i.e., 50 mg/L in this case). For PFRs, the reactor concentration decreases from C_{in} (1,500 mg/L) at the inlet to C_{out} (50 mg/L) at the outlet. The average concentration in the PFR (775 mg/L as the arithmetic average or 274 as the geometric average) is much larger than 50 mg/L (i.e., that of the CFSTR), which makes the reaction rate much higher. Consequently, the required residence time would be much shorter.

Example 11.9: A Reactor with Second-Order Kinetics (PFR)

A biological reactor is used to treat wastewater containing 1,500 mg/L of glucose ($C_6H_{12}O_6$). The required effluent glucose concentration is 50 mg/L. From a bench-scale study, the rate equation was found to be:

$$\gamma = -0.25C^2 \text{ in mg/L/hr}$$

Assuming the reactor behaves as a PFR, determine the required residence time.

Solution:

Strategy: It is a second-order reaction, and the reaction rate constant is equal to 0.25/[(mg/L)(hr)].

Inserting the known values into (11.15),

$$\frac{C_{out}}{C_{in}} = \frac{50}{1{,}500} = \frac{1}{1+(0.25\tau)(1{,}500)}$$

$$\rightarrow \tau = 0.077 \text{ hour } (= 4.6 \text{ minutes})$$

Discussion:

Similar to that in Example 11.8, for the same initial concentration and reaction rate constant, the required residence time to achieve the specified final concentration is 4.6 minutes for a PFR, which is much shorter than that for a CFSTR, 139 minutes (as shown in Example 11.7).

11.4 SIZING REACTORS

Once the reactor type is selected and the required residence time to achieve the desired removal is determined, sizing a reactor is straightforward. The longer the required residence time, the larger the reactor would be for a given flow rate.

For flow reactors such as CFSTRs and PFRs, the *residence time* (or the *hydraulic detention time*), τ, can be defined as:

$$\tau = \frac{V}{Q} \tag{11.20}$$

where V = the volume of the reactor and Q = the flow rate. Consequently, the required reactor volume can be found as:

$$V = \tau \times Q \tag{11.21}$$

For a PFR, by definition, each fluid parcel would spend the same amount of time flowing through the reactor. On the other hand, for a CFSTR, most fluid parcels would flow through the reactor in a shorter or longer time than the average residence time. Therefore, the value of τ in Eq. (11.20) is the average hydraulic detention time for a CFSTR.

For a batch reactor, the residence time calculated by using Eqs. (11.9), (11.10), or (11.11) is the actual time needed for reaction to accomplish. For design of batch reactors, an engineer needs to take the time needed for loading, reaction, unloading, and idle into consideration.

Example 11.10: Sizing a Batch Reactor

A biological reactor is used to treat wastewater containing 1,500 mg/L of glucose ($C_6H_{12}O_6$). The required effluent glucose concentration is 50 mg/L. From a bench-scale study, the rate equation was found to be:

$$\gamma = -0.25C \text{ in mg/L/min}$$

The reactor is operated in a batch mode and the time required for loading and unloading for each batch is 1 hour. The influent flow rate is 100 gallons/minute.

a. What is the total time needed for each batch?
b. Size the batch reactor for this project.

Solution:

Strategy: It is a first-order reaction, and the reaction rate constant is equal to 0.25/min.

a. Inserting the known values into Eq. (11.9),

$$\frac{C_t}{C_o} = e^{-kt} = \frac{50}{1,500} = e^{-0.25t}$$

$\rightarrow t = \underline{13.6 \text{ minutes}}$

The total time needed for each batch of operation

= reaction time + time for loading and unloading = 13.6 + 60 = 74 minutes

b. The required reactor volume, $V = Q \times t$ (from Eq. 11.21)

= (100 gallon/min) × (13.6 minutes) = 1,360 gallons

Discussion:

The total operation time for each batch (74 minutes) is longer than five times of the required reaction time (13.6 minutes). This means that a minimum of six reactors with a size of 1,360 gallons are needed in this case. These reactors would be operated in different phases (i.e., loading, reaction, and unloading) so that the influent flow would not be interrupted.

Example 11.11: Sizing a CFSTR and a PFR

A biological reactor is used to treat wastewater containing 1,500 mg/L of glucose ($C_6H_{12}O_6$). The required effluent glucose concentration is 50 mg/L. From a bench-scale study, the rate equation was found to be:

$$\gamma = -0.25C \text{ in mg/L/min}$$

The influent flow rate is 100 gallon/minute. The reactor can be operated as a CFSTR or as a PFR.

a. What is the required reactor volume if it is operated as a CFSTR?
b. What is the required reactor volume if it is operated as a PFR?

Solution:

a. From Example 11.6, the residence time to achieve the required removal if the reactor is operated in a CFSTR mode = 116 minutes. Thus, the required reactor volume

= (100 gallon/min) × (116 minutes) = <u>11,600 gallons</u>

b. From Example 11.8, the residence time to achieve the required removal if the reactor is operated in a PFR mode = 13.6 minutes. Thus, the required reactor volume

= (100 gallon/min) × (13.6 minutes) = <u>1,360 gallons</u>

Discussion:

1. As mentioned, the design equations for batch reactors and PFRs are essentially the same. The required reaction times for these two reactors are the same, at 13.6 minutes. However, more reactors are needed to accommodate the time needed for loading and unloading, if the reactors are operated in a batch mode.
2. To achieve the same conversion, the size of the PFR, 1,360 gallons, is much smaller than 11,600 gallons for the CFSTR.
3. It should be noted that both PFRs and CFSTRs are "ideal" reactors; the basic assumption for CFSTRs is that all the reactor content is totally mixed, while that for PFRs is no mixing in the longitudinal direction. In actual applications, reactors are designed and operated in either a CFSTR mode or in a PFR mode; however, they could not be 100% CFSTR or 100% PFR in reality The main advantage of having a CFSTR is that it can dampen the influent concentrations (because the concentration in the reactor is theoretically the same as the effluent concentration). When fluctuations of influent concentrations are of concern (e.g., biological processes), CFSTRs would be a good choice. The main advantage of a PFR is that each water parcel that enters the reactor will theoretically receive the same hydraulic detention time. When a more consistent detention time is needed for all water parcels (e.g., water disinfection using chlorine or UV), PFRs would be a better choice.

11.5 REACTOR CONFIGURATIONS

In engineering applications, it is more common to have a few smaller reactors than to have one large reactor for the following reasons:

- flexibility (to handle fluctuations of flow rate)
- maintenance consideration
- a higher removal efficiency

Common reactor configurations include arrangement of reactors in series, in parallel, or a combination of both.

11.5.1 Reactors in Series

For *reactors in series*, the flow rates to all the reactors are the same and equal to the influent flow rate to the first reactor (Figure 11.4). The first reactor, with a volume V_1, will reduce the influent COC concentration of C_0 to yield an effluent concentration of C_1. The effluent concentration from the first reactor becomes the influent

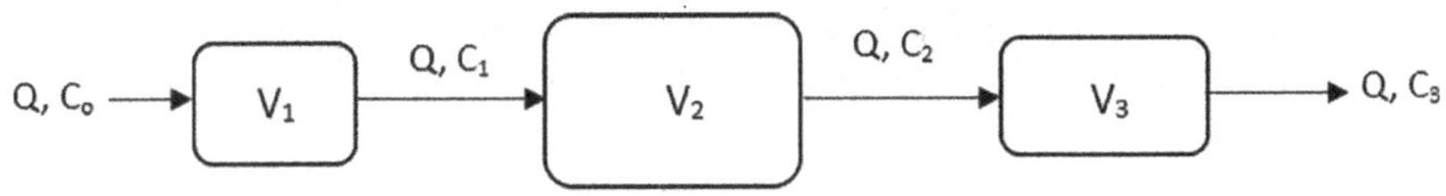

FIGURE 11.4 Three reactors in series.

concentration in the second reactor. Consequently, the effluent concentration from the second reactor (C_2), becomes the influent concentration to the third reactor. More reactors can be added in series until the effluent concentration from the last reactor in series meets the requirement.

For CFSTRs, a few small reactors in series will yield a lower final effluent concentration than a large reactor with the same total volume (this will be illustrated in an example later). For three CFSTRs arranged in series, the effluent concentration from the third reactor can be determined from the COC concentration in the influent as:

$$\frac{C_3}{C_o} = \left(\frac{C_3}{C_2}\right)\left(\frac{C_2}{C_1}\right)\left(\frac{C_1}{C_o}\right) = \left(\frac{1}{1+k_3\tau_3}\right)\left(\frac{1}{1+k_2\tau_2}\right)\left(\frac{1}{1+k_1\tau_1}\right) \quad (11.22)$$

For n identical CFSTRs in series, the final effluent concentration (C_n) can be found as:

$$\frac{C_n}{C_o} = \left[\frac{1}{1+k\tau}\right]^n \quad (11.23)$$

For three PFRs arranged in series, the effluent concentration from the third reactor can be determined from the COC concentration in the influent as:

$$\frac{C_3}{C_o} = \left(\frac{C_3}{C_2}\right)\left(\frac{C_2}{C_1}\right)\left(\frac{C_1}{C_o}\right) = \left(e^{-k_3\tau_3}\right)\left(e^{-k_2\tau_2}\right)\left(e^{-k_1\tau_1}\right) = e^{-(k_3\tau_3+k_2\tau_2+k_1\tau_1)} \quad (11.24)$$

For n identical PFRs in series, the final effluent concentration (C_n) can be found as:

$$\frac{C_n}{C_o} = e^{-n(k\tau)} \quad (11.25)$$

Example 11.12: CFSTRs in Series

A wastewater stream ($Q = 10$ ft³/min) containing 1,000 mg/L of glucose is to be treated biologically. The treatment goal is to reduce the concentration to <100 mg/L. The reaction is first-order with a rate constant 0.1/min, as determined from a bench-scale study.

Four different configurations of bioreactors operated in the CFSTR mode are considered. Determine the final effluent concentration from each of these arrangements and if it meets the clean-up requirement:

a. one 400-ft³ reactor
b. two 200-ft³ reactors in series
c. one 100-ft³ reactor followed by one 300-ft³ reactor
d. one 300-ft³ reactor followed by one 100-ft³ reactor

Solution:

a. For the 400-ft³ reactor, the residence time = V/Q = 400 ft³/(10 ft³/min) = 40 minutes

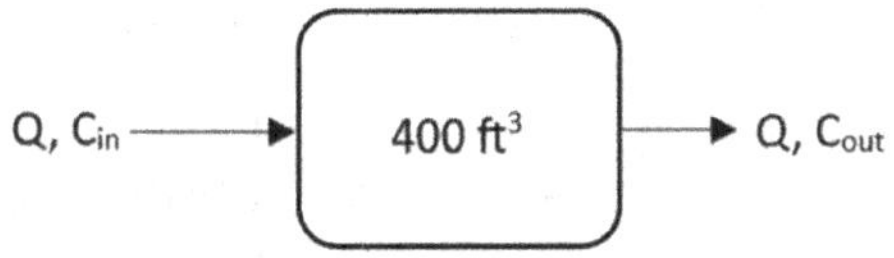

Use Eq. (11.13) to find the final effluent concentration:

$$\frac{C_{\text{out}}}{C_{\text{in}}} = \frac{1}{1+k\tau} = \frac{C_{\text{out}}}{1{,}000} = \frac{1}{1+(0.1)(40)}$$

$C_{\text{out}} = \underline{200\ \text{mg/L}}$ (>100 mg/L)

b. For two 200-ft^3 reactors, the residence time = V/Q = 200 ft^3/(10 ft^3/min) = 20 minutes each.

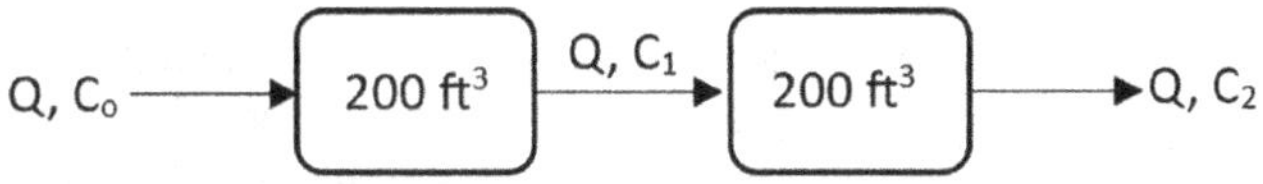

Use Eq. (11.23) to find the final effluent concentration:

$$\frac{C_2}{C_o} = \left[\frac{1}{1+k\tau}\right]^2 = \frac{C_2}{1{,}000} = \left[\frac{1}{1+(0.1)(20)}\right]^2$$

$C_2 = \underline{111.1\ \text{mg/L}}$ (>100 mg/L)

c. The residence time of the first reactor = 100 ft^3/(10 ft^3/min) = 10 minutes; and the residence time of the second reactor = 300 ft^3/(10 ft^3/min) = 30 minutes.

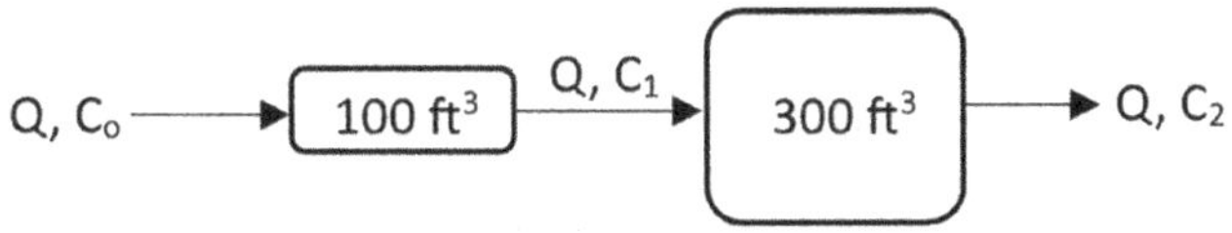

Use Eq. (11.22) to find the final effluent concentration:

$$\frac{C_2}{C_o} = \left(\frac{C_2}{C_1}\right)\left(\frac{C_1}{C_o}\right) = \left(\frac{1}{1+k\tau_2}\right)\left(\frac{1}{1+k\tau_1}\right) = \frac{C_2}{1{,}000} = \left(\frac{1}{1+(0.1)(10)}\right)\left(\frac{1}{1+(0.1)(30)}\right)$$

$C_2 = \underline{125\ \text{mg/L}}$ (>100 mg/L)

d. The residence time of the first reactor = 30 minutes; and the residence time of the second reactor = 10 minutes.

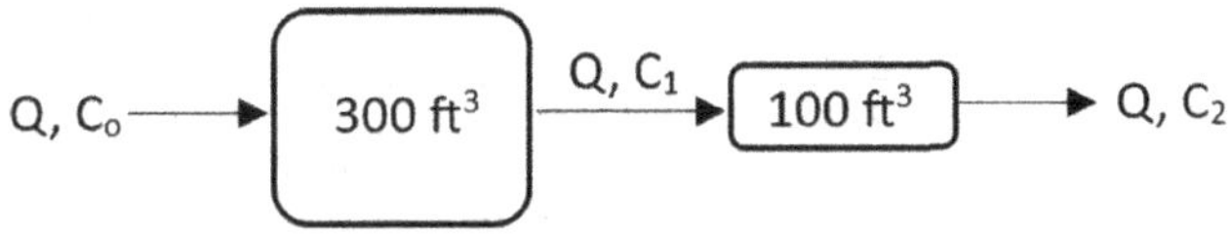

Use Eq. (11.22) to find the final effluent concentration:

$$\frac{C_2}{C_o}=\left(\frac{C_2}{C_1}\right)\left(\frac{C_1}{C_o}\right)=\left(\frac{1}{1+k\tau_2}\right)\left(\frac{1}{1+k\tau_1}\right)=\frac{C_2}{1{,}000}=\left(\frac{1}{1+(0.1)(30)}\right)\left(\frac{1}{1+(0.1)(10)}\right)$$

$C_2 = \underline{125\ \text{mg/L}}$ (>100 mg/L)

Discussion:

1. The total volume of the reactor(s) for each of the four configurations is 400 ft³.
2. The effluent concentration from the first configuration (one large reactor) is the highest. Having a series of smaller CFSTRs in series will always be more efficient than having a large CFSTR. A PFR can be viewed as an infinite series of small CFSTRs, and a PFR is always more efficient than a CFSTR of equal size.
3. For the configurations of having two small reactors in series, the setup of two equal-size reactors yields the lowest effluent concentration.
4. For two reactors of different sizes, the sequence of the reactors does not affect the final effluent concentration, provided the rate constants in the reactors are the same.

Example 11.13: PFRs in Series

A wastewater stream ($Q = 10$ ft³/min) containing 1,000 mg/L of glucose is to be treated biologically. The goal is to reduce the concentration to <100 mg/L. The reaction is first-order with a rate constant 0.1/min, as determined from a bench-scale study.

Four different configurations of bioreactors operated in the PFR mode are considered. Determine the final effluent concentration from each of these arrangements and if it meets the clean-up requirement:

a. one 400-ft³ reactor
b. two 200-ft³ reactors in series
c. one 100-ft³ reactor followed by one 300-ft³ reactor
d. one 300-ft³ reactor followed by one 100-ft³ reactor

Solution:

a. For the 400-ft³ reactor, the residence time = V/Q = 400 ft³/(10 ft³/min) = 40 minutes

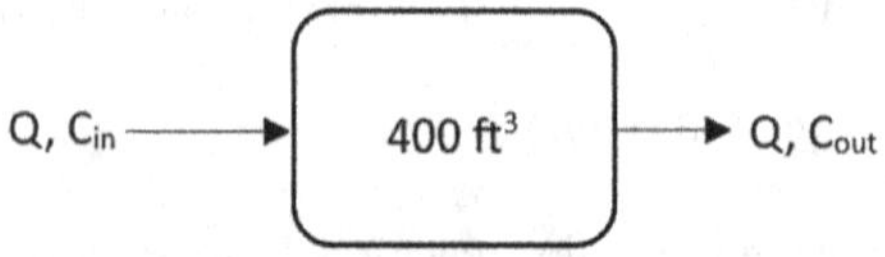

Use Eq. (11.17) to find the final effluent concentration:

$$\frac{C_{\text{out}}}{C_{\text{in}}}=e^{-k\tau}=\frac{C_{\text{out}}}{1{,}000}=e^{-(0.1)(40)}$$

$C_{\text{out}} = \underline{18.3\ \text{mg/L}}$ (<100 mg/L)

b. For two 200-ft^3 reactors, the residence time = V/Q = 200 ft^3/(10 ft^3/min) = 20 minutes each.

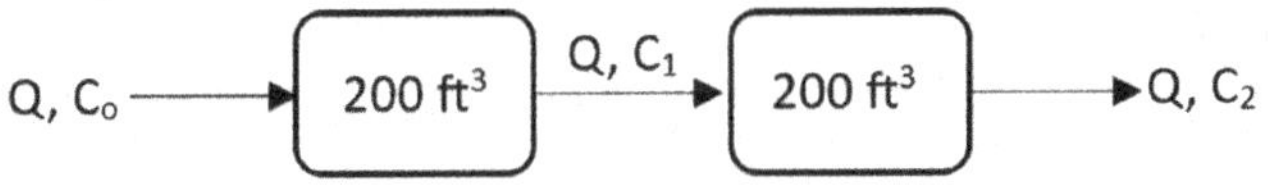

Use Eq. (11.25) to find the final effluent concentration:

$$\frac{C_n}{C_o} = e^{-n(k\tau)} = \frac{C_2}{1{,}000} = e^{-(2)(0.1\times 20)}$$

$$C_2 = \underline{18.3\ \text{mg/L}}\ (<100\ \text{mg/L})$$

c. The residence time of the first reactor = 100 ft^3/(10 ft^3/min) = 10 minutes; and the residence time of the second reactor = 300 ft^3/(10 ft^3/min) = 30 minutes.

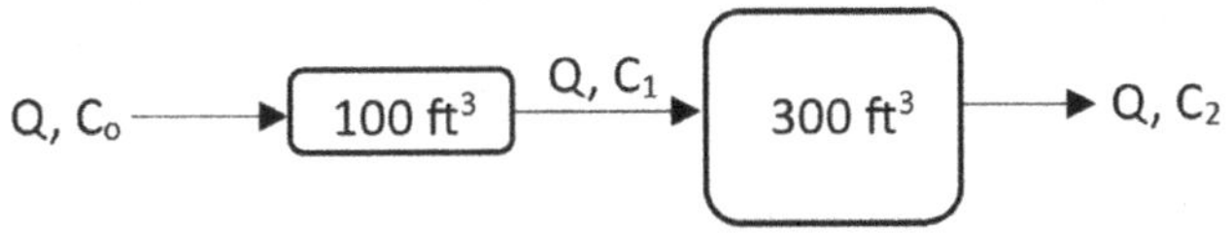

Use Eq. 11.24 to find the final effluent concentration:

$$\frac{C_2}{C_o} = \left(\frac{C_2}{C_1}\right)\left(\frac{C_1}{C_o}\right) = e^{-(k\tau_2+k\tau_2)} = \frac{C_2}{1{,}000} = e^{-(0.1\times 10+0.1\times 30)}$$

$$C_2 = \underline{18.3\ \text{mg/L}}\ (<100\ \text{mg/L})$$

d. The residence time of the first reactor = 30 minutes; and the residence time of the second reactor = 10 minutes.

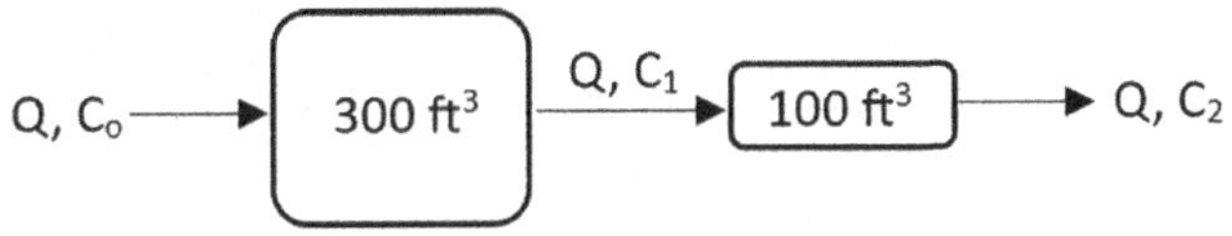

Use Eq. (11.24) to find the final effluent concentration:

$$\frac{C_2}{C_o} = \left(\frac{C_2}{C_1}\right)\left(\frac{C_1}{C_o}\right) = e^{-(k\tau_2+k\tau_2)} = \frac{C_2}{1{,}000} = e^{-(0.1\times 30+0.1\times 10)}$$

$$C_2 = \underline{18.3\ \text{mg/L}}\ (<100\ \text{mg/L})$$

Discussion:

1. The total volume of the reactor(s) for each of the four configurations is 400 ft^3.
2. The effluent concentrations from all four different configurations are the same.
3. The effluent concentration of PFRs is much lower than those of CFSTRs (Example 11.12).

Example 11.14: CFSTRs in Series

A wastewater stream ($Q = 10$ ft^3/min) containing 1,000 mg/L of glucose is to be treated biologically. The goal is to reduce the concentration to <100 mg/L. A CFSTR reactor with 30 minutes of residence time can only reduce the concentration to 200 mg/L. Assuming that it is a first-order reaction, can three smaller reactors (10 minutes of residence time each) in series reduce the glucose concentration to below 100 mg/L?

Solution:

The reaction rate constant is not given, use the given data to estimate its value.

a. Use Eq. (11.13) to find the rate constant:

$$\frac{C_{out}}{C_{in}} = \frac{1}{1+k\tau} = \frac{200}{1{,}000} = \frac{1}{1+(k)(30)}$$

$$\rightarrow k = 0.133/\text{min}$$

b. For three small reactors in series,

$$\frac{C_3}{C_o} = \left[\frac{1}{1+k\tau}\right]^3 = \frac{C_3}{1{,}000} = \left[\frac{1}{1+(0.133)(10)}\right]^3$$

$$\rightarrow C_{out} = \underline{79.1 \text{ mg/L}} \ (<100 \text{ mg/L})$$

Discussion:

This example again demonstrates that three smaller CFSTRs can do a better job than a larger CFSTR with an equivalent total volume. However, three reactors may require a larger capital investment and higher O&M costs.

Example 11.15: PFRs in Series

A wastewater stream ($Q = 10$ ft^3/min) containing 5,000 mg/L of glucose is to be treated biologically. The goal is to reduce the concentration to <100 mg/L. A PFR reactor with 20 minutes of residence time can only reduce the concentration to 1,000 mg/L. Assuming that it is a first-order reaction, how many reactors

(20 minutes of residence time) in series reduce the glucose concentration to below 100 mg/L?

Solution:

a. Use Eq. (11.17) to find out the reaction rate constant:

$$\frac{C_{out}}{C_{in}} = e^{-k\tau} = \frac{1{,}000}{5{,}000} = e^{-(k)(20)}$$

$$k = 0.08/\text{min}$$

b. Use Eq. (11.25) to find the number of PFRs (n) needed:

$$\frac{C_n}{C_o} = e^{-n(k\tau)} = \frac{100}{5{,}000} = e^{-(n)(0.08\times 20)}$$

$$\rightarrow n = \underline{2.45}$$

Three PFRs, each with 20 minutes of residence time, would be needed.

Discussion:

For PFRs in series, we can also determine the total residence time needed to reduce the final concentration to 100 mg/L first, and then determine the number of PFRs needed. Use Eq. (11.17) to find out the required residence time:

$$\frac{C_{out}}{C_{in}} = e^{-k\tau} = \frac{100}{5{,}000} = e^{-(0.08)(\tau)}$$

$$\tau = \underline{48.9 \text{ minutes}}$$

→ Three PFRs of 20-minute residence time are needed.

11.5.2 Reactors in Parallel

For *reactors in parallel*, the reactors share the same influent (the influent is split and fed to the reactors). The flow rate to each reactor in parallel can be different; however, the influent concentrations to all the reactors in parallel are the same. The sizes of the reactors may not be the same, and the effluent concentrations from the reactors can be different (Figure 11.5). In Figure 11.5, the following mass balance equations are valid:

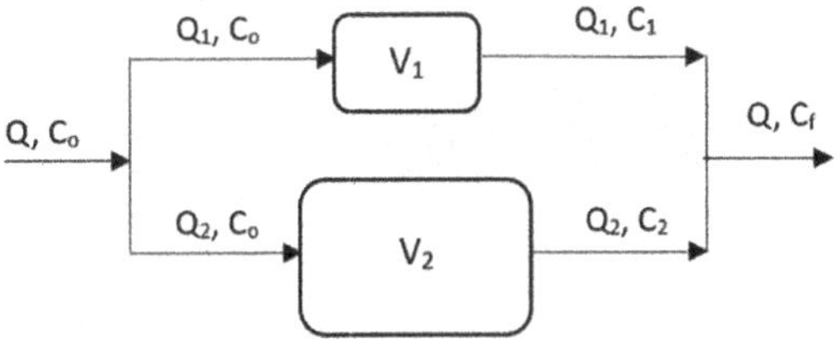

FIGURE 11.5 Two reactors in parallel.

$$Q = Q_1 + Q_2 \tag{11.26}$$

$$C_f = \frac{Q_1C_1 + Q_2C_2}{Q_1 + Q_2} \tag{11.27}$$

Reactors in-parallel configurations are often used for the following cases: (i) a single reactor cannot handle the flow rate, (ii) the total influent rate fluctuates significantly, or (iii) the reactors require frequent off-line maintenance.

Example 11.16: CFSTRs in Parallel

A wastewater stream ($Q = 10$ ft^3/min) containing 1,000 mg/L of glucose is to be treated biologically. The goal is to reduce the concentration to <100 mg/L. The reaction is first-order with a rate constant 0.1/min, as determined from a bench-scale study.

Four different configurations of bioreactors operated in the CFSTR mode are considered. Determine the final effluent concentration from each of these arrangements and if it meets the clean-up requirement:

a. one 400-ft^3 reactor
b. two 200-ft^3 reactors in parallel (each receives 5 ft^3/min flow)
c. one 100-ft^3 reactor and one 300-ft^3 reactor in parallel (each receives 5 ft^3/min flow)
d. one 100-ft^3 reactor ($Q = 2.5$ ft^3/min) and one 300-ft^3 reactor in parallel ($Q = 7.5$ ft^3/min)

Solution:

a. For the 400-ft^3 reactor, the residence time = V/Q = 400 ft^3/(10 ft^3/min) = 40 minutes

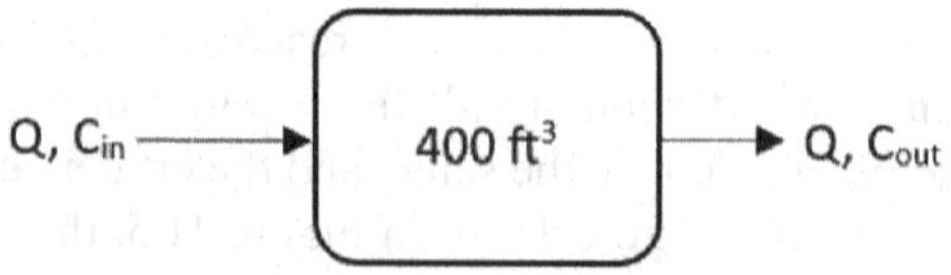

Use Eq. (11.13) to find the final effluent concentration:

$$\frac{C_{out}}{C_{in}} = \frac{1}{1+k\tau} = \frac{C_{out}}{1{,}000} = \frac{1}{1+(0.1)(40)}$$

$$C_{out} = \underline{200 \text{ mg/L}} \ (>100 \text{ mg/L})$$

b. For two 200-ft^3 reactors, the residence time = V/Q = 200 ft^3/(5 ft^3/min) = 40 minutes each.

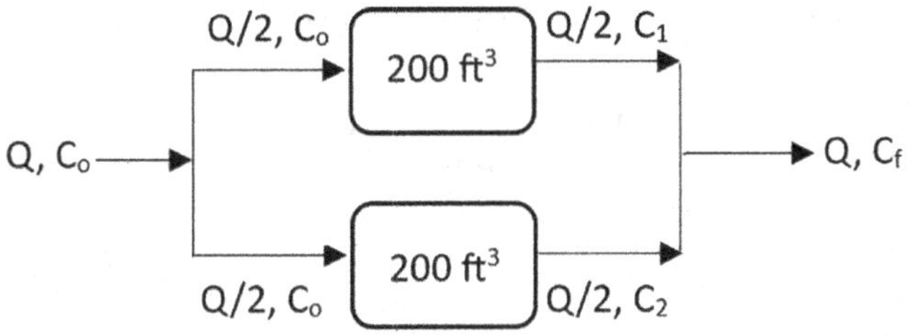

Use Eq. (11.13) to find the C_1 and C_2 (both reactors have the same residence time of 40 minutes):

$$\frac{C_{\text{out}}}{C_{\text{in}}} = \frac{1}{1+k\tau} = \frac{C_{\text{out}}}{1{,}000} = \frac{1}{1+(0.1)(40)}$$

$C_1 = C_2 = C_f =$ 200 mg/L (>100 mg/L)

c. The residence time of the first reactor = 100 ft³/(5 ft³/min) = 20 minutes; and the residence time of the second reactor = 300 ft³/(5 ft³/min) = 60 minutes.

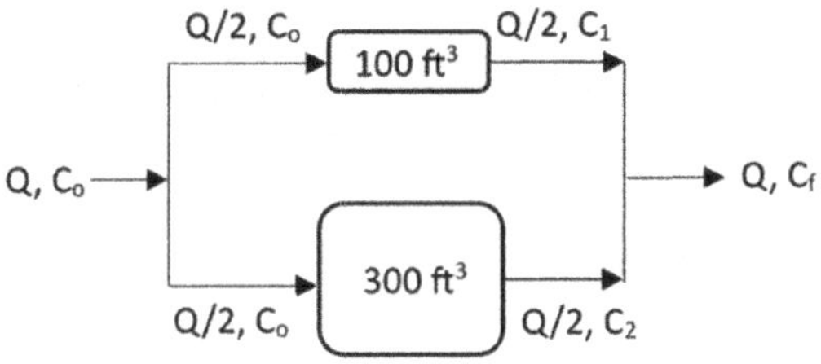

Use Eq. (11.13) to find the values of C_1 and C_2:

$$\frac{C_1}{1{,}000} = \frac{1}{1+(0.1)(20)} \rightarrow C_1 = 333 \text{ mg/L}$$

$$\frac{C_2}{1{,}000} = \frac{1}{1+(0.1)(60)} \rightarrow C_2 = 143 \text{ mg/L}$$

Use Eq. (11.27) and the values of C_1 and C_2 to find C_f:

$$C_f = \frac{Q_1C_1 + Q_2C_2}{Q_1 + Q_2} = \frac{(5)(333)+(5)(143)}{(5+5)} = 238 \text{ mg/L}$$

$C_f =$ 238 mg/L (>100 mg/L)

d. The residence time of both reactors = 40 minutes:

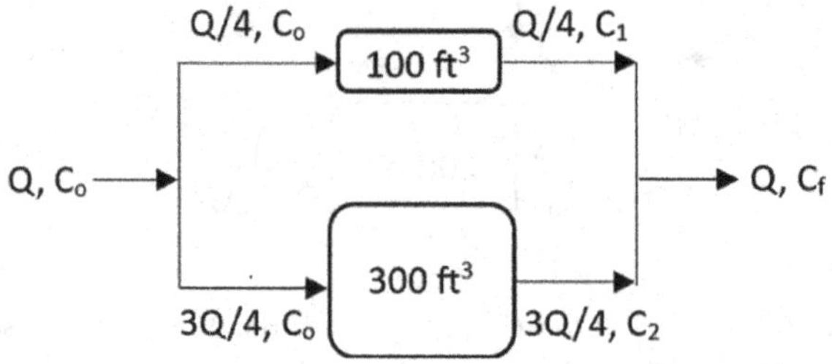

$$C_1 = C_2 = C_f = \underline{200 \text{ mg/L}} \ (>100 \text{ mg/L})$$

Discussion:

1. The total volume of the reactor(s) for each of the four configurations is 400 ft^3.
2. The effluent concentrations from all the configurations exceed the clean-up level. The configurations (a), (b), and (d) have the same effluent concentrations because the residence times of all reactors are identical. The effluent concentration from configuration (c) is the worst among the four.
3. To split the flow into reactors in parallel with the same residence time does not have any impact on the final effluent concentration, as shown in cases (a), (b), and (d).

Example 11.17: PFRs in Parallel

A wastewater stream (Q = 10 ft^3/min) containing 1,000 mg/L of glucose is to be treated biologically. The goal is to reduce the concentration to <100 mg/L. The reaction is first-order with a rate constant of 0.1/min, as determined from a bench-scale study.

Four different configurations of bioreactors operated in the PFR mode are considered. Determine the final effluent concentration from each of these arrangements and if it meets the clean-up requirement:

a. one 400-ft^3 reactor
b. two 200-ft^3 reactors in parallel (each receives 5 ft^3/min flow)
c. one 100-ft^3 reactor and one 300-ft^3 reactor in parallel (each receives 5 ft^3/min flow)
d. one 100-ft^3 reactor (Q = 2.5 ft^3/min) and one 300-ft^3 reactor in parallel (Q = 7.5 ft^3/min)

Solution:

a. For the 400-ft^3 reactor, the residence time = V/Q = 400 ft^3/(10 ft^3/min) = 40 minutes

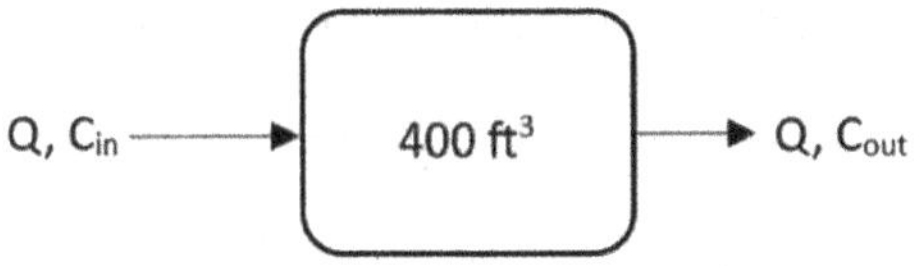

Use Eq. (11.17) to find the final effluent concentration:

$$\frac{C_{out}}{C_{in}} = e^{-k\tau} = \frac{C_{out}}{1{,}000} = e^{-(0.1)(40)}$$

$$C_{out} = \underline{18.3 \text{ mg/L}} \ (<100 \text{ mg/L})$$

b. For two 200-ft³ reactors, the residence time = V/Q = 200 ft³/(5 ft³/min) = 40 minutes each.

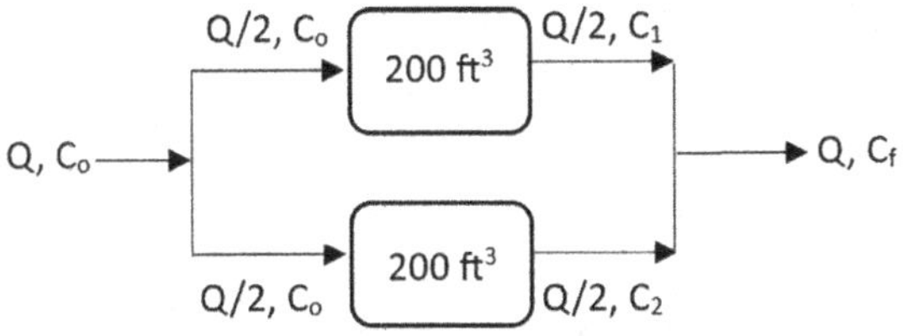

Use Eq. (11.17) to find the C_1 and C_2 (both reactors have the same residence time of 40 minutes):

$$C_1 = C_2 = C_f = \underline{18.3 \text{ mg/L}} \ (<100 \text{ mg/L})$$

c. The residence time of the first reactor = 100 ft³/(5 ft³/min) = 20 minutes; and the residence time of the second reactor = 300 ft³/(5 ft³/min) = 60 minutes.

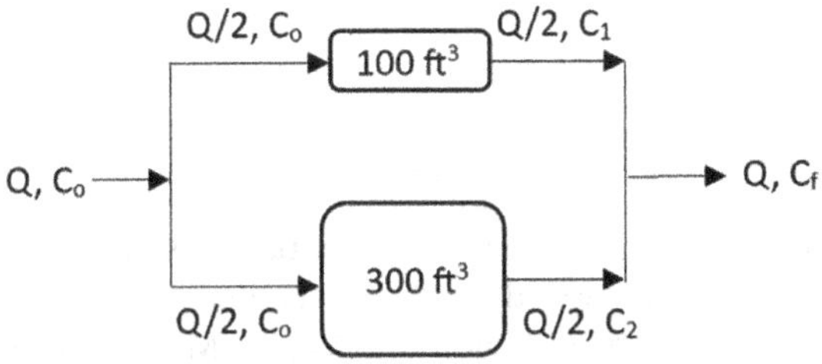

Use Eq. (11.17) to find the values of C_1 and C_2:

$$\frac{C_{out}}{1{,}000} = e^{-(0.1)(20)} \rightarrow C_1 = 135 \text{ mg/L}$$

$$\frac{C_{out}}{1{,}000} = e^{-(0.1)(60)} \rightarrow C_1 = 2.5 \text{ mg/L}$$

Use Eq. (11.27) and the values of C_1 and C_2 to find C_f:

$$C_f = \frac{Q_1C_1 + Q_2C_2}{Q_1 + Q_2} = \frac{(5)(135)+(5)(2.5)}{(5+5)} = 69 \text{ mg/L}$$

C_f = 69 mg/L (<100 mg/L)

d. The residence time of both reactors = 40 minutes:

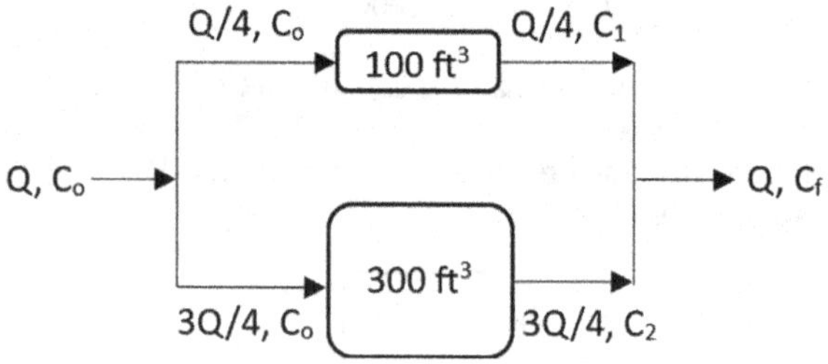

$C_1 = C_2 = C_f$ = 18.3 mg/L (>100 mg/L)

Discussion:

1. The total volume of the reactor(s) for each of the four configurations is 400 ft³.
2. The configurations (a), (b), and (d) have the same effluent concentrations because the residence times of all reactors are identical. The effluent concentration from configuration (c) is the worst among the four.
3. To split the flow into reactors in parallel with the same residence time does not have any impact on the final effluent concentration, as shown in cases (a), (b), and (d).
4. Examples 11.16 and 11.17 illustrate (i) the importance of flow distribution to reactors in parallel, and (ii) PFRs are more efficient than CFSTRs provided the influent concentrations, reaction rate constant, and the residence time are the same.

EXERCISE QUESTIONS

1. A glass bottle containing 600 mL of chloroform (specific gravity = 1.483) was accidentally left uncapped in a poorly ventilated room (5 m × 8 m × 3 m) over a weekend (T = 20°C). On the following Monday, it was found that 20% of chloroform had volatilized. An exhaust fan (Q = 20 m³/min) was turned on to vent the fouled air out of the laboratory.
   a. What is the original chloroform concentration before ventilation, in mg/m³ and in ppm?
   b. How long will it take to reduce the concentration down below the Occupational Health and Safety Association's (OSHA's) short-term exposure limit (STEL) of 2 ppmV?
2. A batch reactor is to be designed to treat wastewater containing 10 mg/L of toluene. The required reduction of toluene is 90%.

a. The rate constant is 0.5 min^{-1}, what is the required reaction time for this batch reactor?
b. What is the required reaction time if the desired final concentration is 400 ppb?

3. A batch reactor was installed to remediate water contaminated with o-xylene. A trial run was conducted with an initial o-xylene concentration of 20 mg/L. After 1 hour of batch-wise operation, the concentration dropped to 4.0 mg/L. However, it is required to reduce the concentration down to 1.0 mg/L. Assuming the reaction is first-order, determine the required residence time to achieve the final o-xylene concentration of 1.0 mg/L.
4. An in-vessel bioreactor is designed to remediate soil impacted by phenol. A bench-scale batch- reactor was set up to determine the order and the rate constant of the reaction. The following concentrations of cresol in the batch reactor at various times were observed and recorded as:

Time (hours)	Phenol Concentration (mg/kg)
0	1,000
0.5	750
1	550
2	300
5	50

a. Use these data to illustrate that it is a first-order reaction.
b. What is the value of the rate constant?

5. A batch reactor is to be designed to treat water contaminated with 40 mg/L of ethylbenzene. The required reduction of ethylbenzene is 80%. The rate constant is 0.6[(mg/L)(hr)]$^{-1}$, what is the required residence time for the batch reactor?
6. A biological reactor is used to treat wastewater containing 1,000 mg/L of glucose. The required effluent glucose concentration is 40 mg/L. The rate equation was found to be:

$$\gamma = -\,0.20\,C \quad \text{in mg/L/min}$$

The content in the reactor is fully mixed. Assuming the reactor behaves as a CFSTR, determine the required residence time.
7. A biological reactor is used to treat wastewater containing 1,000 mg/L of glucose. The required effluent glucose concentration is 40 mg/L. The rate equation was found to be:

$$\gamma = -\,0.20\,C^2 \quad \text{in mg/L/min}$$

The content in the reactor is fully mixed. Assuming the reactor behaves as a CFSTR, determine the required residence time.
8. A biological reactor is used to treat wastewater containing 1,000 mg/L of glucose. The required effluent glucose concentration is 40 mg/L. The rate equation was found to be:

$$\gamma = -0.20\,C \quad \text{in mg/L/min}$$

Assuming the reactor behaves as a PFR, determine the required residence time.

9. A biological reactor is used to treat wastewater containing 1,000 mg/L of glucose. The required effluent glucose concentration is 40 mg/L. The rate equation was found to be:

$$\gamma = -0.20\,C^2 \quad \text{in mg/L/min}$$

Assuming the reactor behaves as a PFR, determine the required residence time.

10. A biological reactor is used to treat wastewater containing 1,000 mg/L of glucose. The required effluent glucose concentration is 40 mg/L. The rate equation was found to be:

$$\gamma = -0.20\,C \quad \text{in mg/L/min}$$

The reactor is operated in a batch mode and the time required for loading and unloading of the slurry for each batch is 1 hour. The influent flow rate is 100 gallons/minute.

a. What is the total time needed for each batch?
b. Size the batch reactor for this project.

11. A biological reactor is used to treat wastewater containing 1,000 mg/L of glucose. The required effluent glucose concentration is 40 mg/L. The rate equation was found to be:

$$\gamma = -0.20\,C \quad \text{in mg/L/min}$$

The influent flow rate is 100 gallons/minute. The reactor can be operated as a CFSTR or as a PFR.

a. What is the required reactor volume if it is operated as a CFSTR?
b. What is the required reactor volume if it is operated as a PFR?

12. A wastewater stream ($Q = 1.0$ m³/min) containing 1,000 mg/L of glucose is to be treated biologically. The goal is to reduce the concentration to <100 mg/L. The reaction is first-order with a rate constant of 0.2/min, as determined from a bench-scale study.

Four different configurations of bioreactors operated in the CFSTR mode are considered. Determine the final effluent concentration from each of these arrangements and if it meets the clean-up requirement:

a. one 60-m^3 reactor
b. two 30-m^3 reactors in series
c. one 20-m^3 reactor followed by one 40-m^3 reactor
d. one 40-m^3 reactor followed by one 20-m^3 reactor

13. A wastewater stream ($Q = 1.0$ m^3/min) containing 1,000 mg/L of glucose is to be treated biologically. The goal is to reduce the concentration to <100 mg/L. The reaction is first-order with a rate constant of 0.2/min, as determined from a bench-scale study.

Four different configurations of bioreactors operated in the PFR mode are considered. Determine the final effluent concentration from each of these arrangements and if it meets the clean-up requirement:

a. one 60-m^3 reactor
b. two 30-m^3 reactors in series
c. one 20-m^3 reactor followed by one 40-m^3 reactor
d. one 40-m^3 reactor followed by one 20-m^3 reactor

14. A wastewater stream ($Q = 1.0\ m^3/min$) containing 1,000 mg/L of glucose is to be treated biologically. The goal is to reduce the concentration to <100 mg/L. A CFSTR reactor with 15 minutes of residence time can only reduce the concentration to 250 mg/L. Assuming that it is a first-order reaction, can three smaller reactors (5 minutes of residence time each) in series reduce the glucose concentration to below 100 mg/L?
15. A wastewater stream ($Q = 1.0\ m^3/min$) containing 6,000 mg/L of glucose is to be treated biologically. The goal is to reduce the concentration to <100 mg/L. A PFR reactor with 5 minutes of residence time can only reduce the concentration to 1,000 mg/L. Assuming that it is a first-order reaction, how many reactors (5 minutes of residence time each) in series reduce the glucose concentration to below 100 mg/L?
16. A wastewater stream ($Q = 1.0\ m^3/min$) containing 1,000 mg/L of glucose is to be treated biologically. The goal is to reduce the concentration to be <100 mg/L. The reaction is first-order with a rate constant of 0.2/min, as determined from a bench-scale study.

    Four different configurations of bioreactors operated in the CFSTR mode are considered. Determine the final effluent concentration from each of these arrangements and if it meets the clean-up requirement:

    a. one 40-m^3 reactor
    b. two 20-m^3 reactors in parallel (each receives 0.5 m^3/min flow)
    c. one 10-m^3 reactor and one 30-m^3 reactor in parallel (each receives 0.5 m^3/min flow)
    d. one 10-m^3 reactor ($Q = 0.25\ m^3/min$) and one 30-m^3 reactor in parallel ($Q = 0.75\ m^3/min$)
17. A wastewater stream ($Q = 1.0\ m^3/min$) containing 1,000 mg/L of glucose is to be treated biologically. The goal is to reduce the concentration to <100 mg/L. The reaction is first-order with a rate constant of 0.2/min, as determined from a bench-scale study.

    Four different configurations of bioreactors operated in the PFR mode are considered. Determine the final effluent concentration from each of these arrangements and if it meets the clean-up requirement:

    a. one 40-m^3 reactor
    b. two 20-m^3 reactors in parallel (each receives 0.5 m^3/min flow)
    c. one 10-m^3 reactor and one 30-m^3 reactor in parallel (each receives 0.5 m^3/min flow)
    d. one 10-m^3 reactor ($Q = 0.25\ m^3/min$) and one 30-m^3 reactor in parallel ($Q = 0.75\ m^3/min$)

12 Water Quality Parameters

12.1 INTRODUCTION

The Earth system, that we live in, is made of four sub-systems: biosphere, hydrosphere, geosphere, and atmosphere. *Biosphere* can be defined as the total sum of living things/organisms. *Hydrosphere* is the sum of all water bodies of the Earth, including surface water, groundwater, and ice sheets/caps. *Geosphere* refers to the solid portion of the Earth [Note: *Lithosphere* refers to the hard and rigid outer layer of the geosphere; while *pedosphere* is often defined as the uppermost of the Earth's surface where the soil-formation processes are still alive]. *Atmosphere* is the air and its content surrounding the entire Earth. These four sub-systems are not independent; interactions among them are occurring continuously. This chapter and the next two chapters (i.e., the last three chapters) of this book will go over essential topics related to main components of the hydrosphere, pedosphere, and atmosphere; they are related to water, soil, and air, respectively.

Sources of water can be classified into surface water (i.e., rivers, streams, lakes, reservoirs, seawater, and ice caps) and groundwater. We need water for daily uses such as drinking, cooking, farming, irrigation, industrial uses, etc. Water is also used for transportation, recreation, and electricity generation. Many plants and animals live in water bodies. Different water bodies have their specific characteristics, while a specific water use may require the water having presence or absence of some specific characteristics. Water treatment may be required to bring a specific water source to meet certain requirement(s); for example, protection of human health is of the utmost concern of potable water. After our use of water, wastewater would be generated and often discharged into receiving water bodies. Wastewater discharge or water reclamation often requires the removal of undesirable substances, that are produced or added during water use, to protect the aquatic species and humans that are using the receiving water bodies as well as to provide quality water for water reuses.

12.2 RELEVANT REGULATIONS

There are many federal, state, and local regulations for different types/uses of water. The focus here is on the quality of potable water and ambient water bodies.

12.2.1 Safe Drinking Water Act

Over 92% of Americans receive their potable water from community water systems and they are subjected to safe drinking water standards specified in the Safe Drinking Water Act (SDWA). The SDWA rules include guidelines for drinking water quality as well as water testing methods and schedules. The National Primary Drinking Water Regulations (NPDWRs) set mandatory water quality standards for drinking water

DOI: 10.1201/9781003502661-12

TABLE 12.1
USEPA's Secondary MCLs (USEPA, 2023b)

Contaminant	Secondary MCL	Noticeable Effects above the Secondary MCL
Aluminum	0.02–0.2 mg/L	Colored water
Chloride	250 mg/L	Salty taste
Color	15 color units	Visible tint
Copper	1.0 mg/L	Metallic taste; blue-green staining
Corrosivity	Non-corrosive	Metallic taste; corroded pipes/fixtures staining
Fluoride	2.0 mg/L	Tooth discoloration
Foaming agents	0.5 mg/L	Frothy, cloudy; bitter taste; odor
Iron	0.3 mg/L	Rusty color; sediment; metallic taste; reddish or orange staining
Manganese	0.05 mg/L	Black to brown color; black staining; bitter metallic taste
Odor	3 TON (threshold odor number)	"Rotten-egg", musty or chemical smell
pH	6.5–8.5	Low pH: bitter metallic taste; corrosion High pH: slippery feel; soda taste; deposits
Silver	0.1 mg/L	Skin discoloration; graying of the white part of the eye
Sulfate	250 mg/L	Salty taste
Total dissolved solids (TDS)	500 mg/L	Hardness; deposits; colored water; staining; salty taste
Zinc	5 mg/L	Metallic taste

contaminants, called "*maximum contaminant levels*" (MCLs) which are established to protect human health. An MCL is the maximum allowable amount of a contaminant in drinking water (USEPA, 2023a). Section 2.1 in Chapter 2 provides a list of compounds having MCLs.

In addition, the National Secondary Drinking Water Regulations (NSDWRs) set non-mandatory water quality standards for 15 contaminants. The "*secondary maximum contaminant levels*" (SMCLs) are established as guidelines to assist public water systems in managing their drinking water for aesthetic considerations, such as taste, color, and odor. These contaminants are not considered to present a risk to human health at the SMCLs. There are a variety of potential effects related to these secondary contaminants; and the effects can be grouped into three categories: aesthetic, cosmetic, and technical effects (USEPA, 2023b) – see Table 12.1.

12.2.2 Clean Water Act

The Clean Water Act (CWA) requires the USEPA to develop *water quality criteria* (WQC) for surface water to determine if the water has become unsafe for humans and wildlife. The forms of WQC can be numeric or narrative. There are five types of WQC; they are (i) *human health criteria* – how much a specific chemical can be present before it is likely to adversely impact human health; (ii) *aquatic life criteria* – how much a specific chemical can be present before it is likely to adversely impact plant and animal life; (iii) *biological criteria* – indications of how healthy water bodies are, based on the types and concentrations of organisms present; (iv) *nutrient criteria* – limits on both total nitrogen and total phosphorus to help prevent eutrophication and proliferation of harmful algal blooms in rivers and streams, lakes and reservoirs, and estuaries and coastal areas, and (v) *microbial/recreational criteria* – limits of certain organisms and their associated toxins in water for safe recreational activities such as swimming (USEPA, 2017).

USEPA's *Priority Pollutants* are a set of chemical pollutants that they regulate, and for which they have developed analytical test methods. The current list has 126 Priority Pollutants. These are not the only pollutants regulated in the CWA programs; however, the list is an important starting point for the USEPA to consider, for example, in developing national discharge standards (e.g., Effluent Guidelines) or in national permitting programs (e.g., NPDES) (USEPA, 2023c & 2023d). The 126 Priority Pollutants are tabulated in Section 2.1 of Chapter 2.

12.2.3 Water Quality Parameters

We have discussed some important properties of water such as hydrogen bonding, heat capacity, amphoteric nature, density, viscosity, and surface tension in previous chapters. They are important characteristics of water and have impacts on the water quality.

Water quality is a measure of the conditions relative to the requirements of a specific need or use. To evaluate the quality, many water quality parameters need to be quantified and compared to the requirements to that need or use. For example, the MCLs are the limits of the water quality parameters that need to be met for potable uses. If one or more water quality parameters exceed the regulatory limits, the use of the water or the discharge of water will not be allowed. More or additional treatment would be needed to make those parameters permissible.

Water quality parameters can be categorized into three groups: physical, chemical, and biological; they will be discussed in the subsequent sections.

12.3 PHYSICAL WATER QUALITY PARAMETERS

Important physical water quality parameters include temperature, solids, turbidity, color, and odor.

12.3.1 Temperature

Temperature (and pH) should be included in any assessment of water quality because it influences many other parameters.

Temperature influences most physical, biological, and chemical processes and ecosystems in aquatic environments. Increased water temperature can result in (EPA, 2021a):

- Increased reaction rates.
- Increased solubility of metals and other toxins in water.
- Possible increased toxicity of some substances to aquatic organisms.

- Decreased dissolved oxygen (DO) available to aquatic life.
- Increased potential for algal bloom.

Temperature will also affect the density and viscosity of water. Extreme temperatures (hot or cold) will affect aquatic life and/or impact negatively on aquatic health. It should be noted that USEPA's description on some ambient WQC can be found in "Water Quality Standards Handbook – Chapter 3: Water Quality Criteria" (USEPA, 2017).

12.3.2 Solids, Electrical Conductivity, and Turbidity

Different types of solids are contained in water; they can be grouped into (i) suspended versus dissolved, (ii) organic versus inorganic, or (iii) settleable versus non-settleable.

Dissolved solid vs. suspended solid. *Dissolved solids* are those that can pass through a filter of openings of ~1.2 micrometers (1.2 µm, 1.2 microns, and 0.00012 cm) or smaller. They include dissolved inorganic ions (e.g., sodium, chloride, calcium, magnesium, nitrate, and phosphorus ions) and dissolved organic compounds. *Suspended solids* are those retained by a 1.2-µm filter; they include both organic (e.g., algae, coffee grounds) and inorganic particulates (e.g., silt and clay). Concentrations of *total solids* (TS), *total dissolved solids* (TDS), and *total suspended solids* (TSS) can be measured by driving off water from corresponding sample through evaporation in an oven (T = 103°C to 105°C). TS is essentially the sum of TDS and TSS as:

$$TS = TDS + TSS \tag{12.1}$$

Volatile solid vs. fixed solids. *Volatile solids* are the portion of the TS that are lost through combustion in a furnace (T = 550°C); while the residual portion is referred to as the *fixed solids*. The terms volatile and fixed can also be applied to TS, TDS, and TSS. Although it is not exactly accurate, *total volatile solids* (TVS) are typically used to represent the total amount of organics in the water; while *total fixed solids* (TFS) represent the total amount inorganic solids in water. The following relationships apply (also see Figure 12.1):

$$TS = TVS + TFS \tag{12.2}$$

$$TVS = VDS + VSS \tag{12.3}$$

$$TFS = FDS + FSS \tag{12.4}$$

$$TDS = VDS + FDS \tag{12.5}$$

$$TSS = VSS + FSS \tag{12.6}$$

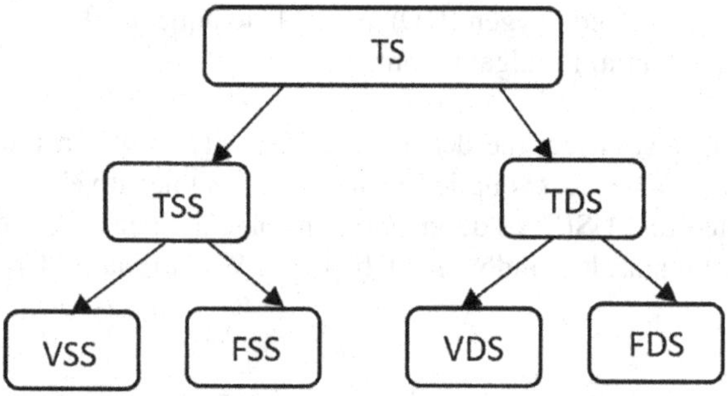

FIGURE 12.1 Types of solids in water.

Concentrations of these solids are typically expressed in mg/L (or in "parts per million"; ppm). If concentrations are relatively low, they can be expressed in μg/L (or in "parts per billion"; ppb) [Note: 1 ppm = 1,000 ppb]. For the cases of relatively high solid concentrations (e.g., in water/wastewater sludges), they can be expressed in "% by weight solid".

Settable Solids. Removal of suspended solids is one of the common practices in water/wastewater treatment. *Sedimentation* (i.e., settling by gravity) is used to remove those suspended solids that can settle to the bottom of the sedimentation tanks (or clarifiers, thickeners, settling tanks) within a reasonable time frame. *Settleable solids* are those solids would settle to the bottom of an Imhoff cone within a specific period, usually 30 or 60 minutes (i.e., V_{30} or V_{60}); and it is expressed in mL/L.

Example 12.1: Types of Solids in Water

Use the information given in the table below to find VSS, FSS, VDS, FDS, TDS, TSS, TVS, TFS and TS of this water sample.

Compound	Concentration (mg/L)	Dissolved?	Volatilize @550°C
Silt/Clay	20	No	No
Salt (NaCl)	300	Yes	No
Coffee grounds	10	No	Yes
Ethanol	120	Yes	Yes

Solution:

a. Volatile suspended solids (VSS) = 10 mg/L (coffee grounds)

b. Fixed suspended solids (FSS) = 20 mg/L (silt/clay)

c. Volatile dissolved solids (VDS) = 120 mg/L (ethanol)

d. Fixed dissolved solids (FDS) = 300 mg/L (salt)

e. TDS = VDS + FDS = 120 + 300 = 420 mg/L (Eq. 12.5)

f. TSS = VSS + FSS = 10 + 20 = 30 mg/L (Eq. 12.6)

g. TVS = VSS + VDS = 10 + 120 = 130 mg/L (Eq. 12.3)

h. TFS = FSS + FDS = 20 + 300 = 320 mg/L (Eq. 12.4)

i. TS = TDS + TDS = 420 + 30 = TVS + TFS = 130 + 320 = 450 mg/L (Eq. 12.1 or Eq. 12.2)

Example 12.2: Total Solid and Volatile Solid Contents of Wastewater Sludge

A laboratory analyzed the solid content of a primary sludge sample. The raw data are shown below:

- Weight of dish = 20.50 g
- Weight of dish + the sludge sample = 140.80 g
- Weight of dish + dry solid = 25.60 g (after dried in an oven at $T = 103°C$)
- Weight of dish + ash = 21.70 g (after a muffle furnace at $T = 550°C$)

What are TS and VS concentrations of this sludge (in %)?

Solution:

a. The weight of the sludge sample = 140.80 − 20.50 = 120.30 g

b. TS = 25.60 − 20.50 = 5.10 g

c. TFS = 21.70 − 20.50 = 1.20 g

d. TVS = TS − TFS = 5.10 − 1.20 = 3.90 g

e. % TS = 5.10 ÷ 120.30 = 4.24%

f. % VS = 3.90 ÷ 120.30 = 3.24%

Discussion:

The solid in this primary sludge is 76.5% (= 3.24% ÷ 4.24%) volatile.

Example 12.3: Concentrations of Solids in a Water Sample

The following test results are for a wastewater sample using a sample size of 200 mL.

- Tare mass of the evaporating dish = 50.6420 g
- Dish + residue after evaporation at 105°C = 50.7170 g
- Dish + residue after ignition at 550°C = 50.6820 g

- Tare of Whatman GF/C filter = 1.5350 g
- Filter + residue after evaporation at 105°C = 1.5590 g
- Filter + residue after ignition at 550°C = 1.5450 g
  a. Determine its TS concentration, in mg/L.
  b. Determine its TDS concentration, in mg/L.
  c. Determine its total volatile solid concentration, in mg/L.

Solution:

a. TS = (50.7170 − 50.6420 g) ÷ 0.20 L = 0.375 g/L = <u>375 mg/L.</u>

b. TSS = (1.5590 − 1.5350 g) ÷ 0.20 L = 0.12 g/L = <u>120 mg/L</u>.

$$\text{TDS} = \text{TS} - \text{TSS} = 375 - 120 = \underline{255\text{ mg/L}}.$$

c. TFS = (50.6820 − 50.6420 g) ÷ 0.20 L = 0.20 g/L = <u>200 mg/L.</u>

$$\text{TVS} = \text{TS} - \text{TFS} = 375 - 200 = \underline{175\text{ mg/L}}.$$

TDS, salinity, and electrical conductivity are water quality parameters related to the amount of dissolved solids in water.

TDS vs. salinity. TDS is the measure of all types of dissolved solid compounds in a water sample; while *salinity* is the measure of dissolved inorganic salt species in a water sample. TDS can include dissolved organic molecular compounds, in addition to dissolved inorganic salt. The commonly used units for salinity are grams of salts/kg of the solution (g/kg) or parts per thousand (ppt). For example, the average salinity of ocean water is around 35 ppt.

TDS vs. electrical conductivity. It would take hours for a laboratory to determine the TDS concentration of a water sample. In water/wastewater treatment, continuously monitoring the TDS level of the flowing water stream is often of necessity. An electronic conductivity meter is often used as an online monitoring tool for TDS. *Electrical conductivity* (EC) of water is directly related to the concentration of dissolved ionized solids in water which can conduct current. The commonly used units for EC are micro-Siemens/cm (μS/cm) and μmho/cm [Note: "mho" is the reciprocal of Ohm in name; and 1 Siemens = 1 mho]. Although EC accounts only for ionic portion of the dissolved solids (i.e., not including the non-ionic dissolved species such as many dissolved organic species), a correlation can often be established for a given type of water. For example, the following relationship was established for the final effluent of a water reclamation plant (WRP) in southern California (Kuo et al., 1994):

$$\text{TDS}\left(\text{in}\frac{\text{mg}}{\text{L}}\right) \sim \left[0.60 \times \text{EC}\left(\text{in}\frac{\mu\text{S}}{\text{cm}}\right)\right] \tag{12.7}$$

For that WRP, the TDS concentrations of the final effluent measured by the on-site laboratory are usually around 600 mg/L and the EC values measured by the online analyzer are typically around 1,000 μS/cm. EC is also commonly used to monitor the water quality of cooling towers to determine if a "blow-down" (i.e., replacement of portion of the recirculating cooling water with fresh water) is necessary.

The secondary MCL of drinking water is 500 mg/L. When the TDS concentration >1,000 mg/L, the water is considered unsuitable for human consumption.

TSS and turbidity are water quality parameters related to the concentrations of suspended solids in water.

TSS vs. turbidity. Similar to the case of TDS, it would take hours for a laboratory to determine the TSS concentration of a water sample. In water/wastewater treatment, continuously monitoring the TSS level of a flowing water stream is also of necessity. *Turbidity* is a measure of water's light-scattering properties. Turbidity is an indication of water clarity which is affected by suspended solids and other substances (such as dissolved organic compounds) present in water. With its online monitoring capability, readings of a turbidity meter are often used as an indicator of TSS concentration of a flowing water stream. The turbidity values are typically expressed in Turbidity Unit (TU) or Nephelometric Turbidity Unit (NTU). In general, higher SS concentrations will incur higher turbidity values. The following correlations between NTU readings and TSS measurements of the secondary and final effluents of several WRPs in southern California were found (Kuo et al., 1994).

$$\text{Secondary Effluent: TSS}\left(\text{in}\,\frac{\text{mg}}{\text{L}}\right) \sim \left[2.4 \times \text{Turbidity}\ (\text{in NTU})\right] \tag{12.8}$$

$$\text{Final Effluent: TSS}\left(\text{in}\,\frac{\text{mg}}{\text{L}}\right) \sim \left[1.5 \times \text{Turbidity}\ (\text{in NTU})\right] \tag{12.9}$$

The main reason for a smaller coefficient (1.5) for the final effluent in Eq. (12.9), when compared to that for the secondary effluent in Eq. (12.8) (2.4), is that the particles in the secondary effluent are mostly bigger than those in the final effluent [Note: The secondary effluents of these WRPs are filtered, disinfected (and dechlorinated) before discharges or reuses]. If two water samples have the same TSS concentration, the water sample containing more particles of larger sizes would not scatter as much light as the one containing more smaller particles.

An increase in turbidity (or SS) can negatively affect aquatic health by (EPA, 2021b):

- Decreasing light penetration into water, and thereby reducing biomass as well as growth rates of aquatic plants.
- Hindering visibility and thereby making it difficult for predators to find prey.
- Affecting the health of aquatic species from clogging fish gills or the filter-feeding systems of other aquatic animals, reducing fish resistance to disease, and altering egg and larval development.

Higher SS concentrations also create challenges to water/wastewater disinfection because particles can shield/block the microorganisms from chemical disinfectants (e.g., chlorine and ozone) as well as UV rays in UV disinfection. Various suspended particles can act as the adsorbent for heavy metals and hydrophobic compounds (e.g., incalcitrant organic compounds) that increases the risk of ingesting these toxic compounds by human as well as aquatic species. Typical turbidity in drinking water would be <1 TU, while the most stringent turbidity requirement for reclaimed water in California Title 22 is 2 TU.

12.3.3 Color and Transmittance/Absorbance

Color. Even pure water is not colorless; it has a slight blue tint. Color of water indicates the presence of organic (e.g., algae or humic substances) and inorganic contaminants. The color strength of water is expressed in *color unit.* Water with one color unit means that its color intensity is equivalent to the color of distilled water containing 1 mg/L platinum in the form of chloroplatinum ion $\left(PtCl_6^{=}\right)$. The current SMCL for color is 15 color units.

Transmittance/absorbance. When light is passed through a water sample, a portion of the light would be absorbed or scattered by substances present in water. *Transmittance* (T) is the ratio of the transmitted light (L_T) to the incident light (L_o) as

$$T\ (\text{in }\%) = \frac{L_T}{L_o} \times 100 \tag{12.10}$$

Absorbance (A) is defined as the amount of light absorbed by the substances present in water. The transmittance and absorbance are related as:

$$A = 2.0 - \log_{10}\left[T(\text{in }\%)\right] \tag{12.11}$$

From Eq. (12.11), water with a transmittance of 100% corresponding to an absorbance of zero and a transmittance of 10% corresponding to an absorbance of 1.0.

UVT and SUVA. The values of transmittance and absorbance of a given water sample would be different for incident lights of different wavelengths. Water and wastewater disinfection using ultraviolet light (UV) is becoming popular. The transmittance of UV light of 254 nanometer (nm) wavelength (UVT_{254}) is critical to the success of UV disinfection because UV wavelength of 254 nm is considered to be the most germicidal. Consequently, *UV light transmittance* (UVT) becomes a critical water quality parameter in UV disinfection. UVT is assumed to mean the water transmittance at 254 nm unless noted otherwise. The UVT values of good secondary effluents are around 70%, while those of good reverse osmosis effluents would be close to 100%.

Specific UV absorbance (SUVA) is defined as the UV absorbance of a water sample at a given wavelength normalized to the concentration of dissolved organic matter (DOM). SUVA analysis has a wide range of applications in water quality. It is used to monitor the quality of drinking water sources with regard to the presence of organic matter and its disinfection byproduct formation potential (DBPFP). For surface water, it can provide information about the sources of organic matter and potential for eutrophication. It can also be used to monitor the efficiency of water/wastewater treatment processes with regard to the removal of organic matter and the DBPFP.

12.3.4 Taste and Odor

Taste and odor (T&O) are human perceptions of water quality. They reflect only aesthetic properties and do not provide direct information about the safety of water. In drinking water, the causes of T&O issues can come from the water fixtures, plumbing, water treatment (e.g., chlorine residual), or water sources (e.g., presence of odorous compounds). Wastewater may contain odorous compounds such as amines, ammonia, mercaptans, and skatole (a foul-smelling constituent of mammalian feces).

Threshold odor numbers (TONs) are whole numbers that indicate how many dilutions it takes to produce odor-free water, and it is calculated by using the following equation:

$$\text{TON} = \frac{\text{mL of sample} + \text{mL of odor free water}}{\text{mL of sample}} \tag{12.12}$$

Example 12.4: Threshold Odor Number

A 40 mL sample of treated wastewater requires 160 mL of distilled water to reduce the odor to a level that is just perceptible. What is the TON of this water sample?

Solution:

Use Eq. (12.12) to find the TON,

$$\text{TON} = \frac{\text{mL of sample} + \text{mL of odor free water}}{\text{mL of sample}} = \frac{40 + 160}{40} = 5$$

Discussion:

The secondary MCL is 3.

12.4 CHEMICAL WATER QUALITY PARAMETERS I – pH, ALKALINITY, HARDNESS, AND DO

To facilitate the presentation, chemical water quality parameters will be presented in three sections. This section talks about pH, alkalinity, hardness, and DO, followed by two subsequent sections which will discuss aggregate organic constituents and some individual inorganic compounds of concern, respectively.

12.4.1 pH

pH (and temperature) should be included in any assessment of water quality because they influence many other parameters.

pH is an important indicator of chemical, physical, and biological changes in a water body, and it plays a crucial role in chemical processes occurring in natural waters. pH is a key factor in water chemistry and toxicity. A change in pH can alter the forms and toxicity of toxic chemicals (see ammonia toxicity later in this section). Heavy metals (e.g., lead, mercury, copper, and arsenic) are generally more soluble at a lower pH; on the other hand, high pHs will incur more precipitation of metals. Elevated concentrations of heavy metals become more toxic to aquatic species due to more absorption into tissues of organisms. In addition, pH also plays a key role in health of aquatic species by affecting biochemical processes and their metabolisms. Most aquatic species have adapted to the natural pH levels in the waterbodies they live in. However, slight changes in pH can have negative impacts on the health of aquatic community. pH in a lake or river often fluctuates daily, with a higher pH during the daytime due to the consumption of CO_2 by aquatic plants and algae (i.e., photosynthesis). At night, aquatic plants and algae respire, giving off CO_2 which would result in lower water pHs (more acidic). The USEPA's recommended water criteria for pH ranges from 6.5 to 9.0, depending on what is protective of aquatic life and the ecosystem (USEPA, 2021c).

The USEPA considers that wastes with a pH ≥12.5, or ≤2 corrosive; and corrosive wastes are regulated as hazardous wastes in the USEPA's regulations. Figure 12.2 shows the pHs of some solutions related to our daily lives.

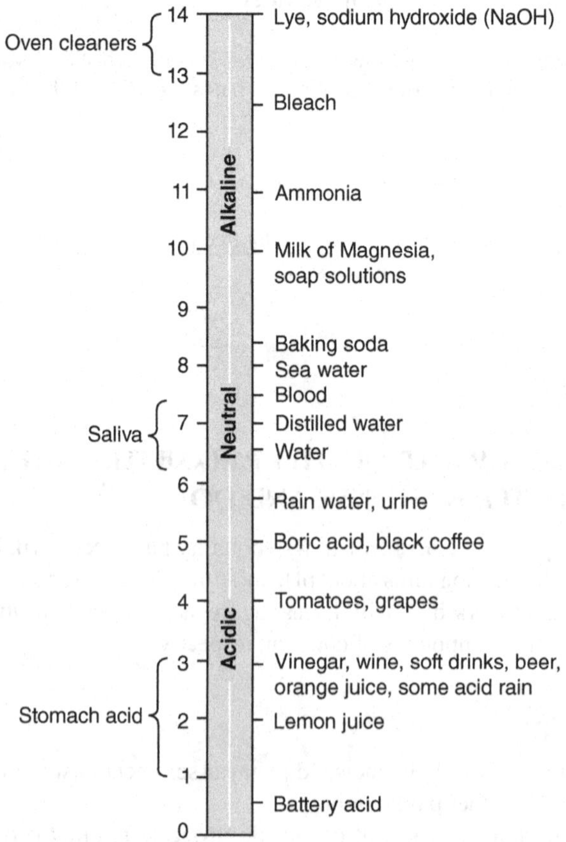

FIGURE 12.2 pH range scale. (EPA, 2006.)

12.4.2 Alkalinity

As mentioned in Chapter 8, a buffer solution contains a weak acid and its conjugate base, or a weak base and its conjugate acid. A buffer solution helps neutralize some of the added acids or bases so that the pH will not change abruptly. In our natural environment, there are many buffer systems. The one that is most important (i.e., the controlling one) is the *carbonate/bicarbonate buffer system.*

The carbonate system in water consists of three carbonate species in equilibrium; they are carbonic acid (H_2CO_3), bicarbonate ions $\left(HCO_3^-\right)$, and carbonate ions $\left(CO_3^=\right)$. For open systems, the aqueous carbonate species can equilibrate with carbon dioxide in the atmosphere as:

$$CO_{2(g)} \overset{H_2O}{\rightarrow} CO_{2(aq)} \tag{8.36}$$

$$CO_{2(aq)} + H_2O_{(l)} \rightarrow H_2CO_{3(aq)} \tag{8.37}$$

$$H_2CO_{3(aq)} \leftrightarrow H^+_{(aq)} + HCO^-_{3\,(aq)} \qquad pK_{a1} \approx 6.3 \tag{8.38}$$

$$HCO^-_{3\,(aq)} \leftrightarrow H^+_{(aq)} + CO^=_{3\,(aq)} \qquad pK_{a2} \approx 10.3 \tag{8.39}$$

In addition, calcium carbonate ($CaCO_3$) is a very common mineral (limestone is one form of calcium carbonate). Dissolution of $CaCO_3$ will add carbonate ions to the aqueous solution as:

$$Ca^{+2}_{(aq)} + CO^=_{3(aq)} \leftrightarrow CaCO_{3(s)} \tag{8.40}$$

Using the relationships shown in Eqs. (8.38) and (8.39), the relative amounts of three carbonate species can be found (see Figure 12.3).

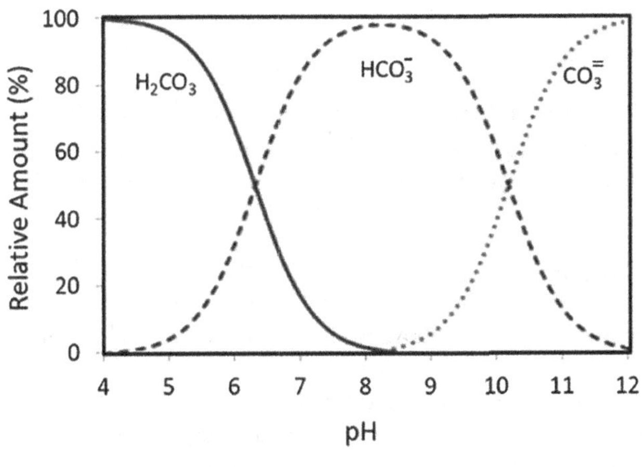

FIGURE 12.3 Relative concentrations of carbonate species vs. pH.

Example 12.5: Carbonate Buffer System

For a surface water system ($T = 25°C$),

a. At what pH, the concentrations of carbonic acid and bicarbonate are equal?
b. At what pH, the concentrations of bicarbonate and carbonate are equal?
c. At what pH, the bicarbonate concentration will be 10 times larger than that of carbonic acid?
d. At what pH, the bicarbonate concentration will be 10 times larger than that of carbonate?
e. Which would be the dominant carbonate species in our ambient water bodies?

Solution:

From Eqs. (8.38) and (8.39), the two acid association equilibrium constants can be written as

$$K_{a1} = 10^{-6.3} = \frac{[H^+][HCO_3^-]}{[H_2CO_3]} \tag{8.41}$$

$$K_{a2} = 10^{-10.3} = \frac{[H^+][CO_3^=]}{[HCO_3^-]} \tag{8.42}$$

a. From Eq. (8.41), $[HCO_3^-] = [H_2CO_3]$ at $pH = 6.3$.

b. From Eq. (8.42), $[HCO_3^-] = [CO_3^=]$ at $pH = 10.3$.

c. From Eq. (8.41), $[HCO_3^-] = 10 \times [H_2CO_3]$ at $pH = 7.3$.

d. From Eq. (8.42), $[HCO_3^-] = 10 \times [CO_3^=]$ at $pH = 9.3$.

e. Bicarbonate will be the dominant species when $6.3 < pH < 10.3$ (pH of most ambient waterbodies fall into this pH range. Also see Figure 12.3 above).

Alkalinity of water is its capacity to neutralize acid. Since each bicarbonate ion can neutralize with one proton and each carbonate ion can neutralize two protons to form carbonic acid, and one hydroxyl ion can neutralize one proton, so alkalinity can be expressed as

$$\text{Alkalinity (in } M) = 2[CO_3^=] + [HCO_3^-] + [OH^-] - [H^+] \tag{8.35}$$

All the concentrations in the above equation are in M, and the minus ahead the proton concentration is because proton is an acid.

In practice, the concentration of alkalinity is often expressed in "mg/L as $CaCO_3$". Since the molecular weight of $CaCO_3 = 100$ and its equivalency $= 2$ (i.e., its equivalent weight $= 50$), the concentrations of the four species involved in alkalinity can be expressed in "mg/L as $CaCO_3$" as

$$\text{Concentration}\left(\frac{\text{mg}}{\text{L}}\text{as CaCO}_3\right)=\text{Concentration}\left(\frac{\text{mg}}{\text{L}}\right)\times\frac{50}{\text{EW of the Species}} \quad (12.13)$$

$$\text{Alkanlity}\left(\text{in}\frac{\text{mg}}{\text{L}}\text{ as CaCO}_3\right)=\left[CO_3^=\right]+\left[HCO_3^-\right]+\left[OH^-\right]-\left[H^+\right] \quad (12.14)$$

Comparing the two equations for alkalinity, a multiply of 2 appears in front of the carbonate concentration in Eq. (8.35), but disappears in Eq. (12.14); it is because concentrations of all species in Eq. (12.14) are all expressed in the equivalency of $CaCO_3$.

Example 12.6: Alkalinity in the Unit of Equivalent Calcium Carbonate

Two water samples were analyzed, and the results are (all the concentrations are in mg/L):

- Sample #1: $Ca^{+2}=70$, $Mg^{+2}=24$, $Na^{+}=36$, $SO_4^{=}=55$, $Cl^{-}=36$, $F^{-}=0.4$, $NO_3^{-}=20$, $HCO_3^{-}=60$ mg/L, and pH = 7.5
- Sample #2: $Ca^{+2}=35$, $Mg^{+2}=12$, $Na^{+}=36$, $SO_4^{=}=55$, $Cl^{-}=36$, $F^{-}=0.4$, $NO_3^{-}=20$, $HCO_3^{-}=30$, $CO_3^{=}=30$ mg/L, and pH = 9.5

What are the alkalinity values (in mg/L as $CaCO_3$) of these two samples?

Solution:

a. Although four terms show up in Eq. (12.14) (i.e., $\left[HCO_3^-\right]$, $\left[CO_3^=\right]$, $[OH^-]$, and $[H^+]$), the concentrations of OH^- and H^+ are usually very small, compared to those of HCO_3^- and $CO_3^=$. Consequently, concentrations of OH^- and H^+ are ignored in most alkalinity calculations.

b. *MW* of $HCO_3^-=61$ and its equivalency = 1 (i.e., equivalent weight = 61); *MW* of $CO_3^==60$ and its equivalency = 2 (i.e., equivalent weight = 60/2 = 30).

c. For sample #1,

$$\text{Alkanlity}\left(\text{in}\frac{\text{mg}}{\text{L}}\text{ as CaCO}_3\right)=\left[HCO_3^-\right]=60\times\frac{50}{(61/1)}=49.2$$

d. For sample #2,

$$\text{Alkanlity}\left(\text{in}\frac{\text{mg}}{\text{L}}\text{ as CaCO}_3\right)=\left[CO_3^=\right]+\left[HCO_3^-\right]$$

$$=30\times\frac{50}{(60/2)}+30\times\frac{50}{(61/1)}=74.6$$

Discussion:

1. The concentrations of carbonate ion and bicarbonate ion of Sample #2 are both 30 mg/L; however, their concentrations are 50 and 24.6 mg/L as $CaCO_3$, respectively.
2. With the same 30 mg/L concentration, carbonate ions contribute almost twice as much to the alkalinity, when compared to that of bicarbonate ion.

12.4.3 Hardness

Hardness is a measure of the mineral content in water. A definition of *hardness* is the concentration of all the multivalent cations (e.g., Ca^{+2}, Mg^{+2}, Cu^{+2}, Fe^{+3}, and Al^{+3}) in water. These dissolved minerals in water will cause problems such as formation of scale deposits on hot surfaces (e.g., calcium carbonate scale on the surface of heat exchangers of household water heaters and industrial boilers) and difficulty in producing lather with soap. Since Ca^{+2} and Mg^{+2} ions have much higher concentrations than the other multivalent cations in ambient waters, we often consider the total hardness in water is the sum of the concentrations of these two ions only as:

$$\text{Total Hardness} = \left[Ca^{+2}\right] + \left[Mg^{+2}\right] \tag{12.15}$$

These dissolved minerals enter water mainly from contact with soil and rock, particularly with limestone deposits (mainly calcium and magnesium carbonate). Consequently, groundwater typically has larger hardness than surface water. There are two types of hardness:

- *Temporary hardness* (or *carbonate hardness*) is hardness associated with carbonate and bicarbonate ions, and it can be removed by boiling.
- *Permanent hardness* (or *non-carbonate harness*) is hardness associated with other anions such as sulfate ions and chlorides, and it will remain after boiling.

Total harness is the sum of these two types of hardness as:

$$\text{Total Hardness} = \text{Carbonate Hardness} + \text{Noncarbonate hardness} \tag{12.16}$$

Similar to alkalinity, concentrations of hardness are expressed in "mg/L as $CaCO_3$" in practice. The concentrations of Ca^{+2} and Mg^{+2} can be converted to $CaCO_3$ equivalent using Eq. (12.13).

Below is a classification of water according to its hardness:

- Soft water: <50 mg/L as $CaCO_3$
- Moderately hard water: 50 to 150 mg/L as $CaCO_3$
- Hard water: 150–300 mg/L as $CaCO_3$
- Very hard water: >300 mg/L as $CaCO_3$

The desirable hardness of potable water is around 100 mg/L as $CaCO_3$.

Example 12.7: Total Hardness

Two water samples were analyzed, and the results are (all the concentrations are in mg/L):

- Sample #1: $Ca^{+2}=70$, $Mg^{+2}=24$, $Na^{+}=36$, $SO_4^{=}=55$, $Cl^{-}=36$, $F^{-}=0.4$, $NO_3^{-}=20$, $HCO_3^{-}=60$ mg/L, and pH = 7.5
- Sample #2: $Ca^{+2}=35$, $Mg^{+2}=12$, $Na^{+}=36$, $SO_4^{=}=55$, $Cl^{-}=36$, $F^{-}=0.4$, $NO_3^{-}=20$, $HCO_3^{-}=30$, $CO_3^{=}=30$ mg/L, and pH = 9.5

What are the total hardness and carbonate hardness concentrations (in mg/L as $CaCO_3$) of these two samples?

Solution:

a. MW of $Ca^{+2}=40$ and its equivalency=2 (i.e., equivalent weight=40/2=20); *MW* of $Mg^{+2}=24.3$ and its equivalency=2 (i.e., equivalent weight=24.3/2).

b. For sample #1,

$$\text{Total Hardness} = \left[Ca^{+2}\right]+\left[Mg^{+2}\right]=70\times\frac{50}{(40/2)}+24\times\frac{50}{(24.3/2)}=274\frac{mg}{L}\text{ as CaCO}_3$$

For sample #2,

$$\text{Total Hardness} = \left[Ca^{+2}\right]+\left[Mg^{+2}\right]=35\times\frac{50}{(40/2)}+12\times\frac{50}{(24.3/2)}=137\frac{mg}{L}\text{ as CaCO}_3$$

c. The alkalinity values of sample #1 and sample #2 are 49.2 and 74.6 mg/L as $CaCO_3$, respectively. Consequently, their carbonate hardness concentrations are 49.2 and 74.6 mg/L as $CaCO_3$, respectively.

12.4.4 Dissolved Oxygen

Nearly all aquatic species need oxygen to survive; in addition, oxygen needs to be present (and provided) in aerobic biological wastewater treatment processes. (e.g., activated sludge processes). The amount of oxygen dissolved in water is usually expressed in DO concentrations (in mg/L).

DO saturation (DO_{sat}) refers to the highest possible DO concentration under those environmental conditions (i.e., temperature, salinity, and atmospheric pressure). For example, DO_{sat} values for fresh water are approximately 10.1, 9.1, 8.3 mg/L at 15°C, 20°C, and 25°C, respectively. It implies that the solubility of oxygen decreases as temperature increases. The DO_{sat} will also decrease as the salinity of water increases. Most aqueous species and plants can grow and reproduce unimpaired when DO concentrations exceed 5 mg/L. If DO concentration of a waterbody drops to 3–5 mg/L, most of aquatic organisms would become stressed (USEPA, 2006).

The primary sources of oxygen in surface waters are transfer of oxygen from the air as well as by plants and algae in the water due to photosynthesis. DO levels in surface waters usually follow a daily cycle. During the day, oxygen is added to the water through photosynthesis. At night, photosynthesis stops, and DO levels drop as oxygen is consumed through respiration. Therefore, DO levels in surface water will likely be the lowest early in the early morning. *Hypoxic* (i.e., low DO concentration) or *anoxic* (virtually no DO) conditions do not support fish or macroinvertebrate populations (USEPA, 2021d).

The level of DO of a given water can also be expressed as *percent saturation* and/or *oxygen deficit*, which are defined below:

$$\text{Percent saturation }(\%)=\frac{DO_{measured}}{DO_{sat}}\times 100 \tag{12.17}$$

$$\text{DO deficit} = DO_{sat}-DO_{measured} \tag{12.18}$$

12.5 CHEMICAL WATER QUALITY PARAMETERS II – INORGANIC COMPOUNDS

In addition to Ca^{+2}, Mg^{+2}, HCO_3^-, and $CO_3^=$, which are related to hardness and alkalinity of water, there are other inorganic species which are of water quality concern. They include chloride, fluoride, sulfate, iron and manganese, metals, and nutrients.

12.5.1 Chloride

Chloride ions (Cl^-) in potable water do not cause any adverse impacts on public health, but high concentrations may cause an unpleasant salty taste for some people. The current SMCL for chloride is 250 mg/L (Table 12.1). The major anthropogenic sources of chloride in surface waters are deicing salts, urban and agricultural runoff, and discharges from municipal wastewater treatment plants, industrial plants, and drilling of oil and gas wells. The USEPA has established ambient aquatic life water quality criteria for chloride (USEPA, 1988).

12.5.2 Fluoride

Fluorides (F^-) are naturally occurring compounds. Low levels of fluoride can help prevention of dental cavities. However, they can result in tooth and bone damages at high concentrations. The USEPA has set an SMCL of 2.0 mg/L for fluoride (Table 12.1). In 2015, the Public Health Service (PHS) replaced the 1962 Drinking Water Standards from 0.7 to 1.2 mg/L to an optimal fluoride concentration of 0.7 mg/L; it is to provide optimal concentration of fluoride in drinking water that provides the best balance of protection from dental caries while limiting the risk of dental fluorosis (ASTDR, 2021).

12.5.3 Sulfate

Sulfate ions $\left(SO_4^=\right)$ are naturally occurring compounds; and they have relatively high concentrations in water and wastewater. The source of sulfate ions in natural water is usually caused by leaching of natural sodium or magnesium sulfate. High sulfate ion concentrations in potable water may create objectionable tastes or unwanted laxative effects. The USEPA has set an SMCL of 250 mg/L for sulfate ions (Table 12.1).

12.5.4 Iron and Manganese

The SMCL for iron (Fe) is 0.3 mg/L and the noticeable effects above this MCL include rusty color, sediment, metallic taste, and reddish or orange staining. The SMCL for manganese (Mn) is 0.05 mg/L and the noticeable effects above this SMCL include black to brown color, black staining, and bitter metallic taste (Table 12.1). When the sources of water for water treatment are from deep groundwater aquifers or deep reservoirs with little or no DO, iron and manganese will be present in more reduced forms as ferrous (Fe^{+2}) and Mn(II) (Mn^{+2}) ions. When these ions are exposed to air or chlorine, they will be oxidized to ferric (Fe^{+3}) and Mn(IV) (Mn^{+4}) ions and form insoluble precipitates as:

$$4\ Fe^{+2} + O_2 + 10\ H_2O \leftrightarrow 4\ Fe(OH)_{3(s)} + 8\ H^+ \tag{12.19}$$

$$2\ Mn^{+2} + O_2 + 2\ H_2O \leftrightarrow 2\ MnO_{2(s)} + 4\ H^+ \tag{12.20}$$

12.5.5 Metals

All metals can be toxic at elevated concentrations, even those that are nutritionally essential for sustaining life. Metal toxicity negatively affects the health of aquatic organisms; for example, metal toxicity (i) decreases abundance and diversity of species, (ii) changes reproduction, juvenile growth, and behavior, and (iii) causes spinal abnormalities, gill damage, and death. In a waterbody, metals are either dissolved or in particulate form. Dissolved metals are those small enough to pass through a 0.45-micron (μm) filter and are more easily absorbed by organisms. Metals can adsorb onto particulates such as clay or organic matter. The dissolved and particulate metals in a waterbody can build up in the tissues of fish and other aquatic organisms over time. *Bioaccumulation* occurs when an organism absorbs, or uptakes, metals more quickly than their body can eliminate them. *Biomagnification* of some metals (such as mercury) can also occur in an aquatic ecosystem; it occurs when concentrations of metals increase from transfer up through the food chain as larger organisms feed on many smaller organisms who have each bioaccumulated metals in their bodies (USEPA, 2021e).

Heavy metal often refers to any metallic chemical element that has a relatively high density and is toxic or poisonous at low concentrations (see also Section 3.2.3). Many of them are on the lists of SDWA's MCLs and CWA's Priority Pollutants; they include antimony (Sb), arsenic (As), barium (Ba), beryllium (Be), cadmium (Cd), chromium (Cr), copper (Cu), lead (Pb), mercury (Hg), nickel (Ni), selenium (Se), silver (Ag), thallium (Tl) and zinc (Zn) (see Section 2.1 and Table 12.1).

12.5.6 Nutrients

Nutrients play a critical role in the biodiversity and healthy functioning of aquatic ecosystems by supporting the growth of aquatic plants and algae that provide food and habitat for fish, shellfish, and smaller organisms. Nitrogen and phosphorus are two of the most important nutrients (USEPA, 2021f).

Nitrogen compounds in water and wastewater can be in the forms of organic nitrogen and inorganic nitrogen. *Organic nitrogen* is nitrogen bonded with carbon and is found in proteins, amino acids, urea, living or dead organisms, and decaying plant materials.; while *inorganic nitrogen* is mainly in the forms of ammonia (NH_3)/ammonium $\left(NH_4^+\right)$, nitrite $\left(NO_2^-\right)$ and nitrate $\left(NO_3^-\right)$. *Total nitrogen* (TN) is the sum of organic nitrogen and *total inorganic nitrogen* (TIN) as:

$$TN = \text{Organic N} + TIN \tag{12.21}$$

The TIN is the sum of the four inorganic species:

$$\text{TIN} = \left[NH_3/NH_4^+\right] + \left[NO_3^-\right] + \left[NO_2^-\right] \tag{12.22}$$

Ammonia (NH_3) in an aqueous solution can be in the form of ammonium ion $\left(NH^+_{4(aq)}\right)$ and $NH_{3(g)}$. Its fractionation is a function of pH as:

$$NH_{3(g)} + H_2O \leftrightarrow NH^+_{(aq)} + OH^-_{(aq)} \quad K_b = 1.8 \times 10^{-5} \text{ @}T = 25°C \tag{12.23}$$

As shown in Eq. (12.23), the reaction will move toward the left at higher pHs; it implies that more ammonia will be in its gaseous form at elevated pHs. That is the reason why ammonia stripping is usually conducted at pH >11 or higher. Ammonia in aquatic environments is toxic to fish and other aquatic organisms. Ammonia in the form of NH_3 is more toxic than in the form of NH_4^+. The ammonia toxicity increases with temperature (because of faster reaction rates) and increases with pH (because more in the form of NH_3). Consequently, the ammonia limit in wastewater discharge is usually more stringent at higher temperatures and pHs.

The primary MCL for nitrate is 10 mg NO_3^--N/L and it is based on the risk of methemoglobinemia (i.e., the blue baby syndrome) in infants younger than 6 months. The primary MCL for nitrite is 1 mg NO_2^--N/L. The MCL limit for the sum of nitrate and nitrate is 10 mg N/L.

In raw wastewater sewage, the nitrogen compounds consist mainly of organic nitrogen compounds, with few or no nitrate and nitrite ions. Most of the organic nitrogen compounds will be converted to ammonia through biological activities, and the process is called *ammonification*. *Total Kjeldahl nitrogen* is the sum of organic nitrogen compounds and ammonia as:

$$\text{TKN} = [\text{Organic nitrogen}] + \left[NH_3/NH_4^+\right] \tag{12.24}$$

In wastewater treatment plants, *nitrification* is a biological process where nitrifying bacteria convert ammonia to nitrite and then to nitrate. It is an aerobic and *autotrophic* (i.e., no carbon source needed) process as:

$$NH_4^+ + 2O_2 \rightarrow NO_3^- + H_2O + 2H^+ \tag{12.25}$$

Denitrification is a biological process where denitrifying bacteria convert nitrate (and nitrite) to nitrogen. It is an anoxic and *heterotrophic* (i.e., carbon source needed) process as:

$$2NO_3^- + \text{organic matter} \rightarrow N_2 + H_2O + CO_2 \tag{12.26}$$

The *anoxic condition* is a form of anaerobic condition, in which nitrate serves as the electron acceptor. Since the four inorganic nitrogen compounds may change their forms from one to another in water/wastewater treatment or in natural systems, their concentrations are usually expressed in the unit of mg N/L in practice. With all the inorganic nitrogen concentrations in mg N/L, conducting a mass balance analysis on these species would be more straightforward, for example. The conversion can be done as:

$$\text{Concentration}\left(\text{in mg }\left(\text{N/L}\right)\right) = \text{Concentration }\left(\text{in}\frac{\text{mg}}{\text{L}}\right)\times\frac{14}{MW} \quad (12.27)$$

Example 12.8: Concentrations of Inorganic Nitrogen Species

Convert the following inorganic nitrogen concentrations to mg N/L: (i) $NO_3^- =$ 45 mg/L; (ii) $NO_2^- = 2.0$ mg/L, and (iii) $NH_4^+ = 5.0$ mg/L.

Solution:

a. *MW* of $NO_3^- = 62$ mg/L; *MW* of $NO_2^- = 46$; and *MW* of $NH_4^+ = 18$.
b. Nitrate concentration $\left(\text{in mg}\left(\text{N/L}\right)\right) = 45\times\frac{14}{62} = 10 \text{ mg } NO_3^-\text{-N/L}$

$$\text{Nitrite concentration}\left(\text{in mg }\left(\text{N/L}\right)\right) = 2.0\times\frac{14}{46} = 0.61 \text{ mg } NO_2^-\text{-N/L}$$

$$\text{Ammonium concentration}\left(\text{in mg }\left(\text{N/L}\right)\right) = 5.0\times\frac{14}{18} = 3.9 \text{ mg } NH_4^+\text{-N/L}$$

Discussion:

1. "14" is the atomic mass of nitrogen (N).
2. "NO_3^--N" means the nitrogen in the form of NO_3^-.
3. This example illustrates the MCL of NO_3^- (10 mg N/L) is equivalent to 45 mg/L of NO_3^-.

Phosphorus in aquatic systems occurs as organic phosphates and inorganic phosphates. Inorganic phosphate is the form that plants can utilize, while animals can use phosphate in the form of inorganic or organic. Phosphorus is usually the limiting nutrient for lake eutrophication.

12.5.7 Other Inorganic Compounds

Other inorganic compounds, that are on the list of SDWA's MCLs and CWA's Priority Pollutants, include (i) cyanides (CN^-), (ii) asbestos, and (iii) radionuclides (alpha/photon emitters, beta photon emitters, radium-226 (Ra-226) and radium 228 (Ra-228), uranium (U)).

12.6 CHEMICAL WATER QUALITY PARAMETERS III – ORGANIC COMPOUNDS

Numerous organic constituents are present in water/wastewater in the forms of dissolved, suspended solids, or colloids. They can be naturally occurring or synthetic. Many of them are toxic at elevated concentrations and are on the lists of SDWA's MCLs and CWA's Priority Pollutants (see Section 2.1).

Identification and measurements of these toxic species are labor-intensive, but necessary. However, not all the organic compounds need to be identified and analyzed frequently. For example, the objective of a biological process in municipal wastewater treatment is to remove the organic compounds as a whole (not much on specific compounds). Instead, aggregate parameters are used to quantify certain types of organics in water/wastewater. The common aggregate parameters include (i) theoretical oxygen demand (ThOD), (ii) total organic carbon (TOC), (iii) biochemical oxygen demand (BOD), (iv) chemical oxygen demand (COD), and (v) volatile organic compounds (VOCs).

12.6.1 Theoretical Oxygen Demand

ThOD is the stoichiometric amount of oxygen needed to oxidize a compound completely to end products. It can be estimated from its chemical formula, as illustrated in Example 12.9.

Example 12.9: Theoretical Oxygen Demand

A wastewater stream contains only 400 mg/L of glucose ($C_6H_{12}O_6$). Determine its ThOD.

Solution:

a. Complete oxidation of glucose with oxygen will form carbon dioxide and water as:

$$C_6H_{12}O_6 + 6O_2 \rightarrow 6CO_2 + 6H_2O$$

b. *MW* of glucose = 180 and *MW* of O_2 = 32

c. $\text{ThOD} = \left(\frac{0.4 \text{ g/L}}{180 \text{ g/mole}}\right)\left(\frac{6}{1}\right)\left(\frac{32{,}000 \text{ mg}}{\text{mole}}\right) = 424.8 \text{ mg } O_2/\text{L}$

Discussion:

1. The (6/1) ratio is the molar (stoichiometric) ratio of oxygen and glucose in the reaction.
2. The "O_2" in the reported ThOD values is often omitted (e.g., 428.8 mg/L for this example).

12.6.2 Total Organic Carbon

Since organic compounds contain carbon in them, TOC is a measure of the total amount of carbon in a water sample. There are various oxidation and detection methods developed to determine the TOC concentrations of water samples. A common approach is high-temperature combustion of water samples in an oxygen-rich environment, in which organic compounds are combusted to produce carbon dioxide. The amount of CO_2 produced is then measured by an analytical instrument. The TOC concentration is often expressed in mg/L, but it is actually in "mg C/L".

12.6.3 Biochemical Oxygen Demand

Biochemical oxygen demand (or *biological oxygen demand*, BOD) measures the amount of oxygen that microorganisms consume while aerobically degrading/decomposing organic matters. It would take a considerable amount of time for microorganisms to completely degrade the organic materials present. The quantity of oxygen used in a specific volume to decompose organics after day t is called BOD_t. For example, BOD_5 concentration of a water sample is determined from measuring the uptake of oxygen in that sample over 5 days at a constant temperature. The amount of oxygen needed to decompose all the biodegradable organics in that sample is called *ultimate BOD* (BOD_u or BOD_L). The relationship among BOD_t, BOD_u, k (the rate constant of biodegradation), and time t is:

$$BOD_t = BOD_u\left(1-e^{-kt}\right) \tag{12.28}$$

The value of the rate constant (k) depends on temperature, type of the organic substances present, and type of microbes exerting the BOD. Since the biodegradation rate will be faster at a higher temperature, the value of the rate constant depends on temperature as:

$$k_T = k_{20}(\theta)^{T-20} \tag{12.29}$$

where k_{20} = rate constant @ T = 20°C, k_T = rate constant at temperature T, and θ = temperature coefficient (typically slightly larger than unity; for example, 1.05). The following general relationships are valid:

$$BOD_3 < BOD_5 < BOD_7 < BOD_u \tag{12.30}$$

$$k_{15°C} < k_{20°C} < k_{25°C} \tag{12.31}$$

$$BOD_{5,\ 15°C} < BOD_{5,\ 20°C} < BOD_{5,\ 25°C} \tag{12.32}$$

$$BOD_{u,\ 15°C} = BOD_{u,\ 20°C} = BOD_{u,25°C} \tag{12.33}$$

The BOD concentrations are often expressed in mg/L, and it is actually in "mg O_2/L". If a water sample has a BOD_5 concentration twice as much as that of another water sample, it means that it has twice as much biodegradable organics in it.

Example 12.10: Biochemical Oxygen Demand

BOD_5 of a wastewater sample is 200 mg/L and the biodegradation rate constant is 0.20/d at 20°C.

a. What would be the value of $BOD_{ultimate}$ at 20°C?
b. What would be the value of BOD_7 at 20°C?
c. What is the biodegradation rate constant at 15°C, if the temperature coefficient is 1.05?
d. What would be the value of $BOD_{ultimate}$ at 15°C?

Solution:

a. Use Eq. (12.28),

$$\text{BOD}_5 = 200 = \text{BOD}_u \left(1 - e^{-(0.20)(5)}\right) \rightarrow \text{BOD}_u = 316.4 \text{ mg/L}$$

b. Use Eq. (12.28) again,

$$\text{BOD}_7 = (316.4)\left(1 - e^{-(0.20)(7)}\right) \rightarrow \text{BOD}_7 = 238.4 \text{ mg/L}$$

c. Use Eq. (12.29),

$$k_{15} = (0.2)(1.05)^{15-20} \rightarrow k_{15} = 0.157 / \text{d}$$

d. Use Eq. (12.28),

$$\text{BOD}_5 = (316.4)\left(1 - e^{-(0.157)(5)}\right) \rightarrow \text{BOD}_u = 172.1 \text{ mg/L}$$

Discussion:

1. The parts (a) & (b) illustrate that Eq. (12.30) is valid: $\text{BOD}_5 < \text{BOD}_7 < \text{BOD}_u$.
2. Part (c) illustrates that Eq. (12.31) is valid: $k_{15°\text{C}} < k_{20°\text{C}}$.
3. Part (d) illustrates that Eq. (12.32) is valid: $\text{BOD}_{5,\,15°\text{C}} < \text{BOD}_{5,\,20°\text{C}}$.
4. Eq. (12.33) $(\text{BOD}_{u,\,15°\text{C}} = \text{BOD}_{u,\,20°\text{C}} = \text{BOD}_{u,25°\text{C}})$ means that BOD_u is a constant, independent of temperature.

12.6.4 Chemical Oxygen Demand

As discussed in the previous section, it would take 5 days to determine the BOD_5 concentration of a water sample; and BOD is a measure of the concentration of biodegradable organic compounds of a water sample. COD is a parameter that measures all the organics (biodegradable and non-biodegradable) that can be degraded in a chemical test using strong oxidizing chemicals (e.g., potassium dichromate, $K_2Cr_2O_7$), strong acid (H_2SO_4)) and elevated T and P. The result is available in less than a few hours. The COD concentration is also expressed in mg/L; but it is actually in "mg O_2/L". For a given sample, its COD value should be ≥ the corresponding BOD value, because some organics present are oxidizable but not biodegradable.

Many water quality reports contain concentrations of TOC, BOD_5, and COD. The value of TOC is usually smaller than those of BOD_5 and COD. The reason is that the unit of TOC is in mg C/L; while those of BOD_5 and COD are in mg O_2/L. As illustrated in the following simplified reaction equation (Eq. 12.34), one mole of C (12 g) will be oxidized by one mole of O_2 (32 g). In other words, it takes 2.67 mg of O_2 to oxidize one mg of C. It implies that the COD concentration of a water sample could be ~3 times of its corresponding TOC concentration.

$$C + O_2 \rightarrow CO_2 \tag{12.34}$$

12.7 COMMON BIOLOGICAL WATER QUALITY PARAMETERS

The presence, condition, and numbers and types of fish, insects, algae, plants, and other organisms provide important information about the health of aquatic ecosystems. Studying these factors as a way of evaluating the health of a waterbody is called *biological assessment*. *Biological criteria* are a way of describing the qualities that must be present to support a desired condition in a waterbody and serve as the standard against which assessment results are compared (USEPA, 2023e). On the other hand, the presence or elevated concentrations of some microbial species may indicate the waterbody has been polluted or the water has not been effectively disinfected.

12.7.1 Types of Microbial Species of Concern

Bacteria. Bacteria are single-cell plants that can ingest food and multiply, provided the conditions (i.e., pH, temperature, DO level, and food supply) are ideal. Higher levels of bacteria are favorable in biological processes to degrade biodegradable organics. However, high levels of bacteria in ambient water can lead to many waterborne diseases.

Algae. Algae are tiny plants composed of photosynthetic pigments (e.g., chlorophyll). They can sustain life by converting CO_2 into more biomass, using the energy from sunlight (i.e., photosynthesis) – as illustrated in Eq. (12.35) below.

$$\text{Photosynthesis: } 6\ CO_2 + 6\ H_2O \overset{\text{sunlight}}{\rightarrow} C_6H_{12}O_6 + 6\ O_2 \qquad (12.35)$$

As shown in Eq. (12.35), the pH and DO concentration of a surface waterbody will increase during the daytime if it contains algae because the photosynthesis process removes CO_2 (an acidic compound) from water and add oxygen into water.

The presence of algae in potable water sources poses T&O problems. In addition, certain species of algae (e.g., blue-green algae) can pose a serious public health risk. On the other hand, algae are essential in wastewater treatment utilizing stabilization/oxidation ponds.

Protozoa. Protozoa are single-celled microscopic animals and consume solid organic particles, bacteria, and algae for food. They are part of microbial systems in the biological treatment of wastewater.

Many protozoa are free-living organisms that can reside in fresh water and pose little or no risk to human health. However, some protozoa are pathogenic to humans. *Cryptosporidium* and *Giardia lamblia* are two protozoa species that present challenges in potable water disinfection, mainly due to their relatively larger sizes (~10 μm).

Viruses. Viruses are tiny biological structures that may be harmful to human health. All viruses need a parasite to survive. Although their concentrations in water or wastewater are relatively low, they can readily pass through filters in water/wastewater treatment. Some of them (e.g., *Adeno virus*) are more resistant to UV disinfection.

12.7.2 Indicator Microorganisms

There are numerous types of microorganisms present in water/wastewater. It would be impractical to analyze all of them. Testing for viruses is very labor-intensive and time-consuming. In addition, specific pathogenic (disease-producing) organisms are not easily identified. For all these reasons, it is necessary to select easily-measured "indicator organisms", whose presence indicates that pathogenic organisms may be present (MN DPH, 2022).

Total coliform, fecal coliform, and *E coli (Escherichia coli)* are all considered indicators of water contaminated with fecal matter. Total coliforms are present throughout the environment; and they are found in soil, water, and human or animal waste. Fecal coliforms are a group within the total coliforms, and they are present in the guts and waste of warm-blood animals. *E. coli* is a specific species of fecal coliforms, and only rare strains are pathogenic. *E.coli* from humans can reach surface water via wastewater treatment plant effluent, broken or leaky sewer pipes, and failing or poorly sited septic systems. *E.coli* from animals can enter waterbodies in stormwater runoff from feedlots, manure storage areas, or areas where there is wildlife (USEPA, 2023g).

12.7.3 Numeration of Bacteria

There are normally two methods to test coliform bacteria in water and wastewater; they are (i) the membrane filter technique and (ii) the multiple-tube fermentation technique.

Membrane-filter technique. The coliform group analyzed in this procedure includes all the organisms that produce a colony, with a golden-green metallic sheen, within 24 hours of inoculation. A pre-determined amount of sample is filtered through a membrane filter which retains the bacteria found in the sample. The filter paper is then placed in a petri dish with a culture medium and then incubated at 35°C for 24 hours. Sheen colonies are then counted under magnification and reported in the unit of colony forming unit (cfu)/100 mL of the original sample (USEPA, 1986a).

Multiple-tube fermentation technique. It is a three-stage procedure in which the results are statistically expressed in terms of the Most Probable Number (MPN). The three stages are (i) the presumptive stage, (ii) the confirmed stage, and (iii) the completed test. For the analysis to be accurate, a five-tube test is required. The results are reported in the unit of MPN/100 mL of the original sample (USEPA, 1986b). The most stringent criterion for water reuse in California is that the total coliform concentration in wastewater discharge should be <2.2 MPN/100 mL.

12.7.4 MCLs of Microorganisms

Cryptosporidium, *Giardia lambia*, *Legionella*, fecal coliform and *E. coli*, total coliform, viruses, heterotrophic plate count (HPC), and turbidity appear on the list of MCLs, with regards to microorganisms.

The main reason for turbidity being on the list is that the presence of suspended solids would have adverse impacts on the effectiveness of disinfection [Note: As mentioned, turbidity is commonly used as an indicator of suspended solid concentrations]. Microorganisms can be shielded by the suspended solids present from chemical disinfectants and these solids can also scatter the UV light if UV disinfection is employed. *Heterotrophs* are a group of microorganisms that use organic carbon sources to grow. Most bacteria found in potable water supplies are heterotrophs. *HPC* is a measure of colony formation on culture media of heterotrophs.

The MCL for total coliform in drinking water is zero. Water systems are required to take samples for total coliforms based on the population served, source type, and vulnerability to contamination. No more than 5% of samples for total coliforms can be positive in 1 month. (For systems that collect fewer than 40 routine samples per month, no more than one sample can be total coliform-positive per month). If a sample tests positive for total coliforms, the system must collect a set of repeat samples within 24 hours and analyze for *E. coli* (Maine DWP, 2023).

Example 12.11: Coliform Concentrations in Wastewater Discharge

The weekly geometric mean of the fecal coliform bacteria density in the effluent shall not exceed 230 MPN/100 mL. The fecal densities measured for 1 week were:

	MPN/100 mL
Monday	110
Tuesday	270
Wednesday	350
Thursday	110
Friday	290
Saturday	430
Sunday	110

What would be the weekly geometric mean of the fecal coliform concentrations for this week (in MPN/100 mL)?

Solution:

a. The geometric mean of n values is the nth root of their multiplication product.

b. Thus, the geometric mean of this data set.

$$= (110 \times 270 \times 350 \times 110 \times 290 \times 430 \times 110)^{1/7} = \underline{206}$$

Discussion:

1. Taking a geometric mean is a way to minimize the impacts of outliers of a data set.
2. The geometric mean of (1, 10, 100) is equal to 10.

REFERENCES

ASTDR (2021). "Fluorides, Hydrogen Fluoride, and Fluorine", Agency for Toxic Substances and Disease Registry (ASTDR), https://wwwn.cdc.gov/TSP/substances/ToxSubstance.aspx?toxid=38, last updated February 10, 2021.

Kuo, J.F., Chen, C., Stahl, J. and Horvath, R. (1994) "Evaluation of Four Different Tertiary Filtration Plants for Turbidity Control", *Water Environ. Res.* V. 66, No. 7, p. 879–86.

Maine DHHS (2023), "Coliform Bacteria", https://www.maine.gov/dhhs/mecdc/environmental-health/dwp/consumers/coliformBacteria.shtml, Maine Department of Health and Human Services (DHHS), last updated February 10, 2023.

MN DPH (2022). "Coliform Bacteria", https://www.health.state.mn.us/communities/environment/water/factsheet/coliform.html,Minnesota Department of Health (MNDPH), last updated December 8, 2022.

USEPA (1986a) "SW-846 Test Method 9132 - Total Coliform: Membrane-Filter Technique", US Environmental Protection Agency, https://www.epa.gov/sites/default/files/2015-12/documents/9132.pdf

USEPA (1986b) "SW-846 Test Method 9131 - Total Coliform, Multiple Tube Fermentation Technique", US Environmental Protection Agency, https://www.epa.gov/sites/default/files/2015-12/documents/9131.pdf.

USEPA (1988), "Ambient Aquatic Life Water Quality Criteria for Chloride", EPA/440/5-88-001, US Environmental Protection Agency, Office of Water.

USEPA (2006), "Voluntary Estuary Monitoring Manual - Chapter 11: pH and Alkalinity", EPA-842-B-06-003, US Environmental Protection Agency, Office of Water.

USEPA (2017), "Water Quality Standards Handbook - Chapter 3: Water Quality Criteria", EPA/823/B17/001, US Environmental Protection Agency, Office of Water.

USEPA (2021a). "Factsheet on Water Quality Parameters - Temperature", 841F21007A, US Environmental Protection Agency.

USEPA (2021b). "Factsheet on Water Quality Parameters - Turbidity", EPA 841F21007D, US Environmental Protection Agency.

USEPA (2021c). "Factsheet on Water Quality Parameters - pH", EPA 841F21007C, US Environmental Protection Agency.

USEPA (2021d). "Factsheet on Water Quality Parameters - Dissolved Oxygen", EPA 841F21007B, US Environmental Protection Agency.

USEPA (2021e). "Factsheet on Water Quality Parameters - Metals", EPA 841F21007J, US Environmental Protection Agency.

USEPA (2021f). "Factsheet on Water Quality Parameters - Nutrients", EPA 841F21007G, US Environmental Protection Agency.

USEPA (2023a). "National Primary Drinking Water Regulations", US Environmental Protection Agency, https://www.epa.gov/ground-water-and-drinking-water/national-primary-drinking-water-regulations, last updated January 9, 2023.

USEPA (2023b). "Secondary Drinking Water Standards: Guidance for Nuisance Chemicals", US Environmental Protection Agency, https://www.epa.gov/sdwa/secondary-drinking-water-standards-guidance-nuisance-chemicals, last updated February 14, 2023.

USEPA (2023c). "Effluent Guidelines", US Environmental Protection Agency, https://www.epa.gov/eg, last updated June 21, 2023.

USEPA (2023d). "Summary of the Clean Water Act", US Environmental Protection Agency, https://www.epa.gov/laws-regulations/summary-clean-water-act, last updated June 22, 2023.

USEPA (2023e). "Biological Water Quality Criteria", US Environmental Protection Agency, https://www.epa.gov/wqc/biological-water-quality-criteria, last updated October 2, 2023.

USEPA (2023g). "*E.coli (Escherichia coli)*", EPA 841F21007F, US Environmental Protection Agency, https://www.epa.gov/wqc/biological-water-quality-criteria.

EXERCISE QUESTIONS

1. Use the information given in the table below to find VSS, FSS, VDS, FDS, TDS, TSS, TVS, TFS and TS of this water sample (in mg/L).

Compound	C (mg/L)	Dissolved?	Volatilize @550°C
Silt/Clay	10	No	No
Salt (NaCl)	400	Yes	No
Coffee grounds	20	No	Yes
Methanol	200	Yes	Yes

2. A laboratory analyzed the solid content of a primary sludge sample and the raw data are given below:
   - Weight of dish = 20.0 g
   - Weight of dish + the sludge sample = 180.0 g
   - Weight of dish + dry solid = 50.0 g (after dried in an oven @ T = 103°C)
   - Weight of dish + ash = 25.0 g (after a muffle furnace @ T = 550°C)

   What is TS and VS concentrations of this sludge (in %)?
3. The following test results are for a wastewater sample using a sample size of 250 mL.

Tare mass of the evaporating dish	50.6000 g
Dish + residue after evaporation at 105°C	50.7500 g
Dish + residue after ignition at 550°C	50.7000 g
Tare of the Whatman GF/C filter	1.5400 g
Filter + residue after evaporation at 105°C	1.5700 g
Filter + residue after ignition at 550°C	1.5500 g

   a. Determine its TS concentration, in mg/L.
   b. Determine its TDS concentration, in mg/L.
   c. Determine its total volatile solid concentration, in mg/L.
4. A 20 mL sample of treated wastewater requires 180 mL of distilled water to reduce the odor to a level that is just perceptible. What is the TON of this water sample?
5. For a surface water system (T = 25°C),
   a. At what pH, the concentrations of carbonic acid and bicarbonate are equal?
   b. At what pH, the concentrations of bicarbonate and carbonate are equal?
   c. At what pH, the bicarbonate concentration will be 100 times larger than that of carbonic acid?
   d. At what pH, the bicarbonate concentration will be 100 times larger than that of carbonate?
   e. Which would be the dominant carbonate species if water pH = 5.0, 8.0, and 11.0, respectively?

6. Two water samples were analyzed, and the results are (all the concentrations are in mg/L):
   - Sample #1: $Ca^{+2}=80$, $Mg^{+2}=48$, $Na^{+}=36$, $SO_4^{=}=55$, $Cl^{-}=36$, $F^{-}=0.4$, $NO_3^{-}=20$, $HCO_3^{-}=60$ mg/L, and pH = 7.5
   - Sample #2: $Ca^{+2}=80$, $Mg^{+2}=48$, $Na^{+}=36$, $SO_4^{=}=55$, $Cl^{-}=36$, $F^{-}=0.4$, $NO_3^{-}=20$, $HCO_3^{-}=30$, $CO_3^{=}=30$ mg/L, and pH = 9.5

   a. What are the alkalinity values of these two samples, in M?
   b. What are the alkalinity values of these two samples, in mg/L as $CaCO_3$?
7. Two water samples were analyzed, and the results are (all the concentrations are in mg/L):
   - Sample #1: $Ca^{+2}=80$, $Mg^{+2}=48$, $Na^{+}=36$, $SO_4^{=}=55$, $Cl^{V}=36$, $F^{-}=0.4$, $NO_3^{-}=20$, $HCO_3^{-}=60$ mg/L, and pH = 7.5
   - Sample #2: $Ca^{+2}=80$, $Mg^{+2}=48$, $Na^{+}=36$, $SO_4^{=}=55$, $Cl^{-}=36$, $F^{-}=0.4$, $NO_3^{-}=20$, $HCO_3^{-}=30$, $CO_3^{=}=30$ mg/L, and pH = 9.5

   a. What are the total hardness concentrations of these two samples, in mg/L as $CaCO_3$?
   b. Using the alkalinity values obtained in Question #6, what are the carbonate and non-carbonate hardness concentrations of these two water samples, in mg/L as $CaCO_3$?
8. Convert the following inorganic nitrogen concentrations to mg N/L: (i) $NO_3^{-}=10$ mg/L; (ii) $NO_2^{-}=10$ mg/L, and (iii) $NH_4^{+}=10$ mg/L.
9. A wastewater stream contains 200 mg/L of methanol and 20 mg/L of ethanol. Determine its ThOD (in mg O_2/L).
10. BOD_5 of a wastewater sample is 300 mg/L and the biodegradation rate constant is 0.25/d @20°C.

    a. What would be the value of $BOD_{ultimate}$ @T = 20°C?
    b. What would be the value of BOD_3 @T = 20°C?
    c. What is the biodegradation rate constant @T = 25°C, if the temperature coefficient is 1.06?
    d. What would be the value of $BOD_{ultimate}$ @T = 25°C?
    e. What would be the value of BOD_3 @T = 25°C?

13 Chemistry in Soils

13.1 FUNDAMENTAL PROPERTIES OF SOIL

13.1.1 Phase Diagram

A soil system contains three phases: (i) soil grains, (ii) water, and (iii) air. The relative proportions of these three phases affect the engineering behaviors of soils. A *phase diagram* illustrates the relative proportions of the volume and the mass/weight of these three phases (see Figure 13.1).

The empty space in soil is called *pores* or *void*. The void in a soil mass is either filled by liquid, which is typically water, or by gases, which is typically air. Thus, the volume of the void (V_v) is equal to the sum of the volume of water (V_w) and the volume of air (V_a) as:

$$V_v = V_w + V_a \tag{13.1}$$

If the entire pore space is occupied by water (i.e., absence of air), the soil is *saturated*; otherwise, it is *unsaturated*. If a soil mass is completely dry, its volume/mass of water in the soil matrix would be equal to zero.

The total volume of a soil mass (V_{total}) is the sum of the volume of solid grains (V_s) and the void volume, thus,

$$V_{\text{total}} = V_v + V_s = V_w + V_a + V_s \tag{13.2}$$

Similarly, the total mass of a soil mass (M_{total}) is the sum of the mass of air (M_a), the mass of water (M_w), and the mass of soil grains (M_s). However, for practical applications, the mass of air is relatively small and negligible, thus:

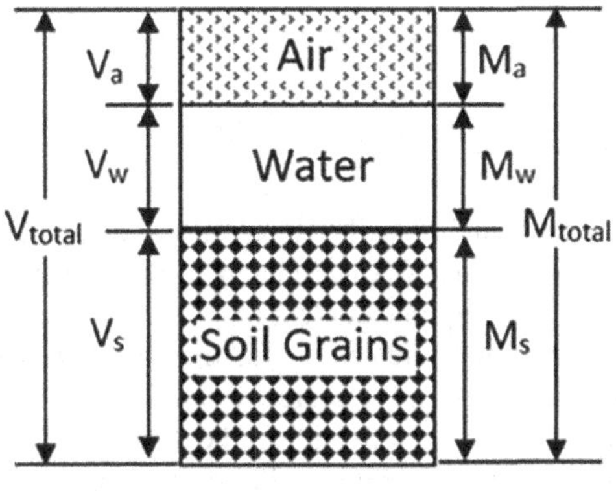

FIGURE 13.1 Phase diagram of soil.

DOI: 10.1201/9781003502661-13

$$M_{total} = M_a + M_w + M_s \cong M_w + M_w \tag{13.3}$$

13.1.2 Porosity and Void Ratio

The void content of a soil sample is often expressed in two simple terms: void ratio (e) and porosity (φ). The *void ratio* is the ratio of the void volume to the volume of the soil grains as:

$$e = \frac{V_v}{V_s} \tag{13.4}$$

The *porosity* is the ratio of the void volume to the total volume of the soil sample as:

$$\varphi \text{ (in \%)} = \frac{V_v}{V_{\text{total}}} \times 100 \tag{13.5}$$

Be noted that the void ratio is usually expressed as a decimal number, and the porosity is expressed in percentage. Values of porosity and void ratio depend on soil composition as well as extent of compaction. Porosity values of soils range from 12% to slightly above 50%, while values of void ratio can be larger than 1.0. The void ratio and porosity are related as:

$$\varphi = \frac{e}{(1+e)} \tag{13.6}$$

13.1.3 Degree of Water Saturation and Water Content

Two simple terms are used to express the amount of water that a soil sample contains; they are water content (w) and degree of saturation (S). The *water content* is the ratio of the mass of water to the mass of the soil grains as,

$$w = \frac{M_w}{M_s} \tag{13.7}$$

The *degree of saturation* is a measure of the void volume that is filled by water as:

$$S \text{ (in \%)} = \frac{V_w}{V_v} \times 100 \tag{13.8}$$

Be noted that the water content is usually expressed as a decimal number, and it is the ratio of the two masses. The natural water content of most soils should be <1.0; however, soils containing a large organic content and very moist can have water content >1.0. In addition, the porosity is often expressed in percentage, and it is a volumetric relationship. Two extreme cases: (i) the degree of saturation of completely dry soils is equal to 0% (and the water content = 0) and (ii) the degree of saturation for saturated soils (e.g., soil below the water table) is equal to 100% because the entire void space is occupied by water.

13.1.4 Specific Weight, Bulk Density, and Unit Weight

Bulk density and unit weight of a soil mass are the mass and the weight per unit volume. *Bulk density* (ρ_b) and *bulk unit weight* (γ) are the mass and the weight of the total soil sample (including water it contains) divided by its total volume, respectively. Thus, the bulk density and bulk unit weight can be found as,

$$\rho_b = \frac{M_{\text{total}}}{V_{\text{total}}} \tag{13.9a}$$

$$\gamma = \frac{W_{\text{total}}}{V_{\text{total}}} \tag{13.9b}$$

Dry bulk density (ρ_d) and *dry unit weight* (γ_d) are the bulk density and the bulk unit weight when the soil is dry, respectively. Common units for bulk density are lb_m/ft^3 and kg/m^3, and those for unit weight are lb/ft^3 and kN/m^3. The bulk density of well-compacted soil is approximately 110 lb/ft^3 or 1,800 kg/m^3 [Note: 62.4 lb/ft^3 = 1,000 kg/m^3]; and the unit weight of well-compacted soil is approximately 110 lb/ft^3 or 18 kN/m^3 [Note: 1 N = 1 $kg{\cdot}m/s^2$].

Specific gravity (*SG*) of a soil grain is the ratio of its density to the density of water, and it tells how many times the soil grain is heavier than water. The SG values of soil grains depend on their types; that of sand is equal to 2.65 and that of organic soils is smaller.

13.1.5 Relationships among Void Ratio, Water Content, SG, and Bulk Density

The void ratio (e), water content (w), degree of water saturation (S), specific gravity of soil grains (SG), dry bulk density (ρ_d) of a soil mass, and density of water (ρ_w) are related as:

$$w = \frac{(e \times S)}{SG} \tag{13.10}$$

$$e = \frac{(SG \times \rho_w)}{\rho_d} - 1 \tag{13.11}$$

Example 13.1: Degree of Water Saturation

The results of a feasibility study indicate that the soil in a stockpile is suitable for on-site above-ground bioremediation. Estimate the amount of nutrient solution needed for the first spray. Use the following information in your calculation:

a. Mass of excavated soil in the soil pile = 150 tons
b. Bulk density of soil = 90 lb/ft^3
c. Soil porosity = 40%

d. Initial degree of water saturation = 15%
e. Desired degree of water saturation after the first spray = 60%.

Solution:

a. Mass of excavated soil = 180 tons × 2,000 lb/ton = 360,000 lb

b. Volume of the excavated soil = 360,000 lb ÷ 90 lb/ft³ = 4,000 ft³

c. Volume of water needed = (Volume of soil) × (porosity) × (S_{final} − $S_{initial}$)

$$= (4{,}000\,\text{ft}^3)(40\%)(60\% - 15\%) = \underline{720\,\text{ft}^3} = \underline{5{,}400\text{ gallons}}.$$

Discussion:

1. 1 ton = 2,000 lb in the US customary units; 1 ton = 1,000 kg (= 2,200 lb) in SI.
2. Make-up water may be needed as the remediation project proceeds.

Example 13.2: Void Ratio and Porosity

Given a sandy soil sample with a dry density equal to 110 lb/ft³ (pcf) and SG of soil grains equal to 2.65, determine (a) the void ratio and (b) the porosity of this soil sample.

Solution:

a. Use Eq. (13.11) to find the void ratio,

$$e = \frac{(SG \times \rho_w)}{\rho_d} - 1 = \frac{(2.65 \times 62.4)}{110} - 1 = 0.50$$

b. Use Eq. 13.6 to find the porosity,

$$\varphi = \frac{e}{(1+e)} = \frac{0.50}{(1+0.50)} = 33.3\%$$

Example 13.3: Phase Relationships

The weight of a soil sample is 46.0 lb. The volume of the soil sample measured before drying is 0.40 ft³. After the sample was dried out in an oven, its weight became 38.0 lb. The SG of solid is 2.65. Determine (a) water content, (b) unit weight of the moist soil, (c) void ratio, (d) porosity, and (e) degree of saturation.

Solution:

a. Use Eq. (13.7) to find the water content,

$$w = \frac{M_w}{M_s} = \frac{(46.0 - 38.0)}{(38.0)} = 0.21$$

b. Use Eq. (13.9) to find the bulk density,

$$\rho_b = \frac{M_{\text{total}}}{V_{\text{total}}} = \frac{46.0}{0.4} = 115\ \text{lb/ft}^3$$

c. Volume of water $(V_w) = M_w \div \rho_{\text{water}} = (46.0 - 38.0)/(62.4) = 0.13\ \text{ft}^3$

$$\text{Volume of the solid } (V_s) = M_s \div \rho_{\text{solid}} = (38.0)/(62.4 \times 2.65) = 0.23\ \text{ft}^3$$

$$\text{Void volume } (V_v) = V_{\text{total}} - V_{\text{solid}} = 0.40 - 0.23 = 0.17\ \text{ft}^3$$

$$e = V_v \div V_s = (0.17)/(0.23) = 0.74$$

d. $\varphi = \dfrac{e}{(1+e)} = \dfrac{0.74}{(1+0.74)} = 42.5\%$

e. $S\ (\text{in }\%) = \dfrac{V_w}{V_v} \times 100 = \dfrac{0.13}{0.17} \times 100 = 76.5\%$

Discussion:

The degree of saturation can also be found by using Eq. (13.10).

13.2 SOIL CLASSIFICATION SYSTEMS

Soil texture affects many soil properties, such as structure/porosity, water infiltration rate, water holding capacity, and chemistry. There are several systems/methods used to classify soils. The three most commonly used systems are (1) the American Association of State Highway and Transportation Officials (AASHTO) soil classification system, (2) the United States Department of Agriculture (USDA) soil classification system, and (3) the Unified Soil Classification System (USCS). The AASHTO system was developed specifically for highway construction. The focus of this book will be on the other two systems: the USDA soil classification system and the USCS.

13.2.1 USDA Soil Classification System

The USDA soil classification system was developed for agricultural purposes. Soil textures are classified by the fractions of sand, silt, and clay in a soil. Clay particles are defined as those smaller than 0.002 mm in diameter (i.e., <2 μm). Silt particles are from 0.002 to 0.05 mm in diameter (i.e., 2–50 μm). Sand ranges from 0.05 to 2.0 mm.

The system defines 12 major soil texture classifications; they are sand, loamy sand, sandy loam, loam, silt loam, silt, sandy clay loam, clay loam, silty clay loam, sandy clay, silty clay, and clay. Classifications are typically named for the primary constituent particle size or a combination of the most abundant particle sizes (e.g., sandy clay, silty clay). Loams are soils having roughly equal proportions of sand, silt, and/or clay in a soil sample (Figure 13.2). It should be noted that these classes are descriptive and not directly related to engineering properties.

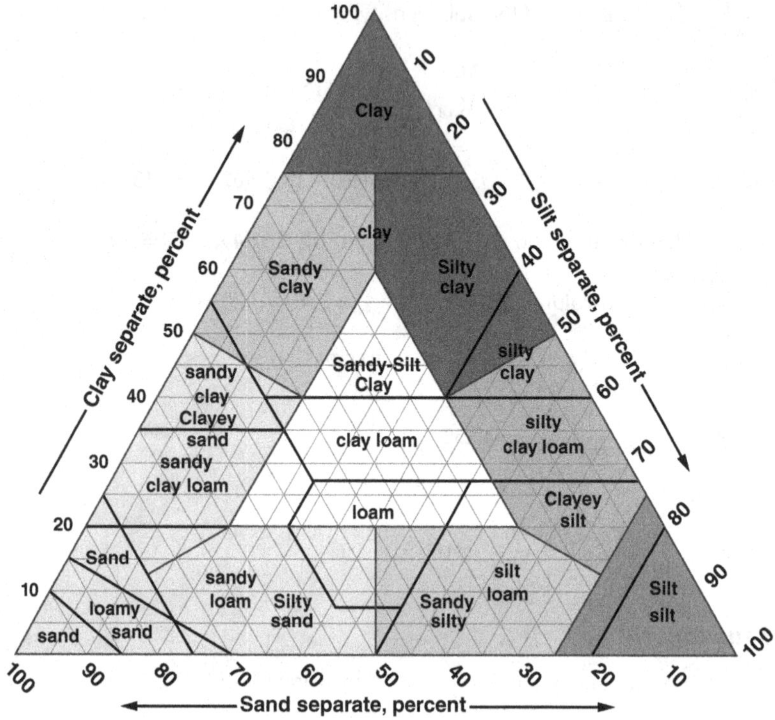

FIGURE 13.2 USDA soil classification system. (USDA, 2012.)

Example 13.4: USDA Soil Classification System

a. What is the USDA texture for a soil having 40% sand, 40% silt, and 20% clay?
b. What is the USDA texture for a soil having 30% sand, 35% silt, and 35% clay?
c. What is the USDA texture for a soil having 65% sand, 20% silt, and 15% clay?

Solution:

Read from Figure 13.2,

a. Loam
b. Clay loam.
c. Sandy loam.

13.2.2 Atterberg Limits

Before we discuss the USCS, we need to know about the Atterberg limits. As water content increases, fine-grained soils (i.e., silt and clay) go through four distinct states:

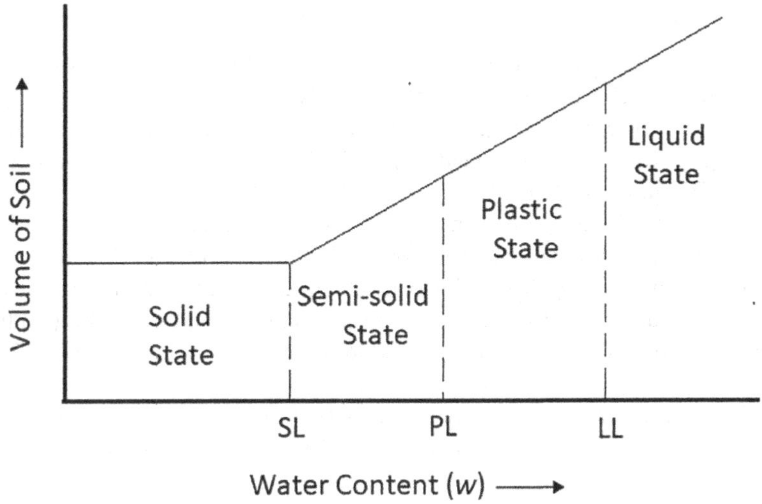

FIGURE 13.3 States of soil vs. Atterberg limits.

solid, semi-solid, plastic, and liquid. Soil in each state exhibits significant changes in strength, behavior, and consistency; and consequently, their engineering properties. The *Atterberg limit tests* determine the shrinkage limit (SL), plastic limit (PL), and liquid limit (LL). These three limits are water contents at the points where the states of the soil change (Figure 13.3).

- ***Liquid limit.*** LL is the water content at which the soil changes from a plastic state to a liquid state.
- ***Plastic limit.*** PL is the water content at which the soil changes from a plastic state to a semi-solid state.
- ***Shrinkage limit.*** SL is the water content where further loss of moisture does not cause a decrease in the volume of the soil.

It is apparent that the following relationship exists:

$$\text{SL} < \text{PL} < \text{LL} \tag{13.12}$$

The Atterberg limits can be used to distinguish between silt and clay and to distinguish between different types of silts and clay (see the USCS in the next section). *Plasticity index* (PI) is the difference between the LL and PL for a given fine-grained soil as:

$$\text{Plasticity Index (PI)} = \text{Liquid Limit (LL)} - \text{Plastic Limit (PL)} \tag{13.13}$$

Soils with a larger PI value indicate it has a higher clay content.

13.2.3 The USCS

The USCS is used to describe the texture and grain size of a soil sample. Unconsolidated materials are represented by a two-letter symbol, based on (i) the type of materials (i.e., gravel (*G*), sand (*S*), silt (*M*), clay (*C*), organic (*O*)), (ii) grading (i.e., well-graded (*W*), poorly graded (*P*)), and (iii) plasticity (i.e., high plasticity (*H*) with LL > 50% and low plasticity (*L*) with LL <50%). For example, SC represents clayed sand, SM represents silty sands, SW represents well-graded sands, CH materials consist of clay with high plasticity, and OH materials consist of organic clays of medium to high plasticity. Major divisions of the USCS are (i) coarse-grained soils (i.e., gravels and sands), (ii) fine-grained soils (i.e., silts and clays), and (iii) highly organic soils. More details on the classifications are (Caltrans, 2022; USDA, 2012):

- Coarse-grained soils (>50% retained on the #200 sieve (0.075 mm opening))
  - Gravels (≥50% of the coarse fraction retained on the #4 sieve (4.75 mm opening)
  - Sands (≥50% of the coarse fraction passes the #4 sieve
- Fine-grained soils (>50% passes on the #200 sieve (0.075 mm opening)
  - Silts and clays with liquid limit >50%
  - Silts and clays with liquid limit ≤50%
- Highly organic soils

A *plasticity chart* is used to differentiate the plasticity and organics characteristics of fine-grained soils, based on their LL and PI (Figure 13.4).

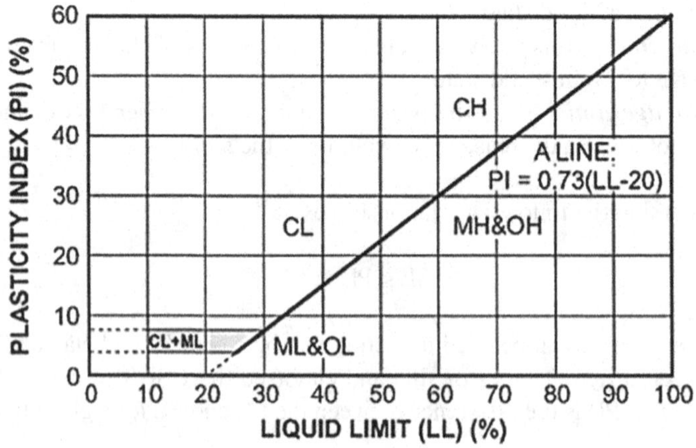

FIGURE 13.4 Plasticity chart. (Caltrans, 2023).

13.3 SOIL CHEMISTRY

Soil chemistry deals with chemical composition, chemical properties, and chemical reactions of soils. The soil phase diagram simply shows three basic phases: air, water, and soil grains. Microscopically a soil system is more complicated. The soil grains are usually heterogenous mixtures of inorganic and organic solids of different types and sizes. In addition to microorganisms, various chemical compounds (organic or inorganic) are contained in all three phases. These three phases are interconnected. For example, volatile organic compounds (VOCs) would travel between the air void and the soil moisture through dissolution and volatilization; and those in the soil moisture would also travel to and from the soil grains through adsorption and desorption.

13.3.1 Chemical Composition of Soils

The solid portion of soil includes inorganic minerals and organic matter. Soil organic matter can be grouped into three major types: (1) plant residues and living microorganisms; (2) active soil organic matter, referred to as *detritus*; and (3) stable soil organic matter, referred to as *humus*. Environmental engineers' concerns with regard to soil organic matter include (i) the presence of organic matter will enhance the adsorption of contaminants onto soils which make soil remediation more challenging; (ii) the presence of proper soil microorganisms may incur natural biodegradation of organic contaminants in soil; and (iii) humic substances are commonly known as disinfection byproduct (DBP) precursors in chlorine disinfection. Typical soils may contain a few percentages of organics by weight.

The mineral components of soils are derived from their parental rocks. Common elements in these minerals include oxygen, hydrogen, silicon, aluminum, iron, sodium, potassium, calcium, and others. The major groups include (i) silicates (e.g., quartz (SiO_2) and clay minerals such as kaolinite ($Si_4Al_4O_{10}(OH)_8$), (ii) oxides and hydroxides (e.g., gibbsite ($Al(OH)_3$), hematite (Fe_2O_3)), and (iii) carbonates and sulfates (e.g., calcite ($CaCO_3$) and gypsum ($CaSO_4{\cdot}2H_2O$)).

Phosphorus, nitrogen, and some trace metals are also present in soils because they are essential for plants. However, many trace metals become toxic above certain concentrations.

13.3.2 Chemical Properties of Soils

Important chemical properties of soils include soil pH, cation exchange capacity, electrical conductivity, and sodium adsorption ratio.

Soil pH. Soil pH is an important aspect of soil chemistry because it affects the availability of nutrients to plants as well as the activities of microorganisms in soil. Most plants grow better in soils with neutral to slightly acidic pH.

Acidic soil pH will enhance the dissolution of minerals and leaching of metals into soil solutions (e.g., groundwater).

Cation exchange capacity. *Cation exchange capacity* (CEC) is the maximum quantity of cations that a soil can hold, in equilibrium with the soil solution, at a given pH. It is often expressed in meq/100 g (i.e., milliequivalents of element per 100 grams of dry soil). Most of the soil's CEC occurs on its clay and organic content. CEC is used as a measure of soil fertility, nutrient retention capacity, as well as the capacity to protect underlying aquifers from contamination of cations (e.g., heavy metals). Soils with a large CEC could inhibit the migration of heavy metals through vadose zones and/or groundwater aquifers.

Electrical conductivity. *Electrical conductivity* (EC) is the ability of a material to conduct an electrical current. The EC of soil is primarily used to assess salt concentration in soil. Electrokinetic remediation is one of the emerging technologies for soil remediation. Effectiveness of electrokinetic remediation depends significantly on the EC of the soil to be remediated, and it is often more applicable to clayey soils because of their larger EC values.

Sodium adsorption ratio. *Sodium adsorption ratio* (SAR) is a measure of the concentration of sodium ion (Na^{+}) relative to those of calcium ion (Ca^{+2}) and magnesium ion (Mg^{+2}) in a soil solution as:

$$SAR = \left\{ \frac{[Na^{+}]}{([Ca^{+2}] + [Mg^{+2}])/2} \right\}^{1/2} \tag{13.14}$$

Soils with a large SAR may have increased dispersion of clay particles and organic matters, which may result in reduced hydraulic conductivity.

13.3.3 Chemical Reactions in Soil

When "foreign" organic and inorganic compounds/substances enter a soil system (e.g., subsurface soil or a groundwater aquifer) consisting of air, moisture, minerals, and soil organic matter, many chemical reactions/processes can occur; they include:

- dissolution, hydrolysis, precipitation, complexation, and oxidation/reduction in water
- adsorption/desorption between water and soil grains/soil organic matter
- volatilization/absorption between air and water

In addition, biological processes may also occur with the presence of microorganisms. All these processes may change the physical and chemical properties of soil due to the changes in pH and dissolved oxygen, for example. The change in pH would affect the dissolution of minerals as well as leaching of metals from the minerals. It would also affect the behaviors of colloidal particles (e.g., clay) in water. Understanding these processes is important (i) to understand the fate and transport of chemicals through a soil system and/or a groundwater aquifer, and (ii) to select a proper alternative for remediating contaminated soil and/or groundwater.

13.4 CLAY

Although clay minerals may only be a small fraction of a soil mass, they often require special attention because they significantly affect the engineering properties/behaviors of soils. It is mainly due to (i) their small sizes, (ii) platy shapes, (iii) molecular structure, (iv) water-bearing lattice, and (v) surface charges.

13.4.1 Types of Clay

Clay minerals are predominantly hydrous aluminum silicates or more rarely, hydrous magnesium or iron silicates. They are composed of layers of (i) silicon and oxygen (i.e., the silica layer) and (ii) aluminum and oxygen. They are categorized into three principal groups: kaolinites, montmorillonites, and illites (USDA, 2012). Figure 13.5 illustrates the platy structure of three groups of clay.

- ***Kaolinite.*** Kaolinite consists of two-layer molecular sheets: one of silica and one of alumina (1:1 ratio). The sheets are firmly bonded together with no variations in distance between them. Consequently, the sheets do not take up water.
- ***Montmorillonite.*** Montmorillonite consists of three-layer molecular sheets: two layers of silica and one layer of alumina (a 2:1 ratio). The molecular sheets are weakly bonded, permitting water and associated chemicals to enter between the sheets. Consequently, they are subject to considerable expansion when saturated, and shrinkage when drying.

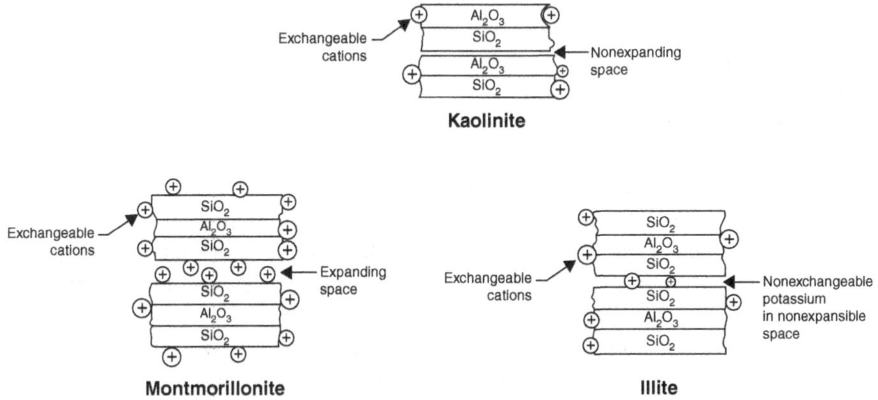

FIGURE 13.5 Types and structures of clay minerals. (USDA, 2012.)

- ***Illite.*** Illite has the same molecular structure as montmorillonite, but it has stronger molecular bonding, resulting in less expansion and shrinkage properties.

13.4.2 Properties of Clay Minerals

Clay fraction in soil has significant impacts on the fate and transport of chemicals in the environment because of the following special properties:

- ***Large specific surface area.*** *Specific surface area* (SSA) is defined as the surface area of particles per unit mass, typically in m^2/g. Particles with large specific areas have more surface area per unit mass available for reactions/processes to take place on their surface (e.g., adsorption). Clay minerals have large SSAs because of their small sizes (<2 μm) and platy shape. The SAA of expanding clay mineral (e.g., montmorillonite) can be a few hundreds of m^2/g and that of non-expanding clay (e.g., kaolinite) is also relatively large, at tens of m^2/g.
- ***Large water holding capacity.*** Due to the layer structure of clay minerals, if water is added to dry clay minerals, water can enter the interlayer space causing swelling or expansion of the interlayer space. Sodium bentonite (a type of montmorillonite) has the ability of absorbing water 4–5 times its own weight and it can swell 5–15 times its dry volume at saturation. Bentonite pellets are often added to the annular space between the borehole and the casing of groundwater monitoring wells to isolate the screened intervals of the wells.
- ***High density of surface charges and CEC.*** As mentioned, clay minerals are made of sheets of silica and alumina. Silicon (Si) in the silica sheet has four electrons to share, while aluminum (Al) in the alumina sheet has three electrons to share. Sometimes an aluminum atom may accidentally replace a silicon atom in the silica sheet (or replacement of Al^{+3} by Mg^{+2}) in its formation; consequently, this renders an unmatched negative charge on the clay. Clay minerals typically carry a relatively high density of net negative charges. That is the main reason why clay minerals possess large CEC values.
- ***Strong adsorptive capacity.*** The clay minerals are strong adsorbents because of their large SSA and high density of negative surface charge. The soil organic matter has a strong tendency of getting adsorbed onto the clay fraction of soil, and this turns the soil into an even stronger adsorbent for many organic and inorganic compounds. Once contaminants (organic or inorganic) enter the subsurface, many of them will get adsorbed onto clay; this creates complexation for remediation if removal of the contaminants is the remedial objective.
- ***Low hydraulic conductivity.*** Due to the small particle sizes and the large water-retaining capability, hydraulic conductivity of clayey formation is very low. A compacted clay liner is often used as the bottom liner of modern

landfills to protect underlying groundwaters. The hydraulic conductivity of the compacted clay liner is typically $\leq 10^{-7}$ cm/s.

Soil vapor extraction (SVE) is commonly used in soil remediation. If the contaminated zone in subsurface containing clayey layers, SVE may not be effective because less/no extracted air will flow through the clay layers which are less permeable and may also have higher contaminant concentrations. Migration of fine particles in a groundwater aquifer may plug its pores so that the hydraulic conductivity of the aquifer will decrease, which may be a concern for in-situ groundwater remediation.

- ***Catalytic capability.*** Due to their large SSA and high surface charge density, clay minerals may serve as catalysts for reactions occurring in natural systems as well as in industrial processes.

REFERENCES

Caltrans (2022). "*Soil and Rock Logging, Classification and Presentation Manual*", California Department of Transportation (Caltrans), https://dot.ca.gov/-/media/dot-media/programs/engineering/documents/geotechnical-services/202212-lm-soilandrocklogging-manualrev70-a11y.pdf.

Caltrans (2023). "*Unified Soil Classification System*", California Department of Transportation, https://dot.ca.gov/-/media/dot-media/programs/maintenance/documents/office-of-concrete-pavement/pavement-foundations/uscs-a11y.pdf.

USDA (2012). "*Chapter 3 - Engineering Classification of Earth Materials*" Part 631, National Engineering Handbook, United States Department of Agriculture (USDA), https://directives.sc.egov.usda.gov/OpenNonWebContent.aspx?content=31847.wba, last updated January 2012.

EXERCISE QUESTIONS

1. The results of a feasibility study indicate that the soil in a stockpile is suitable for on-site above-ground bioremediation. Estimate the amount of nutrient solution needed for the first spray. Use the following assumptions in your calculation:
   a. Volume of excavated soil in the soil pile = 200 yd^3
   b. Bulk density of soil = 90 lb/ft^3
   c. Soil porosity = 45%
   d. Initial degree of water saturation = 10%
   e. Desired degree of water saturation after the first spray = 65%.
2. Given a sand sample with a dry density equal to 100 lb/ft^3 (pcf) and the SG of soil grains equal to 2.80, determine (a) the void ratio and (b) the porosity of this soil sample.
3. The weight of a soil sample is 50.0 lb. The volume of the soil sample measured before drying is 0.50 ft^3. After the sample was dried out in an oven, its weight became 40.0 lb. The SG of solid is 2.80. Determine (a) water content, (b) unit weight of the moist soil, (c) void ratio, (d) porosity, and (e) degree of saturation.

4. a. What is the USDA texture for a soil having 25% sand, 25% silt, and 50% clay?
   b. What is the USDA texture for a soil having 25% sand, 50% silt, and 25% clay?
   c. What is the USDA texture for a soil having 50% sand, 25% silt, and 25% clay?

14 Chemistry in the Atmosphere

14.1 OUR ATMOSPHERE

14.1.1 Composition

Atmosphere of the Earth is a mixture of many different gases held by the gravitational force. Major components of dry air include nitrogen (78.08% by volume or by mole), oxygen (20.95%), and argon (0.93%). Carbon dioxide (CO_2), the major greenhouse gas (GHG), accounts for most of the balance (0.04%) with other gases in trace amounts. Water vapor concentrations vary at locations. Our atmosphere also contains particulate matter (PM).

14.1.2 Structure

Earth's atmosphere is about 480 km (300 miles) in thickness. Atmospheric pressure decreases as the altitude increases due to the decreasing mass of air above. The atmosphere can be divided into four layers based on its temperature: troposphere, stratosphere, mesosphere, and thermosphere. Beyond the atmosphere, it is the exosphere. Figure 14.1 illustrates the temperature profile of the atmosphere.

Troposphere, the lowest part of the atmosphere, starts from the Earth's surface to 8–14.5 km high (5–9 miles). It is the part of the atmosphere that we live in. Its thickness is less than 3% of the entire atmosphere; however, it contains about 75% of all the atmospheric mass and almost all the water vapor molecules. In the troposphere, its temperature decreases with altitude. The bottom part of the troposphere is called the *planetary boundary layer* in which the air motion is affected by the Earth's surface. The air in the troposphere will move horizontally and vertically and redistribute heat, moisture, and other constituents, including air pollutants, along with its movement. The upper limit of the troposphere is called the *tropopause* in which the temperature is relatively constant. Clouds and weather generally occur below the tropopause in the troposphere. The airlines often fly around the tropopause.

The *stratosphere* starts above the troposphere and ends about 50 km (31 mi) from the Earth's surface, and it contains much of the ozone in the atmosphere. With regard to air pollution, the troposphere is the layer we have most of our daily activities as well as almost all the weather in; while the ozone layer, which absorbs harmful ultraviolet (UV) radiation from the Sun, is within the stratosphere. The temperature in this layer increases with altitude because of the absorption of solar UV radiation.

The layer above the stratosphere is the *mesosphere*, in which its temperature decreases with height, reaching the minimum temperature of about 185 K at the

DOI: 10.1201/9781003502661-14

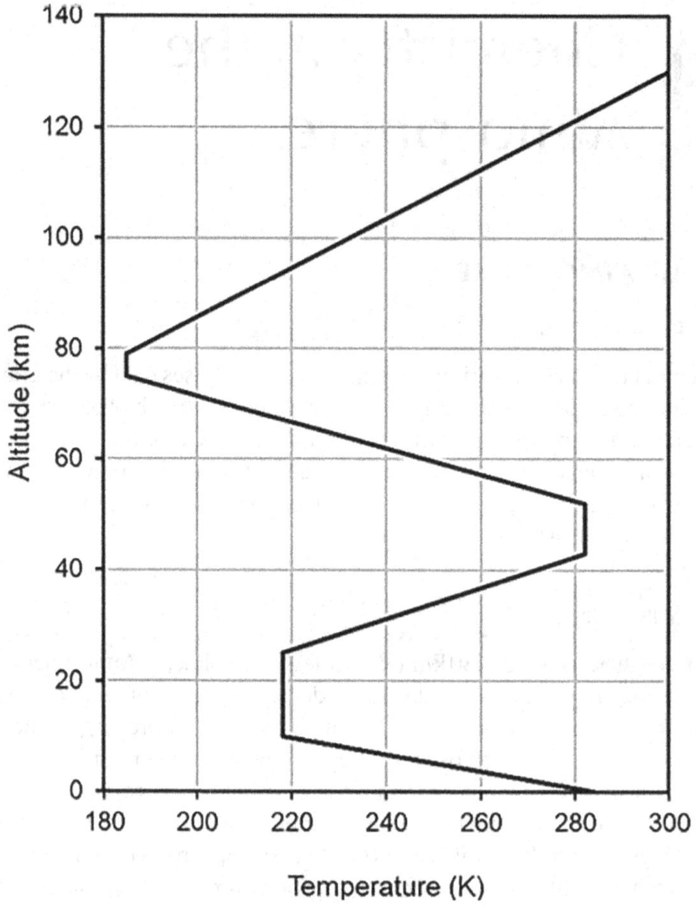

FIGURE 14.1 Temperature profile of the atmosphere.

mesopause from the Earth's surface. The *thermosphere* is the top layer of the atmosphere. Its temperature increases with altitude because of the absorption of solar UV and X-ray radiation. The region of the atmosphere above about 80 km (50 mi) is also called the *ionosphere* since the energetic solar radiation would knock electrons off molecules and atoms in this region.

14.2 AMBIENT AIR QUALITY

14.2.1 Composition of Our Ambient Air

Air is a mixture of gases. *Gas* is a state of matter that has no fixed shape and volume. *Vapor* refers to a gas phase where the same substance also exists in its liquid and/or solid state under that condition. For example, we say that our ambient air contains nitrogen gas and water vapor. It is because no liquid or solid nitrogen coexists with the nitrogen gas, while water and/or ice are present with water vapor under ambient conditions.

In addition to gases and vapors, air also contains PM which includes suspended solid and liquid, such as dust, soot, smoke, and liquid droplets as well as living matters such as pollen, bacteria, spores, mold, and fungus. *Dust* is a loose term applied to solid particles, predominantly larger than colloids and capable of temporary suspension in air or other gases. Dusts do not diffuse but settle under the influence of gravity. *Soot* is particles of amorphous carbon and tars generated from incomplete combustion of hydrocarbons.

An *aerosol* is a mixture of fine PM (liquid or solid) and air/gas. Aerosols can be natural (e.g., fog and geyser steam) or artificial (e.g., haze). *Fog* is a loose term applied to visible aerosols in which the dispersed phase is liquid, mainly formed from condensation. *Smoke* is an aerosol which is a visible mixture of gases and fine particles generated from combustion; the particles are present in sufficient quantity to be observable independently of the presence of other solids. *Smog* is a term derived from smoke and fog, and it was first used around 1950 to describe the combination of smoke and fog in London. *Photochemical smog* is air contamination caused by photochemical reactions among nitrogen dioxide and hydrocarbons by the action of sunlight. Ozone is its main component.

14.2.2 Criteria Pollutants

The Clean Air Act (CAA) is the federal law that defines the USEPA's responsibilities for protecting and improving the nation's air quality and the stratospheric layer. It regulates air emissions from stationary and mobile sources. Among other things, the CAA authorizes the USEPA to establish National Ambient Air Quality Standards (NAAQS) to protect public health and public welfare and to regulate emissions of hazardous air pollutants (HAPs).

The criteria pollutants included in the NAAQS are:

1. Carbon monoxide
2. Nitrogen dioxide (NO_2)
3. Sulfur dioxide (SO_2)
4. Ozone (O_3)
5. Particulate matter (PM)
6. Lead (Pb)

The air quality index (AQI) is the USEPA's index for reporting air quality, with values ranging from 0 to 500. The larger the AQI value, the greater the level of air pollution (and the greater the health concern). An AQI value of 100 generally corresponds to an ambient air concentration that equals to the level of the short-term NAAQS. AQI values of ≤ 100 are considered satisfactory.

14.2.3 Hazardous Air Pollutants

The CAA also requires the USEPA to regulate toxic air pollutants (also known as "*air toxics*" or "*hazardous air pollutants* (HAPs)) from large industrial facilities. The original list of HAPs included 189 pollutants, and it has been modified to 188 HAPs – see Chapter 2. Many of the HAPs are volatile organic compounds.

14.2.4 Volatile Organic Compounds

Volatile organic compounds (VOCs) are present in water, soil, and air; and many of them are toxic. There are different definitions of VOCs, and sometimes they are confusing.

The general definition of VOCs is "organic compounds whose composition makes it possible to evaporate under normal atmospheric conditions of temperature and pressure. Since the volatility of a compound is higher if its boiling point is lower, the boiling point is part of some definitions of VOCs. For example, VOCs are defined by the European Union as "*A VOC is any organic compound having an initial boiling point≤250°C, measured at standard atmospheric pressure of 101.3 kPa (= 1 atm)*" (USEPA, 2023a).

In the United States, emissions of VOCs to ambient air are regulated mostly to prevent the formation of ozone, a constituent of photochemical smog (see next section). However, only some VOCs are considered "reactive" enough to be of concern. VOCs that are non-reactive or of negligible reactivity to form ozone are exempt from the definitions of VOCs by the USEPA in its regulations. In addition, some states have their own definitions and lists of the exempted compounds (USEPA, 2023a). A complete list of the exempted compounds can be found on the USEPA's website. The notable ones include methane, ethane, 1,1,1-TCA, acetone, some chlorofluorocarbon compounds (CFCs), and some perfluorocarbon compounds (PFCs) (USEPA, 2023b).

14.2.5 Chemistry of Photochemical Smog

Ozone (O_3) can be "good" or "bad" for human health and the environment, depending on where it is in the atmosphere. Ozone in the stratosphere is considered "good" because it protects living things from harmful ultraviolet (UV) radiation from the Sun. Ground-level ozone (i.e., ozone in the troposphere) is considered bad because it can trigger various health problems and adverse impacts on the environment.

Tropospheric (or ground-level) *ozone* is mostly not emitted directly into the atmosphere, but it is created by photochemical reactions between nitrogen dioxide (NO_2) and VOCs in the presence of sunlight. The major sources of NO_2 and VOCs include industrial facilities and electric utilities, motor vehicles, gasoline vapors, and chemical solvents.

Figure 14.2 illustrates concentration profiles of NO, NO_2, hydrocarbon (HC), O_3, and the reaction products in ambient air during morning rush hours resulting from automobile emissions. As shown, times of the concentration peaks are in the following sequence: NO → NO_2 → O_3 → the reaction products. Below is a simplified explanation of the chemistry of this smog formation.

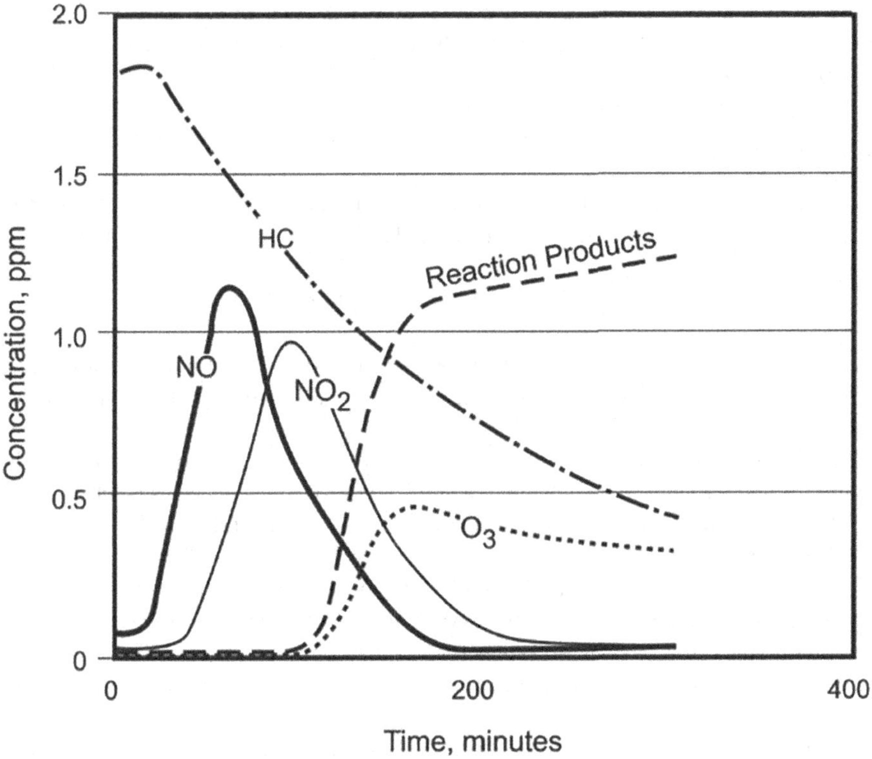

FIGURE 14.2 A typical photochemical profile. (EPA, 2014.)

Nitrogen oxide (NO) emitted from mobile (and stationary) combustion sources is rapidly converted to NO_2 as:

$$NO + \frac{1}{2}O_2 \rightarrow NO_2 \tag{14.1}$$

Nitrogen dioxide (NO_2), those newly formed and those directly from the tailpipes, can dissociate by sunlight photolysis to form NO and atomic oxygen (O) by absorbing photons of energy ($h\nu$) as,

$$NO_2 + h\nu \rightarrow NO + O \tag{14.2}$$

The atomic oxygen produced can then react rapidly with oxygen to form ozone with various hydrocarbons working as *catalysts* [Note: A catalyst is a substance that speeds up the chemical reaction rate without being consumed in the reaction] as:

$$O + O_2 \xrightarrow{\text{Hydrocarbons}} O_3 \tag{14.3}$$

The main source of these hydrocarbons is VOCs present in the atmosphere. They participate in the reaction by absorbing the excess vibrational energy and stabilizing the ozone molecules.

14.2.6 Acid Rains

Normal rain is slightly acidic and has pH of about 5.6 because of the dissolution of carbon dioxide into it to form carbonic acid (H_2CO_3) as:

$$CO_{2(g)} + H_2O \leftrightarrow CO_{2(aq)} + H_2O \leftrightarrow H_2CO_3 \tag{14.4}$$

Acid rain, or *acid deposition*, includes any form of precipitation with acidic components (i.e., sulfuric acid or nitric acid) that fall from the atmosphere to the ground in a wet or dry form. *Wet deposition* is the sulfuric and nitric acids formed in the atmosphere fall to the ground in the form of rain, snow, fog, or hail. Acidic particles and gases in the atmosphere can also deposit to the surfaces (such as water bodies, vegetation, and buildings) in the absence of moisture as *dry deposition* (USEPA, 2023c).

The acidic components in the acid rain come from SO_2 and/or NO_x in the atmosphere. They react with water, oxygen to form sulfuric and nitric acids, then mix with water and other materials before deposit onto the ground. Since wind can blow them over a long distance, acid rain can be a regional and even an international problem.

Formation of nitric acid (HNO_3) from NO_x, oxygen, and water can be illustrated as:

$$NO_{(g)} + \frac{1}{2}O_{2(g)} \rightarrow NO_{2(g)} \tag{14.5}$$

$$3NO_{2(g)} + H_2O \rightarrow 2HNO_{3(aq)} + NO_{(g)} \tag{14.6}$$

Formation of sulfuric acid (H_2SO_4) from SO_2, oxygen, and water can be illustrated as:

$$SO_{2(g)} + \frac{1}{2}O_{2(g)} \rightarrow SO_{3(g)} \tag{14.7}$$

$$SO_{3(g)} + H_2O \rightarrow H_2SO_{4(aq)} \tag{14.8}$$

Acid rain usually has a pH between 4.2 and 4.4, but it can be lower.

14.3 REACTIONS IN STRATOSPHERE

14.3.1 Light Radiation

Electromagnetic radiation is the flow of energy at the speed of light, in the form of waves, through free space or a medium, in the combination of electric and magnetic fields. Electromagnetic waves are classified according to their wavelength (λ) or frequency (ν). They are related to the speed of light (C) as:

$$\nu = \frac{C}{\lambda} \tag{14.9}$$

where C = speed of light = 3.00×10^8 m/s and frequency is typically in s^{-1} (i.e., Hertz or Hz). The electromagnetic waves include microwaves, infrared (IR), visible light, ultraviolet (UV), X-ray, and γ-ray, in the order of decreasing wavelengths.

Planck's equation relates the energy of a photon (E) with its wavelength (λ) or frequency (ν) as:

$$E = h\nu = h\left(\frac{C}{\lambda}\right) \tag{14.10}$$

where h = Planck's constant = 6.626×10^{-34} J·s. As shown in Eq. (14.10), a photon with a shorter wavelength will have a larger frequency and possesses a larger energy.

The wavelength of visible light is from ~400 to ~700 nanometer (nm); the wavelength of the red light is the longest and that of the violet light is the shortest. Ultraviolet light (UV) has wavelength shorter than that of the violet light. The UV region covers the wavelength range of 100–400 nm and is divided into three bands: (i) UVA (315–400 nm), (ii) UVB (280–315 nm), and (iii) UVC (100–280 nm). Short-wavelength UVC is the most damaging type of UV radiation. However, it is completely filtered by the atmosphere and does not reach the Earth's surface (WHO, 2023). For UV disinfection, the wavelength of 253.6 nm is considered as the most germicidal wavelength. Infrared radiation (IR), also known as thermal radiation, has wavelengths above the red visible light and between 780 nm and 1 mm; and it is divided into three bands: (i) IR-A (780–1,400 nm), (ii) IR-B (1,400–3,000 nm) and (iii) IR-C, also known as the far-IR (3 μm–1 mm).

Example 14.1: Energy of Photons

a. Light with a wavelength of 500 nm is green. Determine the frequency and energy of this green light photon.
b. What is the energy of a photon with a wavelength of 254 nm?
c. What is the energy of one mole of photons with a wavelength of 254 nm?

Solution:

a. From Eq. (14.9),

$$\nu = \left(\frac{C}{\lambda}\right) = \left(\frac{3.00 \times 10^8 \text{ m/s}}{500 \times 10^{-9} \text{ m}}\right) = 6.0 \times 10^{14} \frac{1}{\text{s}}$$

From Eq. (14.10),

$$h\nu = \left(6.626 \times 10^{-34} \text{ J} \cdot \text{s}\right)\left(6.0 \times 10^{14} \frac{1}{\text{s}}\right) = 3.97 \times 10^{-19} \frac{\text{J}}{\text{photon}}$$

b. From Eq. (14.10),

$$hv = \left(6.626\times10^{-34}\,\text{J}\cdot\text{s}\right)\left(\frac{3.00\times10^{8}\,\text{m/s}}{254\times10^{-9}\,\text{m}}\right) = 7.83\times10^{-19}\,\frac{\text{J}}{\text{photon}}$$

$$hv = \left(7.83\times10^{-19}\,\frac{\text{J}}{\text{photon}}\right)\left(6.022\times10^{23}\,\frac{\text{photons}}{\text{mole}}\right) = 471{,}300\ \text{J}$$

Discussion:

1. The energy of 254-nm light is approximately twice of that of 500-nm green light.
2. One mole of UV_{254} photons possesses a relatively large energy of 471,300 J.

14.3.2 Ozone Chemistry in the Stratosphere

Photochemical reactions start with the absorption of a photon of energy (hv) by an atom, molecule, ion, or free radical. Equation (14.11) illustrates the first step of a photochemical reaction of a species A:

$$A + hv \rightarrow A^{*} \tag{14.11}$$

where A^* is the excited state of the species A. This excited species may reduce its energy level through dissociation, fluorescence, reaction with another species, or collisional deactivation by transferring energy to another energy-absorbing species (USEPA, 2002). The energy for atmospheric photochemical reactions comes from the Sun. Light intensity is a critical parameter for photochemical reactions.

Stratospheric zone (i.e., the "good" ozone") occurs naturally in the upper atmosphere, 6–30 miles (10–50 km) above the Earth's surface. It forms a protection layer that shields us from harmful UV rays in the sunlight. As a natural process, ozone molecules are constantly formed and destroyed in the stratosphere; and the total amount remained relatively constant in the recent past.

Stratospheric ozone is naturally formed in photochemical reactions involving sunlight and oxygen molecules. In the first step, sunlight breaks apart one oxygen molecule (O_2) into two atomic oxygen (O) (Eq. 14.12). In the second step, each atomic oxygen combines with another oxygen molecule to form an ozone molecule (Eq. 14.13). The overall reaction is that three oxygen molecules react with sunlight to form two ozone molecules (Eq. 14.14).

$$O_2 + hv \rightarrow 2\ O \tag{14.12}$$

$$O + O_2 \rightarrow O_3 \tag{14.13}$$

$$3O_2 + hv \rightarrow 2O_3 \tag{14.14}$$

The ozone layer in the stratosphere absorbs the UVB portion of the sunlight. UVB has been linked to many harmful effects, including skin cancers, cataracts, and harm to some crops and marine life. The stratospheric ozone layer is the Earth's "sunscreen" (USEPA, 2023d).

The total amount has remained relatively stable until recently. Scientific evidence has shown that the ozone shield has been depleted beyond the natural process since the early 1970s. This beneficial ozone is being gradually destroyed by manmade chemicals. At locations where the protective ozone layer has been significantly depleted (e.g., over the Antarctic), it is called a "hole" in the ozone layer and the problem is referred to as *ozone depletion*. This allows more UV radiation to reach the Earth's surface and leads to increased incidences of skin cancer, cataracts, and other health problems.

Ozone can be destroyed by radicals such as hydroxyl (·OH), nitric oxide (NO·), chlorine (Cl·), and bromine (Br·) radicals. In the stratosphere, the hydroxyl and nitric oxide radicals are naturally occurring, while levels of chlorine and bromine radicals have significantly increased due to human activities. *Ozone-depleting substances* (ODS) are compounds that can travel to the stratosphere, due to its relative inertness in the troposphere, and release chlorine or bromine radicals by solar UV radiation to destroy ozone. ODS that may release chlorine radicals include chlorofluorocarbons (CFCs), hydrochlorofluorocarbons (HCFCs), carbon tetrachloride (CCl_4), and 1,1,1-trichloroethane ($C_2H_3Cl_3$). Those that can release bromine radicals include bromofluorocarbons (halons) and bromomethane (CH_3Br) (USEPA, 2023d).

The discussion below uses trichloroflurometHane ($CFCl_3$, CFC-11) and simplified reactions to illustrate ozone depletion by ODS. $CFCl_3$ can be broken down by solar UV radiation to release chlorine radicals as:

$$CFCl_3 + h\upsilon \rightarrow Cl\cdot + \cdot CFCl_2 \qquad (14.15)$$

The chlorine radical can destruct an ozone molecule and form chlorine monoxide radical (ClO·):

$$Cl\cdot + O_3 \rightarrow ClO\cdot + O_2 \qquad (14.16)$$

ClO can then react with atomic oxygen (O), formed from the reactions among solar UV, ozone and oxygen molecules as:

$$ClO\cdot + O \rightarrow Cl\cdot + O_2 \qquad (14.17)$$

The net reaction of two reactions shown immediately above is:

$$O_3 + O \rightarrow 2O_2 \qquad (14.18)$$

Chlorine radical works as a catalyst for ozone destruction. In this way, one chlorine atom/radical can participate in many cycles to destroy many ozone molecules. With the presence of these substances in the stratosphere, ozone is being destroyed more quickly than it is naturally created.

14.4 GREENHOUSE GASES

14.4.1 Greenhouse Gases

Nearly all the Earth's energy comes from the Sun. The climate on Earth is determined by the balance between the energy received from the Sun and the energy emitted back to the outer space from the Earth and its atmosphere. Gases in the atmosphere that trap some of the outgoing energy, retaining heat in our atmosphere, are called *greenhouse gases* (GHGs). Without this *greenhouse effect*, the average surface temperature of the Earth will be −18°C (−0.4°F). Water vapor is the major GHG. However, human activities are changing the Earth's natural greenhouse effect with a dramatic increase in the release of other GHGs into the atmosphere. These GHGs are now believed to be the cause of global warming and climate change.

The GHGs of concern are:

- Carbon dioxide (CO_2)
- Methane (CH_4)
- Nitrous oxide (N_2O)
- Fluorinated gases: hydrofluorocarbons (HFCs), perfluorocarbons (PFCs), sulfur hexafluoride (SF_6), and nitrogen trifluoride (NF_3). Fluorinated gases (especially HFCs) are sometimes used as substitutes for stratospheric ODS (USEPA, 2023e).

14.4.2 Global Warming Potential

Global warming refers to the recent and the ongoing rise in the global average temperature near the Earth's surface. It has risen by 1.5°F (0.83°C) over the past century and is project to rise further. Although climate change may be caused by factors other than global warming, it is now a common belief that the global warming is mostly caused by the increasing concentrations of GHGs in the atmosphere. Small changes in the average Earth's temperature can translate to significant and potentially adverse shifts in weather and climate. Climate change can affect human health, cause changes to forests and other ecosystems, impact our energy supply, and cause severe weather events.

Each of these GHGs can stay in the atmosphere for different durations, ranging from a few years to thousands of years. Due to their long stay, they are well mixed in the atmosphere so that their concentrations are relatively the same all over the world. Different GHGs can have different effects on global warming. Two key factors for the differences are their ability to absorb energy ("*radiative efficiency*") and their "lifetime" in the atmosphere. By taking these into consideration, *global warming potential* (GWP) was developed to allow comparisons of the global warming impacts of different GHGs. A gas with a larger GWP would absorb more energy per unit mass than those with smaller GWPs, and it thus would contribute more to global warming on a per-unit-mass basis. The GWP of CO_2 is unity, and it serves as the yardstick.

The United States primarily uses the 100-year GWP, which is based on the energy by a gas absorbed over a period of 100 years, as a measure of the relative impact of

TABLE 14.1
Lifetime and GWP of Greenhouse Gases

	Lifetime in Atmosphere (yr)	GWP
CO_2	Varies	1
CH_4	12	28
N_2O	114	265
Flourinated Gases		
PFCs	2,600–50,000	up to 11,100
HFCs	up to 270	up to 12,400
NF_3	740	16,100
SF_6	3,200	23,500

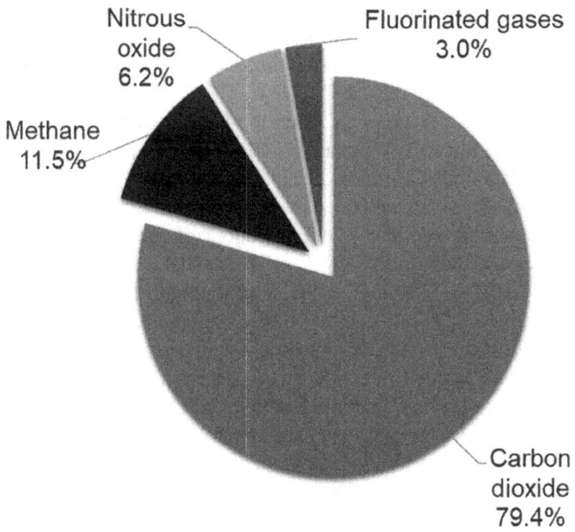

FIGURE 14.3 Percentages of total US GHG emissions of 2021.

different GHGs. However, it can be different in some metrics used by the scientific community. Using a 20-year GWP would prioritize gases with shorter lifetimes than CO_2 in the atmosphere. For example, the GWP values for CH_4 are 84–87 and 28–36 for the 20-year and 100-year cycles, respectively, because of its shorter lifetime. On the other hand, the 100-year GWPs of fluorinated gases would be larger than their 20-year GWPs.

By taking the GWPs into account, the total greenhouse gas emission is 6,340 Million Metric Tons of CO_2 equivalent ($MMT_{CO2\text{-}eq}$) in 2021; CO_2 (79.4%), CH_4 (11.5%), N_2O (6.2%), and fluorinated gases (3.0%) (USEPA, 2023e) (see Figure 14.3).

Example 14.2: Global Warming Potentials

A source emits 100 kg of methane per day. Determine its annual emission of methane gas as a greenhouse gas in $MMT_{CO2\text{-}Eq}$, using GWP of methane = 30 (based on a lifetime of 20 years).

Solution:

The annual methane emission, in million metric ton CO_2-equivalent/yr

$$= (100\ \text{kg/d}) \times (365\ \text{d/yr}) \times (\text{ton}/1{,}000\ \text{kg}) \times (30) = 1{,}095\ \text{ton } CO_2\text{-equivalent/yr}$$

$$= 1.095 \times 10^{-3}\ MMT_{CO2\text{-}eq}/\text{yr}.$$

REFERENCES

USEPA (2002). *APTI 482: Sources and Control of Volatile Organic Air Pollutants - Student Manual (3rd edition)*, Prepared by J.W. Crowder for Air Pollution Training Institute, United States Environmental Protection Agency, Research Triangle Park, NC 27711.

USEPA (2003). *APTI 452: Principles and Practices of Air Pollution Control - Student Manual (3rd edition)*, Air Pollution Training Institute, United States Environmental Protection Agency, Research Triangle Park, NC 27711.

USEPA (2023a). "Technical Overview of Volatile Organic Compounds", US Environmental Protection Agency, https://www.epa.gov/indoor-air-quality-iaq/technical-overview-volatile-organic-compounds, last updated March 14, 2023.

USEPA (2023b). "Volatile Organic Compound Exemptions", US Environmental Protection Agency, https://www.epa.gov/ground-level-ozone-pollution/volatile-organic-compound-exemptions, last updated June 16, 2023.

USEPA (2023c). "What is Acid Rain?", US Environmental Protection Agency, https://www.epa.gov/acidrain/what-acid-rain#:~:text=Acid%20rain%20results%20when%20sulfur,form%20sulfuric%20and%20nitric%20acids, last updated June 1, 2023.

USEPA (2023d). "Ozone Layer Protection", US Environmental Protection Agency, https://www.epa.gov/ozone-layer-protection, last updated October 19, 2023.

WHO (2023). "Radiation: Ultraviolet (UV) Radiation", World Health Organization (WHO), https://www.who.int/news-room/questions-and-answers/item/radiation-ultraviolet-(uv)#:~:text=The%20UV%20region%20covers%20the,(100%2D280%20nm).

EXERCISE QUESTIONS

1. a. Determine the frequency and the energy of two light photons: (i) red light with a wavelength of 700 nm and (ii) violet light with a wavelength of 400 nm.
   b. What is the energy of one mole of photons with a wavelength of 400 nm?
2. A source emits 100 kg of carbon dioxide, 10 kg of methane and 1.0 kg nitrous oxide per day. Use the GWP values in Table 14.1 to estimate its total annual greenhouse gas emission in $MMT_{CO2\text{-}Eq}$.

Index

Note: **Bold** page numbers refer to tables; *italic* page numbers refer to figures.

Printed in the United States
by Baker & Taylor Publisher Services